AF512207

INTRODUCTION TO ARTIFICIAL NEURAL NETWORKS

By

GUNJAN GOSWAMI

Assistant Professor, Department of CSE,
JSS Academy of Technical Education,
Noida, U.P.

S.K. KATARIA & SONS®

Educational Publishers

4885/109, Prakash Mahal, Dr. Subhash Bhargav Lane,
Opposite Delhi Medical Association, Daryaganj, New Delhi-110002 (INDIA)
Phone : +91-11-43551243; Mobile: +91-9871775858, 7678425209
e-mail: katariabooks@gmail.com, katariabook@yahoo.com
Website: www.skkatariaandsons.com
(e-books, subscriptions and bulk buying also available on our website)

Published by:

S.K. Kataria & Sons®

4885/109, Prakash Mahal, Dr. Subhash Bhargav Lane,
Opposite Delhi Medical Association, Daryaganj, New Delhi-110002 (INDIA)
Phone: +91-11-43551243; *Mobile:* +91-9871775858, 7678425209
e-mail: katariabook@gmail.com, katariabooks@yahoo.com
Website: www.skkatariaandsons.com

First Edition : 2009, Third Edition: 2012, Reprint: 2026

Laser Typesetting by :
AISA ENTERPRISES, DELHI-110092

Printed at:
New A.S. Offset Printers, Delhi

Preface

We all know that computers are extremely fast and precise at executing the sequence of instructions that have been formulated for them. But, humans are more efficient than computers in some of the applications like speech recognition, face recognition etc. Therefore this book introduces the foundation of artificial neural networks.

Throughout the years, the computational changes have brought growth to new technologies. Such is the case of artificial neural networks, that over the years, they have given various solutions to the industry.

A neural network is a parallel system, capable of resolving paradigms that linear computing cannot. The fundamentals of ANN, the algorithms, examples are included in this book. This is new and fundamental different approach for computing and information processing.

Organization of the book :

The book consists of 10 chapters providing a comprehensive coverage of course as prescribed by M.D.U. for final year students for the course "Artificial Neural Networks". A large number of examples, theoretical exercises and numerical problems have been included in the text.

Chapter 1 presents an overview of Artificial Neural Networks and Biological Neurons. The Historical development of ANN and the various applications of ANN are also described in this chapter.

Chapter 2 presents the most important neuron processing paradigms, the network properties, the learning concepts, learning algorithms, models of ANN, Types of learning. The numerical are discussed with each topic where applicable.

Chapter 3 discusses Single Layer Perceptron Classifiers. In this the main stress is over the feed forward networks. The examples are discussed to make classification more clear.

Chapter 4 describes the derivation of generalized delta learning rule that is based on minimization of steepest descent rule. The error back-propagation algorithm is discussed and illustrated with examples.

Chapter 5 covers the discrete –time and continuous time gradient type networks. The Hopfield networks are discussed in this chapter.

Chapter 6 describes the associative memories. The concept of auto-associative and hetero-associative memories is also discussed in this chapter. The concept of storage and retrieval in Bidirectional memory is also covered in this chapter.

Chapter 7 describes the self organizing networks. These networks have found their applications in ANN. Clustering, Winner-take –all, MAXNET etc. are discussed in this chapter.

Other chapters are all about Fuzzy Logic, Genetic Algorithms. Although a very basic introduction is presented here, but the students will find it useful.

Concluding.....

To end this preface I just really hope you find what you are looking for.

AUTHOR

Acknowledgement

The book itself is an acknowledgement to the intensity, drive and technical competence of many individuals, who have contributed to it.

I am thankful to **Almighty God**, without whose grace nothing is possible for any one. I am very thankful to my friends and students who have inspired me to write this book.

I am grateful to **my family members** for their constant motivation and guidance that has helped me to complete this book. I would also express my gratefulness for their patience and understanding during the time I devoted to writing the book.

I would like to take this opportunity to express my thanks to **Mr. Sanjeev Kumar Kataria** of M/s S.K. Kataria and Sons for publishing this book in such a short time. I am also thankful to **Mr. Sachin** for his assistance.

Gunjan Goswami
Author

Dedicated to

Almighty God

Contents

CHAPTER

1

OVERVIEW OF BIOLOGICAL NEURONS

INSIDE THIS CHAPTER

1.1. BACKGROUND

What is Intelligence? Where does intelligence emerge? There are two ways to answer these questions from computational point of view. One is based on symbolism and the other based on connectionism. The difference between them is that the former approach models the intelligence using symbols and latter using connections and associated weights. Both have achieved their successes in practical applications.

1.1.1. History of Artificial Neural Networks

The history of neural networks that was described above can be divided into several periods:

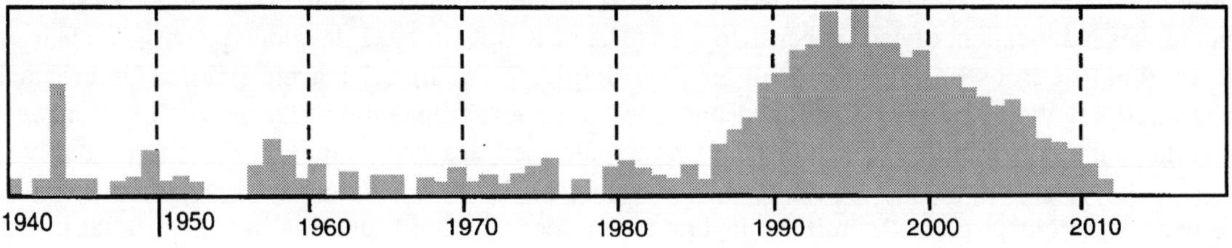

Fig. 1.1

1. First Attempt

There were some initial simulations using formal logic. McCulloch and Pitts (1943) developed models of neural networks based on their understanding of neurology. These models made several assumptions about how neurons worked. Their networks were based on simple neurons which were considered to be binary devices with fixed thresholds. The results of their model were simple logic functions such as "*a* or *b*" and "*a* and *b*". Another attempt was by using computer simulations. Two groups (Farley and Clark, 1954; Rochester, Holland, Haibit and Duda, 1956). The first group (IBM researchers) maintained closed contact with neuroscientists at McGill University. So whenever their models did not work, they consulted the neuroscientists. This interaction established a multidisciplinary trend which continues to the present day.

2. Promising and Emerging Technology

Not only was neuroscience influential in the development of neural networks, but psychologists and engineers also contributed to the progress of neural network simulations. Rosenblatt (1958) shown considerable interest and activity in the field when he designed and developed the Perceptron. The Perceptron had three layers with the middle layer known as the association layer. This system could learn to connect or associate a given input to a random output unit.

Another system was the ADALINE (Adaptive Linear Element) which was developed in 1960 by Widrow and Hoff (of Stanford University). The ADALINE was an analogue electronic device made from simple components. The method used for learning was different to that of the Perceptron; it employed the Least-Mean-Squares (LMS) learning rule.

3. Period of Frustration and Disrepute

In 1969 Minsky and Papert wrote a book in which they generalized the limitations of single layer Perceptron to multilayered systems. In the book they said: "...our intuitive judgment that the extension (to multilayer systems) is sterile". The significant result of their book was to eliminate funding for research with neural network simulations. The conclusions supported the disenchantment of researchers in the field. As a result, considerable prejudice against this field was activated.

4. Innovation

Although public interest and available funding were minimal, several researchers continued working to develop neuromorphically based computational methods for problems such as pattern recognition.

During this period several paradigms were generated which modern work continues to enhance. Grossberg's (Steve Grossberg and Gail Carpenter in 1988) influence founded a school of thought which explores resonating algorithms. They developed the ART (Adaptive Resonance Theory) networks based on biologically plausible models. Anderson and Kohonen developed associative techniques independent of each other. Klopf (A. Henry Klopf) in 1972 developed a basis for learning in artificial neurons based on a biological principle for neuronal learning called heterostasis. Werbos (Paul Werbos 1974) developed and used the back-propagation learning method, however several years passed before this approach was popularized. Back-propagation nets are probably the most well known and widely applied of the neural networks today. In essence, the back-propagation network is a Perceptron with multiple layers, a different threshold function in the artificial neuron, and a more robust and capable learning rule. Amari (A. Shun-Ichi 1967) was involved with theoretical developments: he published a paper which established a mathematical theory for a learning basis (error-correction method) dealing with adaptive pattern classification. While Fukushima (F. Kunihiko) developed a step wise trained multilayered neural network for interpretation of handwritten characters. The original network was published in 1975 and was called the Cognitron.

5. Re-Emergence

Progress during the late 1970s and early 1980s was important to the re-emergence on interest in the neural network field. Several factors influenced this movement. For example, comprehensive books and conferences provided a forum for people in diverse fields with specialized technical languages, and the response to conferences and publications was quite positive. The news media picked up on the increased activity and tutorials helped disseminate the technology. Academic programs appeared and courses were introduced at most major Universities (in US and Europe). Attention is now focused on funding levels throughout Europe, Japan and the US and as this funding becomes available, several new commercial with applications in industry and financial institutions are emerging.

6. Today

Significant progress has been made in the field of neural networks-enough to attract a great deal of attention and fund further research. Advancement beyond current commercial applications appears to be possible, and research is advancing the field on many fronts. Neurally based chips are emerging and applications to complex problems developing. Clearly, today is a period of transition for neural network technology.

1.2. KNOWLEDGE-BASED INFORMATION PROCESSING

Knowledge-based system is a computer program that acquires, represents, and uses knowledge for a specific purpose. The basic structure is shown in the Fig. 1.2, consists of a knowledge base which stores the knowledge and an inference engine which makes inference using the knowledge. However as Feigenbaum indicated, the power of such a system derives from the knowledge it possesses rather than from the inference methods it employs. Still there are some differences between the knowledge-based programming and the conventional programming.

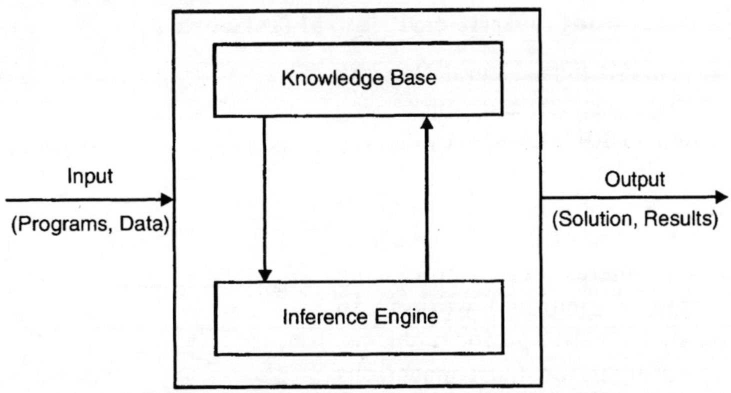

Fig. 1.2. Knowledge-Based Information Processing

Table 1.1. Comparison of Knowledge-based Processing and Conventional Processing.

Knowledge-based Processing	Conventional Processing
1. Declarative knowledge	1. Procedural knowledge
2. Separation control from knowledge	2. Integrating control and knowledge
3. Strategic and heuristic processing	3. Algorithmic processing
4. Symbolic Processing (dominant)	4. Numeric processing (dominant)
5. Explanation capability	5. No Explanation

The Knowledge-Based Algorithm (General View)

Given a problem (initial conditions and goal)

1. The inference engine selects a piece of knowledge from the knowledge base.

2. The inference engine executes the selected knowledge either to transform the goal or to generate a new fact.
3. If the goal (original or transformed) is solved, then exit and succeed. If a certain stopping condition is met such as knowledge available is exhausted but the goal is not solved yet, then exit and fail. Otherwise go to step 1.

1.3. OVERVIEW OF NEURAL NETWORKS

There are number of different answers possible for the question "How to define Neural Networks". The answer may be, "ANN are simply a class of mathematical algorithms and these algorithms produce the solutions to a number of specific problems" or the answer may be "these are synthetic networks that emulate (copy) the biological neural networks".

Many answers are provided by the Researchers. ANN undoubtedly have been biological inspired, but the close correspondence between them and real neural systems is still weak. The definition of ANN and biological Neural networks, the similarities and dissimilarities are explained further in this chapter.

1.3.1. What is Artificial Neural Network ? [MDU, May 2008]

An Artificial Neural Network (ANN) is an information-processing paradigm that is inspired by the way biological nervous systems, such as the brain, process information. It is composed of a large number of highly interconnected processing elements (neurons) working in parallel to solve specific problems. The weights on the connections encode the knowledge of a network. Each neuron has a local memory and the output of each neuron depends upon only the input signals arriving at the neuron and value in neuron's memory. ANNs, **like people**, learn by example. An ANN is configured for a specific application, such as pattern recognition or data classification, through a learning process. Learning in biological systems involves adjustments to the synaptic connections that exist between the neurons. This is true for ANNs as well.

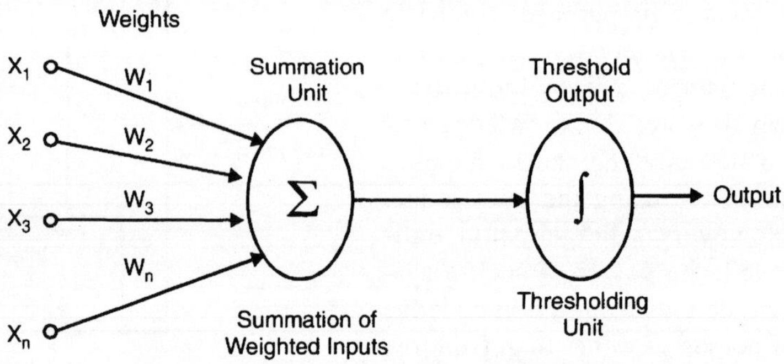

Fig. 1.3. Simple Model of An Artificial Neuron

In many tasks such as recognizing human and understanding speech, current AI systems cannot do better than humans. It is conjectured that the structure of the brain is somehow suited to these tasks and not suited to these tasks such as high speed arithmetic calculation. Rather than by being programmed an ANN can solve the problems simply by example mapping.

The intelligence of a neural network emerges from the collective behavior of neurons, each of which performs only very limited operation. Even though each individual neuron works slowly, they can still quickly find the solution by working in parallel. This fact can explain why humans can recognize a visual scene faster than a digital computer, while an individual brain cell responds much more slowly than a digital cell in VLSI circuit.

Neural Networks are identified by different names such as

(*i*) Artificial Neural Networks

(*ii*) Connectionists Network

(*iii*) Parallel distributed processing systems

1.3.2. Advantages of Neural Network

Neural networks, with their remarkable ability to derive meaning from complicated or imprecise data, can be used to extract patterns and detect trends that are too complex to be noticed by either humans or other computer techniques. A trained neural network can be thought of as an "expert" in the category of information it has been given to analyze. Other advantages include:

(*i*) Its structure massively distributed in parallel. The information processing takes place through the iteration of a great amount of computational neurons, each one of them send exciting or inhibiting signals to other nodes in the network. Differing from other classic Artificial Intelligence methods where the information processing can be considered sequential—this is step by step even when there is not a predetermined order, in the Artificial Neural Networks this process is essentially in parallel, which is the origin of its flexibility. Because the calculations are divided in many nodes, if any of them gets deviated from the expected behaviour it does not affect the behaviour of the whole network.

(*ii*) Its ability to learn and generalize. The ANN have the capability to acquire knowledge from its surroundings by the adaptation of its internal parameters, which is produced as a response to the presence of an external stimulus. The network learns from the examples which are presented to it, and generalizes knowledge from them. The generalization can be interpreted as the property of artificial neural networks to produce an adequate response to unknown stimulus which are related to the acquired knowledge.

These two characteristics for information processing make an ANN able to give solution to complex problems normally difficult to manage by the traditional ways of approximation. Additionally, using them gives the following benefits :

1. No linearity: The answer from the computational neuron can be linear or not. A neural network formed by the interconnection of nonlinear neurons, is in itself nonlinear, a trait which is distributed to the entire network. No linearity is important over all in the cases where the task to develop presents a behaviour removed from linearity, which is presented in most of real situations.

2. Adaptive learning: The ANN is capable of determine the relationship between the different examples which are presented to it, or to identify the kind to which it belong, without requiring a previous model.

3. Self-organization: This property allows the ANN to distribute the knowledge in the entire network structure, there is no element with specific stored information.

4. Fault tolerance: This characteristics is shown in two senses : The first is related to the samples shown to the network, in which case it answers correctly even when the examples exhibit variability or noise; the second, appears when in any of the elements of the network occurs a failure, which does not impossibilitate its functioning due to the way in which it stores information.

1.4. NEURAL NETWORKS VERSUS CONVENTIONAL COMPUTERS

Neural networks take a different approach to problem solving than that of conventional computers. Conventional computers use an algorithmic approach *i.e.,* the computer follows a set of instructions in order to solve a problem. Unless the specific steps that the computer needs to follow are known the computer cannot solve the problem. That restricts the problem solving capability of conventional computers to problems that we already understand and know how to solve. But computers would be so much more useful if they could do things that we don't exactly know how to do.

Neural networks process information in a similar way the human brain does. The network is composed of a large number of highly interconnected processing elements (neurons) working in parallel to solve a specific problem. Neural networks learn by example. They cannot be programmed to perform a specific task. The examples must be selected carefully otherwise useful time is wasted or even worse the network might be functioning incorrectly. The disadvantage is that because the network finds out how to solve the problem by itself, its operation can be unpredictable.

On the other hand, conventional computers use a cognitive approach to problem solving; the way the problem is to solve must be known and stated in small unambiguous instructions. These instructions are then converted to a high-level language program and then into machine code that the computer can understand. These machines are totally predictable; if anything goes wrong is due to a software or hardware fault.

Neural networks and conventional algorithmic computers are not in competition but complement each other. There are tasks are more suited to an algorithmic approach like arithmetic operations and tasks that are more suited to neural networks. Even more, a large number of tasks require systems that use a combination of the two approaches (normally a conventional computer is used to supervise the neural network) in order to perform at maximum efficiency.

Von Neumann Computer vs. Biological Neural System

	Von-Neumann Computer	**Biological Neural System**
processor	complex high-speed one (or a few)	simple low-speed a large number
memory	separate localized noncontent addressable	integrated in the neuron distributed content addressable
computing	centralized sequential strored-programs	distributed parallel self-learning
reliability	vulnerable	robust
expertise	numerical symbolic	perceptual problems manipulations
environment	well-defined	poorly defined unconstrained

Why would anyone want a 'new' sort of computer? (ANN)

What are (everyday) computer systems good at........and not so good at?

Good at (Conventional Computers)	Not so good at (Conventional Computers)
Fast arithmetic	Interacting with noisy data or data from the environment
Doing precisely what the programmer programs them to do	Massive parallelism
	Fault tolerance
	Adapting to circumstances

Where can neural network systems help?

- Where we can't formulate an algorithmic solution.
- Where we can get lots of examples of the behavior we require.
- Where we need to pick out the structure from existing data

1.5. BIOLOGICAL NEURON AND ARTIFICIAL MODELS

1.5.1. How the Information Flow Takes Place in Brain? [MDU, May 2007]

An estimated 10^{11} neurons participating in approximately 10^{15} interconnections over the transmission path that may range for a meter or more gives the human nervous system an amazing complexity. These special kinds of cells also called nerve cells are organized into a very complicated interconnecting network, in the human brain. Each neuron is physically connected to millions of other and share many characteristics however each neuron has unique capability to receive, process and transmit electrochemical signals over the neural pathway that comprises the brain communication system.

The input to the network is provided by sensors receptors. Receptors deliver stimuli both from within the body as well as from sense organs when the stimuli originate from the external world. Stimuli are in the form of electrical impulses that convey the information into the network of neurons. As a result of information processing in the central nervous systems, the effectors are controlled and give the human responses in the form of miscellaneous actions.

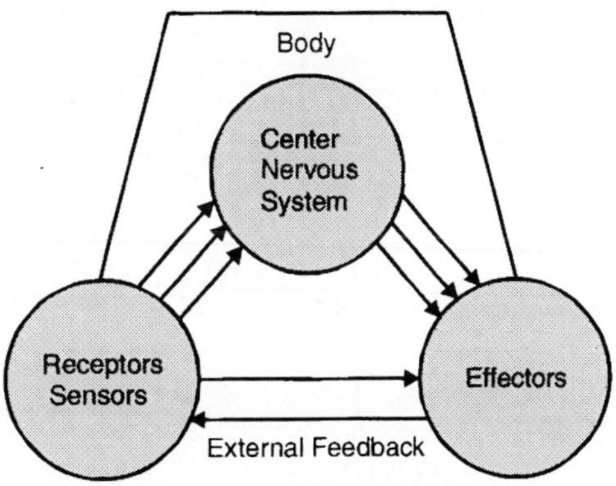

Fig. 1.4. Information Flow in Nervous System

Therefore the three stage system, consisting of receptors, neural network and effectors, in control of the organism and its actions. As it is clear from the Fig. 1.4, the information is processed, evaluated and compared with the stored information in the central nervous system. When necessary, commands are generated there and transmitted to the motor organs. The motor organs are monitored in the central nervous system by the feedback links that verify their actions.

1.5.2. Biological Neuron

[MDU, Dec. 2007, May 2008, Dec. 2008, Dec. 2009, May 2010, Dec. 2010]

The elementary nerve cell called a neuron is the fundamental building block of biological neural network. The three main components of a biological neuron are:

1. A neuron cell body called **soma.**
2. Branching extensions called **dendrites** for receiving input, and
3. An **axon** that carries the neuron's output to the dendrites of other neurons.

In the human brain, a typical neuron collects signals from others through a host of fine structures called *dendrites*. The neuron sends out spikes of electrical activity through a long, thin stand known as an *axon*, which splits into thousands of branches. The axon-dendrite contact organ is called a synapse. At the end of each branch, a structure called a *synapse* converts the activity from the axon into electrical effects that inhibit or excite activity from the axon into electrical effects that inhibit or excite activity in the connected neurons. When a neuron receives excitatory input that is sufficiently large compared with its inhibitory input, it sends a spike of electrical activity down its axon. Learning occurs by changing the effectiveness of the synapses so that the influence of one neuron on another changes. The neuron is able to respond to the total of its inputs aggregated within a short time intervai called period of latent summation.

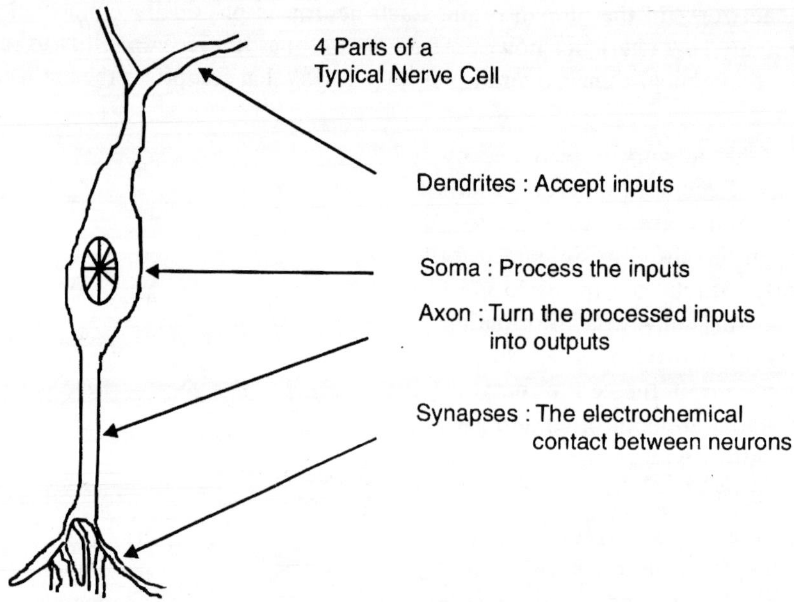

Fig. 1.5. Biological Neuron

Table 1.3. Difference between ANN and Biological Networks.

ANN	Biological Networks
ANNs have usually been limited to 10000 units with hundreds of connections per unit.	The human brain is extremely large for a neural network containing 10^{11} neurons and there exists 10^{15} interconnections.
The memory has an organized behaviour and the data retrieved follows an ordered sequence.	The memory follows a pattern of distributed representation and data retrieval depends on retention capacity of the brain
Learning takes place by a formulated set of rules and hence the system is less faulty. The processing elements receive many signals and sum up the weighted inputs. These networks can be retrained in case of significant damage (*i.e.*, loss of data).	Learning takes place arbitrarily and hence the system is bound for failure. Biological networks are fault tolerable and even in a traumatic loss, other neurons can take over the functions of damaged cells.
Redundancy can increase the reliability of the system, allowing it to function even when some of the neural units are destroyed.	Redundancy is used. When some cells die, the nervous system appears to perform the same. It can counteract source of noise in biological systems.
The strength of neuron depends on whether it is active or inactive and has relatively simpler interconnections.	The strength of neuron depends on active chemicals present and connections are stronger or weaker as a result of structure layer rather than individual synapses.
ANNs focus on learning a single task.	Biological systems have broad capabilities and can address many different types of tasks.
Artificial systems are slow to converge and usually require hundreds or thousands of training presentations for learning to take place	Biological systems have the property of being able to learn with as little as one training presentation. A face that is viewed once, can be recognized again.

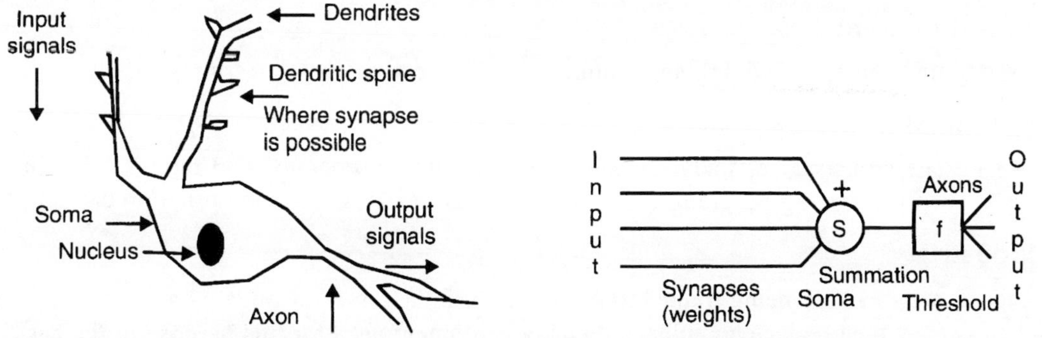

Biological Neuron **Model of Artificial Neuron**

Fig. 1.6. Similarity between Biological Neuron and ANN.

1.6. APPLICATIONS OF ARTIFICIAL NEURAL NETWORKS [MDU, May 2008]

Neural networks have broad applicability to real world business problems. In fact, they have already been successfully applied in many industries.

1. Business Applications

The 1988 DARPA Neural Network Study [DARP88] lists various neural network applications, beginning in about 1984 with the adaptive channel equalizer. This device, which is an outstanding commercial success, is a single- neuron network used in long-distance telephone systems to stabilize voice signals. The DARPA report goes on to list other commercial applications, including a small word recognizer, a process monitor, a sonar classifier, and a risk analysis system.

2. Aerospace

ANN found its application in high performance aircraft autopilot, flight path simulation, aircraft control systems, autopilot enhancements, aircraft component simulation, aircraft component fault detection.

3. Automotive

ANN found application in automobile automatic guidance system, warranty activity analysis.

4. Credit Card Activity Checking

Neural networks are used to spot unusual credit card activity that might possibly be associated with loss of a credit card.

5. Defense

ANNs are used in defence for weapon steering, target tracking, object discrimination, facial recognition, new kinds of sensors, sonar, radar and image signal processing including data compression, feature extraction and noise suppression, signal/image identification.

6. Electronics

In electronics ANNs are used for code sequence prediction, integrated circuit chip layout, process control, chip failure analysis, machine vision, voice synthesis, nonlinear modeling.

7. Entertainment

ANNs are used for providing animation, special effects, market forecasting.

8. Financial

Real estate appraisal, loan advisor, mortgage screening, corporate bond rating, credit-line use analysis, portfolio trading program, corporate financial analysis, currency price prediction.

9. Industrial

Neural networks are being trained to predict the output gasses of furnaces and other industrial processes. They then replace complex and costly equipment used for this purpose in the past.

10. Insurance

ANN found application in policy application evaluation, product optimization, etc.

11. Manufacturing

Manufacturing process control, product design and analysis, process and machine diagnosis, real-time particle identification, visual quality inspection systems, beer testing, welding quality analysis, paper quality prediction, computer-chip quality analysis, analysis of grinding operations, chemical product design analysis, machine maintenance analysis, project bidding, planning and management, dynamic modeling of chemical process system.

12. Medical

ANN are used in medical for breast cancer cell analysis, EEG and ECG analysis, prosthesis design, optimization of transplant times, hospital expense reduction, hospital quality improvement, emergency-room test advisement.

13. Robotics

In Robotics ANN are used for trajectory control, forklift robot, manipulator controllers, vision systems.

14. Speech

ANN have shown remarkable progress in speech recognition, speech compression, vowel classification, text-to-speech synthesis.

15. Telecommunications

Image and data compression, automated information services, real-time translation of spoken language, customer payment processing systems.

16. Transportation

ANN are also helpful in truck brake diagnosis systems, vehicle scheduling, routing systems.

17. Forecasting and Risk Management

NN have the capability to predict the situations from past trends. Therefore, they have found applications in stock market, banking (for document reading, credit application evaluation).

ANN's Application in the Field of Speech and Vision : [MDU, May 2008]

1. Speechreading (Lipreading)

As part of the research program Neuroinformatic the IPVR develops a neural speechreading system as part of a user interface for a workstation. The three main parts of the system include a face tracker (done by Marco Sommerau), lip modeling and speech processing (done by Michael Vogt) and the development and application of SNNS for neural network training (done by Günter Mamier).

Automatic speechreading is based on a robust lip image analysis. In this approach, no special illumination or lip make-up is used. The analysis is based on true color video images. The system allows for realtime tracking and storage of the lip region and robust off-line lip model matching. The proposed model is based on cubic outline curves. **A neural classifier detects visibility of teeth edges and other attributes**. At this stage of the approach **the edge between the closed lips is automatically modeled if applicable, based on a neural network's decision**.

To achieve high flexibility during lip-model development, a model description language has been defined and implemented. The language allows the definition of edge models (in general)

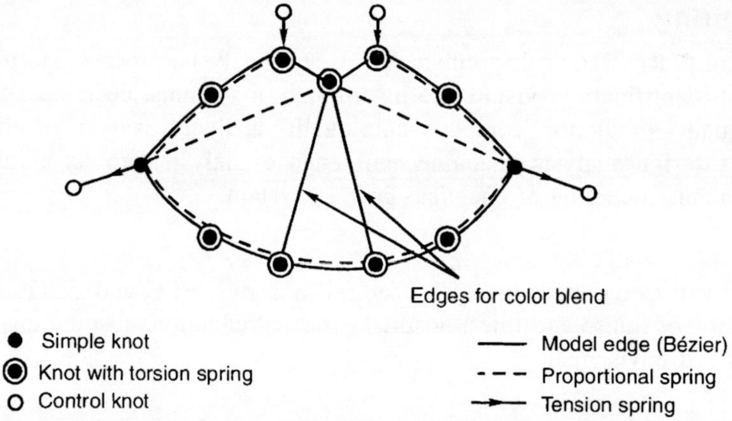

Fig. 1.7. The model interpreter enables a permanent measurement of model knot positions and color blends along model edges during adaptation to an utterance. The resulting parameters may be used for speech recognition tasks in further steps

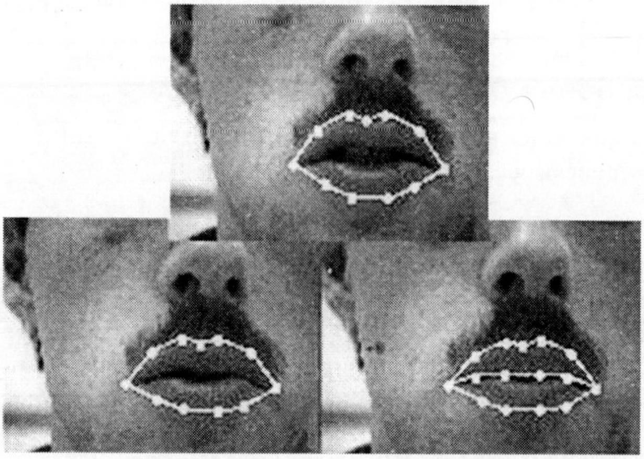

Fig. 1.8. Lipread (190 kb)

based on knots and edge functions. Inner model forces stabilize the overall model shape. User defined image processing functions may be applied along the model edges. These functions and the inner forces contribute to an overall energy function. Adaptation of the model is done by gradient descent or simulated annealing like algorithms. Figure 1.8 shows one configuration of the lip model, consisting of an upper lip edge and a lower lip edge. The model edges are defined by Bezier-functions. Outer control knots stabilize the position of the corners of the mouth.

The Active Vision System

The active vision system consists of a framework for feeding a 2-dimensional image into a recurrent artificial neural network and allowing that network to scan the image. The receptive field, or artificial retina, is a square region composed of cells, or receptors, that read pixel values from the surface. The retine size is considered to be of dimension 5x5.

The retina is able to move across the image vertically and horizontally, as well as zooming in and out. Unlike that system, this one includes the ability to rotate. All images used are in grayscale

TIFF format. Each pixel contains a value between 0 and 255. These are scaled to values between 0 and 1 and input to the neural network (so, 5x5 retina receives 25 pixel inputs). Any portion of the retina that wanders past the image boundary receives pixel inputs of –1.

Also input to the neural network is the retina's current orientation, an "hourglass", and a bias. The orientation consists of the retina's x and y position, angle of rotation, and zoom factor. The active vision system is allotted a certain number of steps for each image evaluation, and the "hourglass" input is the ratio of steps remaining to total steps allocated. The bias input is a constant value of 1. Each time step, neural net outputs are used to update the position and orientation of the retina. The specific movement outputs include change in horizontal location , change in vertical

location, change in rotation, and change in zoom. The fifth output, affinity, represents the confidence the network has that the image contains the target shape.

The network architecture is shown below in Fig. 1.9.

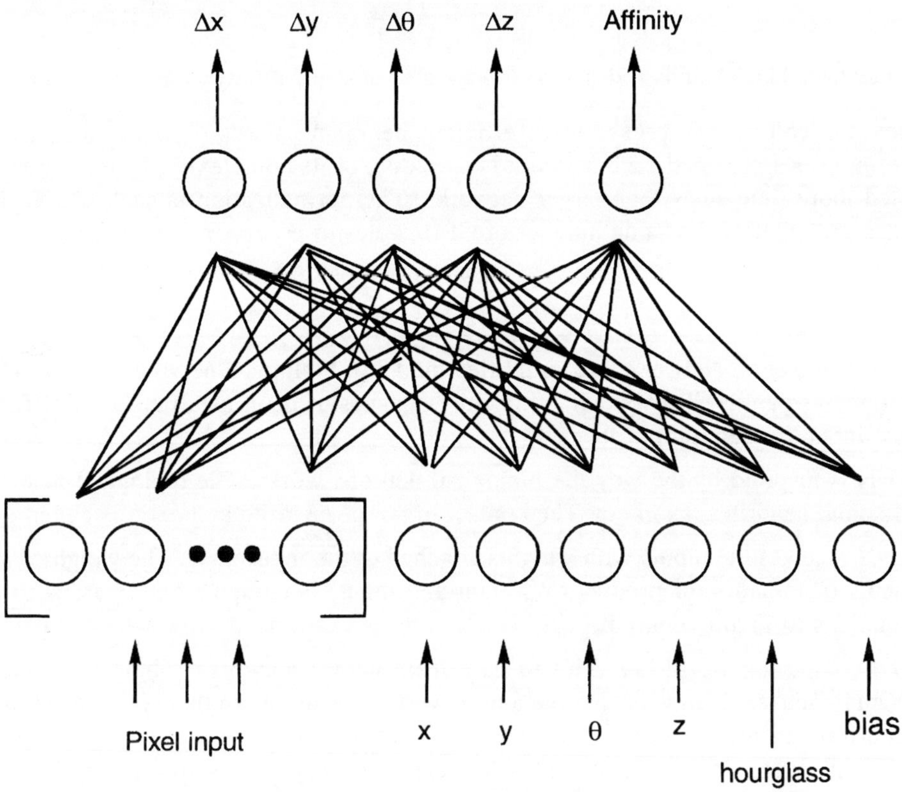

Fig. 1.9. Active Vision Neural Network Initial Architecture

The activation function for the output neurons is a modified hyperbolic tangent (tanh).

Each output ranges from –1 to 1. For movement controls (Δx, Δy, $\Delta\Theta$ and Δz) the neuron's output value is multiplied by the maximum data for that value. The maximum change for location (x and y) each time step is +/– 20 pixels. The maximum change for rotation is 3.6 degrees clockwise or counterclockwise, and the maximum change in zoom is +/– 1%. The affinity output is scaled to a confidence value between 0 and 1.

The network's final judgment regarding the target shape is a product of this affinity, with greater weight given to affinity responses nearer the end of the evaluation. The weighted sum of affinity values is given in the equation below.

$$\frac{\sum_{i=1}^{n}(\text{affinity}_i \, . i^2)}{\sum_{i=1}^{n} i^2}$$

Fig. 1.10. Final Affinity Calculation (*n* = number of steps, affinity $_i$ = affinity at step *i*)

To provide evolutionary pressure for efficient neural networks (*i.e.,* to mitigatc bloat) the number of time steps provided each network is a product of its complexity. Thus, smaller networks are allocated more time steps than larger networks. The normalization is such that each network should use approximately the same number of CPU cycles to process an image fully.

Summary

- Artificial Neural Networks are simplified model of biological nervous system. ANN can be defined as parallel distributed architecture consists of neurons arranged in different layers like input, hidden, output layers.
- ANN is inspired by the way the biological neurons works. The biological neuron consists of soma, dendrites, axon etc. The contact organ of dendrite and axon is called synapse.
- ANN receives the inputs with weights attached to the input links. The weighted sum of the inputs (summation of product of weight and input) is computed and passed to activation functions to give the output.
- The computing world has a lot to gain from neural networks. Their ability to learn by example makes them very flexible and powerful. Furthermore there is no need to devise an algorithm in order to perform a specific task; *i.e.,* there is no need to understand the internal mechanisms of that task. They are also very well suited for real time systems because of their fast response and computational times, which are due to their parallel architecture. Neural networks also contribute to other areas of research such as neurology and psychology. They are regularly used to model parts of living organisms and to investigate the internal mechanisms of the brain.
- *Period of latent summation:* The neuron is able to respond to the total of its inputs aggregated within a short time interval that is called period of latent summation.
- *Refractory period:* It is the time period in which an axon fiber is in a state of complete nonexcitability for a certain time.

➡ ANN is finding its importance in various fields like transportation, pattern matching, voice recognition etc.

Review Questions

1. What is ANN? How ANN is different from Biological Neuron?
2. Explain the similarities and dissimilarities between ANN and the biological neuron. **[MDU, May 2010]**
3. Differentiate between Conventional Processing and knowledge based processing.
4. How information flow takes place in biological neuron? **[MDU, May 2010]**
5. Explain the advantages and disadvantages of ANN.
6. What are the main components of biological neuron?
7. State the conditions where ANN is used effectively.
8. Write the applications of ANN.

❑❑❑

CHAPTER

2

FUNDAMENTAL CONCEPTS OF ARTIFICIAL NEURAL NETWORKS

INSIDE THIS CHAPTER

2.1. INTRODUCTION

ANN typically consists of many hundreds of simple processing units, which are wired, together in a complex communication network. Each unit or node is a simplified model of a simple neuron, which fires if it receives a sufficiently strong input signal from the other node. These nodes can be grouped into different layers. A layer of processing elements makes independent decision or computation on data and passes the result to another layer. The next layer in turn makes its computation on data and passes the result to another layer. Each processing element makes the computation based upon weighted sum of its inputs. The first layer is the input layer and the last layer is called the output layer. The layers that are placed in between are called hidden layers.

Figure 2.1 shows general idea how an artificial neural network works

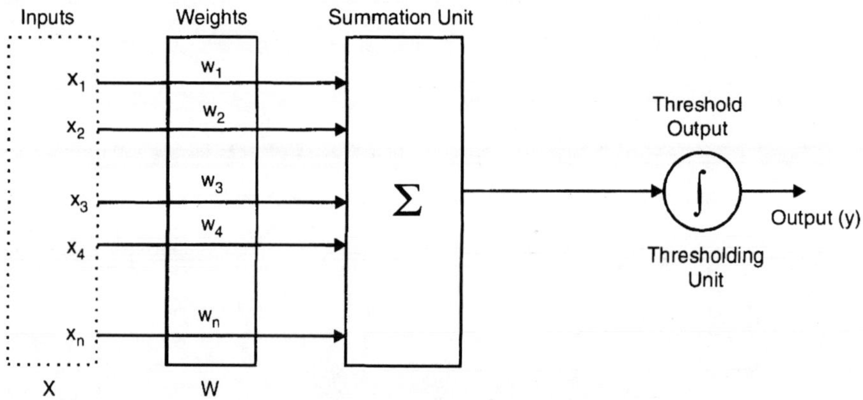

Fig. 2.1. Simple Model of an Artificial Neuron

Here a set of input labeled $x_1, x_2 \ldots x_n$ is applied to the artificial neuron. These inputs collectively referred to as the vector 'X' corresponds to the signals into the synapse of a biological neuron. Each signal is multiplied by an associated weight $w_1, w_2 \ldots w_n$ before it is applied to the summation block. Each weight corresponds to the 'strength' of a single biological synaptic connection. The set of weight is referred collectively as the vector 'W'. The summation block refers to the biological cell body, adds all the weighted inputs algebraically, producing an output that we call SUM. This may compactly state in vector notation as:

Stated in vector notation as:

$$\text{SUM} = X*W$$

or

$$\text{SUM} = x_1*w_1 + x_2*w_2 + x_3*w_3 \ldots x_n*w_n = \sum_{i=1}^{n} X_i W_i \qquad \ldots(2.1)$$

To generate the final output y, the sum is passed onto a non-linear filter ϕ called activation function or transfer function or squash function which releases the output.

$$\therefore \qquad y = \phi\,(\text{Sum})$$

Activation functions are described in more detail in next topic.

A very commonly used activation function is the thresholding function. In this, the sum is compared with a threshold value θ. If the value of sum is $> \theta$, then the output is 1 else it is 0.

$$\phi(\text{Sum}) = \begin{cases} 1 & \text{if sum} > 0 \\ 0 & \text{if sum} \leq 0 \end{cases}$$

2.2. BASIC CONCEPTS OF ARTIFICIAL NEURAL NETWORKS

Artificial neural network has a parallel-distributed architecture with a large no. of nodes and connections. Each connection points from one node to another and is associated with a weight. A simple view of the network structure and behaviors is shown in Fig. 2.2. Construction of neural network involves the following tasks:

- **Determine the network properties:** Network properties includes the network topologies, the type of connections, the order of connections and the weight range.
- **Determine the node properties:** the activation range, and the activation transfer function
- **Determine the system dynamics:** the weight initialization scheme, activation-calculating formula, and the learning rule.

2.2.1. Network Properties

The topology of ANN refers to its framework as well as its interconnection scheme. The framework is often specified by the number of layers and the number of nodes per layer. The presence of weights is compulsory on each connection. The types of layer include:

(*a*) **The Input Layer:** The nodes in this layer are called input units, which encodes the instance present to the network for processing. For example, each input unit may be designated by an attribute value possessed by the instance.

(*b*) **The Hidden Layer:** The nodes in this layer are called hidden units, which are not directly observable and hence hidden. They provide the nonlinearities for the network.

(*c*) **The Output Layer:** The nodes in this layer are called output nodes, which encodes possible concepts or values to be assigned to the instance under consideration *e.g.*, each output unit presents a class of objects.

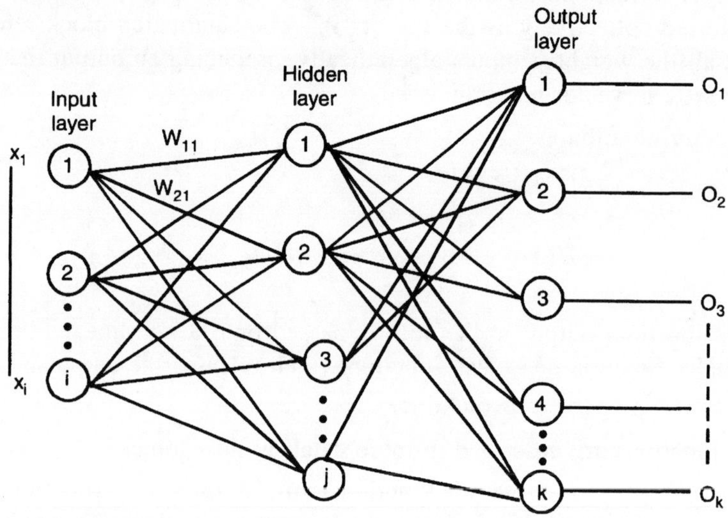

Fig. 2.2. The Neural Network Computation Model

Terms used in ANN

➡ **Input Units:** Input units do not process the information; they simply distribute the information to the other units. According to the interconnection scheme, a network can be either feedforward or recurrent and its connections either symmetrical or asymmetrical as will be discussed later on.

➡ **Weights:** ANN consists of large number of simple processing elements called neurons. These neurons are connected to each other by directed communication links, which are associated with weights.

"Weight is an information used by the neural net to solve a problem".

The weights may be fixed or can take random values. Weights can be set to zero, or can be calculated by some methods. Initialization of weights is an important criteria in Artificial Neural network. From the above Fig. 2.3.

x_1, ... x_i are the input signals

O_1, ... O_k are the output signals

And w_{ji} and w_{kj} are the weights from layer *i*(source) to layer *j*(destination) and from layer *j*(source) to layer *k*(destination) respectively.

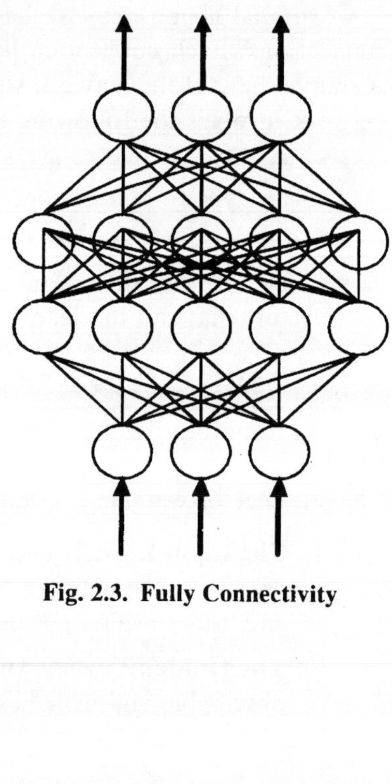

Fig. 2.3. Fully Connectivity

Based on these parameters, the net input "Net" is calculated and

$$\text{Net} = x_1 * w_1 + x_2 * w_2 \ldots + x_n * w_n$$

or

$$\text{Net} = \sum_{i=1}^{n} X_i * W_i$$

Note: Weight is always written as $W_{(\text{dest. source})}$

- **Symmetrical Connections:** If there is a connection pointing from node i to node j, then there is also a connection from node j to node i, and the weights associated with both the connections are equal or $W_{ij} = W_{ji}$.
- **Asymmetrical Connections:** If the connections are not symmetrical, they are called asymmetrical connections.
- **Interlayer Connections:** A connection between the nodes in different layers.
- **Intralayer Connections:** A connection between the nodes within the same layer.
- **Self Connection:** A connection pointing from the node to itself.
- **Connectivity:** It means how the nodes are connected. Full connectivity means that every node in one layer is connected to every node in its adjacent layer. Two types of connectivity fully and sparsely connectivity is shown in Figs. 2.3 and 2.4 respectively.

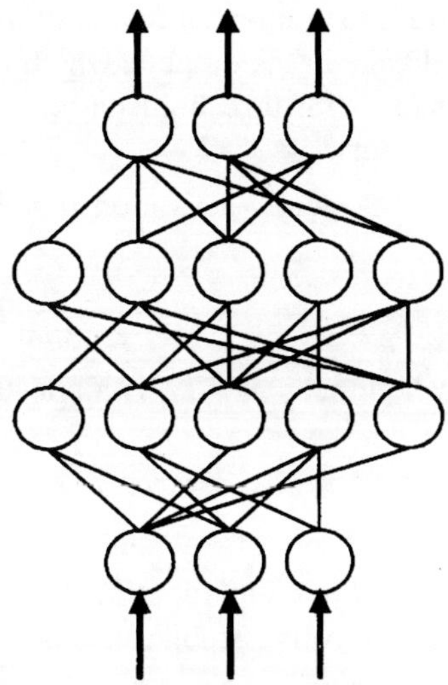

Fig. 2.4. Sparsely Connectivity

- **Threshold:** The threshold 'θ'is a factor which is used in calculating the activation of the given net. Based on the value of threshold the output may be calculated.

  For example $f(\text{net}) = \begin{cases} +1 \text{ if net } > 0 \\ -1 \text{ if net } < 0 \end{cases}$

  The threshold value is defined by the user.
- **Bias:** A bias acts exactly as a weight on a connection from a unit whose activation is always 1. Increasing the bias increases the net input to unit ($b = w_0$).

  The bias improves the performance of the ANN. Similar to initialization of weights, bias should also be initialized either to 0 or to any specified value, based on neural net. If bias is present, then the net input calculated as Net = $\sum_{i=1}^{n} X_i * W_i + b$

  Where Net = Net input

  b = Bias

  X_i = Inputs from neuron i

  W_i = Weight of neuron i to output neurons.

2.2.2. Node Properties

Activation Functions

Activation Functions for the hidden units are needed to introduce non-linearity into the network. Without non-linearity hidden units would not make networks more powerful than just plain Perceptron

(Which do not have any hidden units, just input and output units). The reason is that a linear function of linear functions is again a linear function. However, it is the nonlinearity (*i.e.,* the capability to represent non-linear functions) that makes multilayer networks so powerful. Almost any nonlinear function does the job, except for polynomials. The differential activation functions are useful for solving function which is bounded; the sigmoid functions such as logistic and tanh and the Gaussian function are the most common choices. Functions such as tanh or arctan that produce both positive and negative values tend to yield faster training than functions that produce only positive values such as logistic, because of better numerical conditioning.

For hidden units, sigmoid activation functions are difficult to train because the error function is stepwise constant, hence the gradient either does not exist or is zero, making it impossible to use back propagation or more efficient gradient-based training methods. Sigmoid units are easier to train than threshold units. With sigmoid units, a small change in the weights will usually produce a change in the outputs, which makes it possible to tell whether that change in the weights is good or bad. With threshold units, a small change in the weights will often produce no change in the outputs.

Typically activation functions used are:

$$f(\text{net}) = \frac{2}{1+\exp(-\lambda \text{net})} - 1 \qquad \ldots(i)$$

where net = $w^t X$, λ is a constant.

and
$$f(\text{net}) = \text{sgn}(\text{net}) = \begin{cases} +1, & \text{net} > 0 \\ -1, & \text{net} < 0 \end{cases} \qquad \ldots(ii)$$

Activation function (*i*) and (*ii*) are called bipolar continuous and bipolar binary functions respectively. Bipolar means both positive and negative responses of a neuron.

The unipolar continuous (*iii*) and unipolar binary activation functions (*iv*) can be obtained from the above equation.

$$f(\text{net}) = \frac{1}{1+\exp(-\lambda \text{net})} \qquad \ldots(iii)$$

and
$$f(\text{net}) = \begin{cases} 1 & \text{net} > 0 \\ 0 & \text{net} < 0 \end{cases} \qquad \ldots(iv)$$

The soft limiting activation functions (*i*) and (*iii*) are often called sigmoidal characteristics as opposed too hard-limiting activation functions given in (*ii*) and (*iv*). Hard limiting activation function describe the discrete neuron model. Other activation functions are :

1. Sigmoid Function

This function is a continuous function varies gradually between the values 0 and 1 or –1 and 1 and is given by

$$F(x) = \frac{1}{1+e^{-\alpha x}}$$

where α is the slope parameter, which adjusts the abruptness of the function as it changes between two values.

By analogy to analog electronic system, Activation function is defined as a non-linear gain for the artificial neuron. This gain is calculated by finding the ratio of the change in $F(X)$ to a small change in X. Thus gain is the slope of the wave at a specific excitations level.

Figure 2.5 describes the block labeled F that accepts the SUM output and produces the signal OUT. The SUM signal is produced by an activation function F to produce the neuron output signal, OUT. This may be a simple linear function.

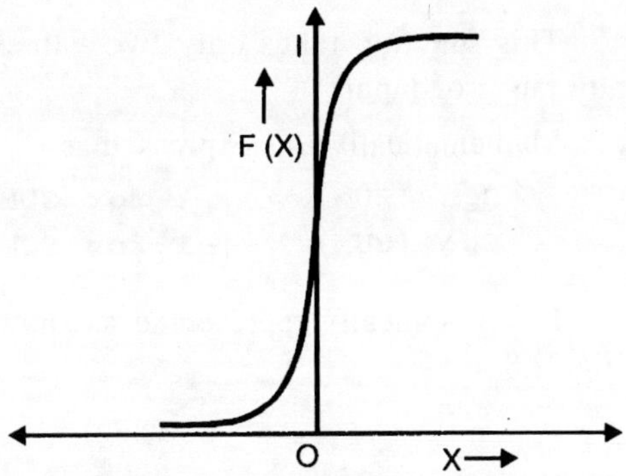

Fig. 2.5. The Sigmoid Function

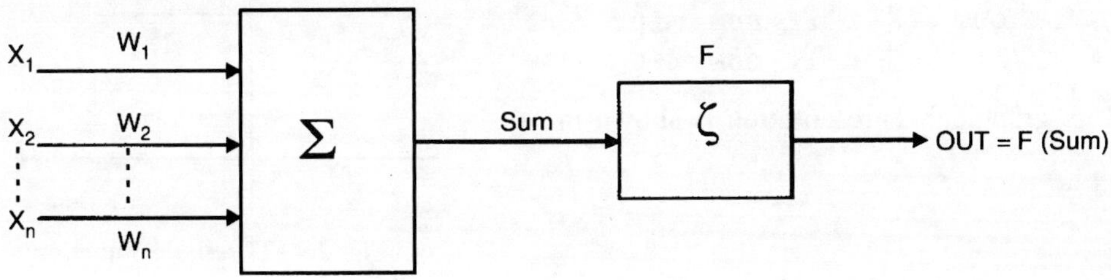

Fig. 2.6 ANN's Working Model.

```
OUT=F  (SUM)
OUT=  { 1 if SUM >T
OUT=  { 0 if otherwise.
```

where K is a constant and T is a threshold constant value.

2. Hyperbolic Tangential Functions

This function can be expressed as

```
OUT = tanh (SUM)
```

This is also S-shaped function, as shown in Fig. 2.7, but it is symmetrical about origin. This function confines the output between 1 and –1.

Or

$$y = \frac{\exp(x)-\exp(-x)}{\exp(x)+\exp(-x)}$$

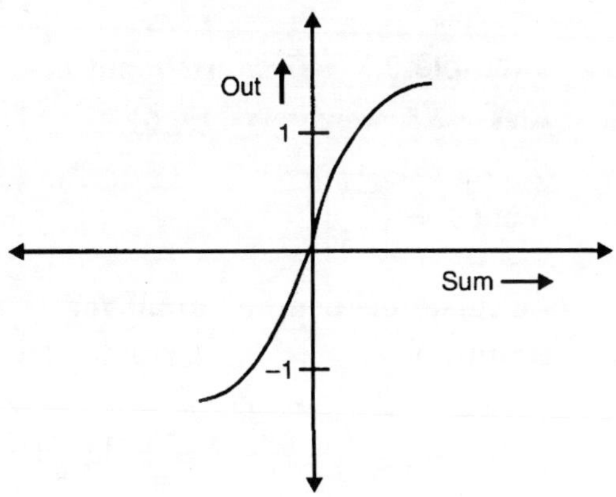

Fig. 2.7. Hyperbolic Tangential Function

3. Signum Function

Signum function can be expressed as

```
OUT= Sgn (SUM)
```

This function takes only two values for any range of inputs.

Mathematically it is expressed as

$$\text{Sgn (SUM)} = \begin{cases} 1 & \text{for SUM} > 0 \\ -1 & \text{for SUM} < 0 \end{cases}$$

It is graphically represented as shown in Fig. 2.8.

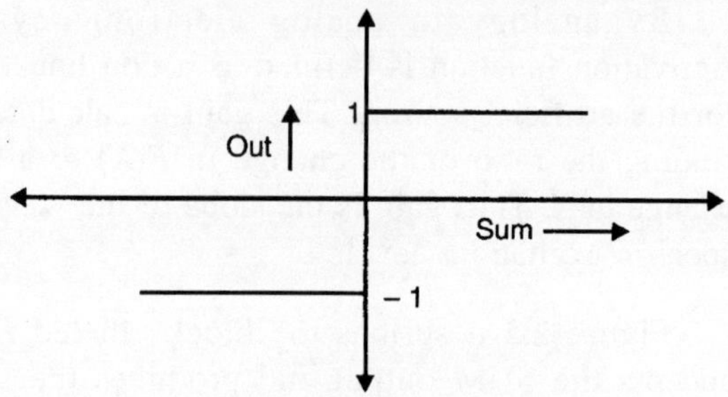

Fig. 2.8. Signum Function

4. Threshold Function

This function output is either 0 or 1 and given by the function as

$$\text{OUT} = \begin{cases} 1 & \text{if SUM} > 0 \\ 0 & \text{if SUM} < 0 \end{cases}$$

And its graphical representation is shown in Fig. 2.9.

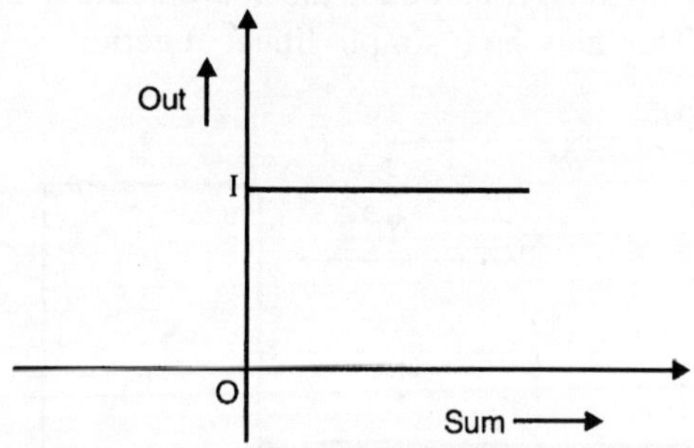

Fig. 2.9. Threshold Function

Example 2.1. The input to the single-input neuron is 3 and its weight is 2.4 and bias is –2. What is the net input to the transfer function?

Solution. Input to the transfer function is =weight* input + bias

$$\text{Therefore input} = (2.4*3) + (-2)$$
$$= 7.2 - 2 = 5.2$$

Example 2.2. Given a 2 input neuron with the following parameters:
Bias = 2.5, weight W= [3, 6]

$$\textbf{Input X} = \begin{bmatrix} -3 \\ 5 \end{bmatrix}$$

Calculate the neuron output for symmetric hard-limiting function.

Solution.

$$\text{Net} = W*X + \text{bias}$$
$$= [3 \quad 6]\begin{bmatrix} -3 \\ 5 \end{bmatrix} + 2.5$$
$$= [-9 + 30] + 2.5$$
$$= [21] + 2.5 = 23.5$$
$$f(\text{net}) = +1$$

because for hard limiting function

$$f(\text{net}) = \begin{cases} +1 & \text{if net} > 0 \\ -1 & \text{if net} < 0 \end{cases}$$

Exercise: Given a 2 input neuron with the following parameters:

Bias = 3.5, weight W= [3, 7]

Input = $\begin{bmatrix} -3 \\ 5 \end{bmatrix}$

Calculate the neuron output for all above activation functions.

2.2.3. System Dynamics

- The weight initialization scheme is specific to the particular neural network model chosen. However, in many cases, initial weights are just randomized to small real numbers.
- The learning rule is one of the most important attributes to specify for a neural network. The learning rule determines how to adapt connection weights in order to optimize the network performance. It indicates how to calculate the weight adjustment during each training cycle. However, the rule is suspended after training is completed.
- When the neural network is used to solve the problem, the solution lies in the activation levels of the output units *e.g.,* suppose a NN is implemented for classifying fruits into lemons, oranges, apples. The network has three output units representing the three kinds respectively. Given an unknown fruit, we want to classify it. So we present the characteristics of the fruit to the network. The information is received by the input layer and propagates further. If the output unit corresponding to the class apple reaches the maximal activation, then the class assigned to the fruit is apple. From this example, the inference behavior of a NN involves how to compute the activation levels across the network. The activation levels of the input units need not be calculated since they are given. Those of hidden and output units are calculated according to the activation function used.

2.3. NEURAL NETWORK ARCHITECTURES

Generally ANN structures can be represented using directed graph. We know that graph consists of set V of vertices and a set E of edges. When a direction is assigned to each edge, then the graph is called directed. The vertices of the graph may represent neurons and the edges represent the weights attached to the synaptic links.

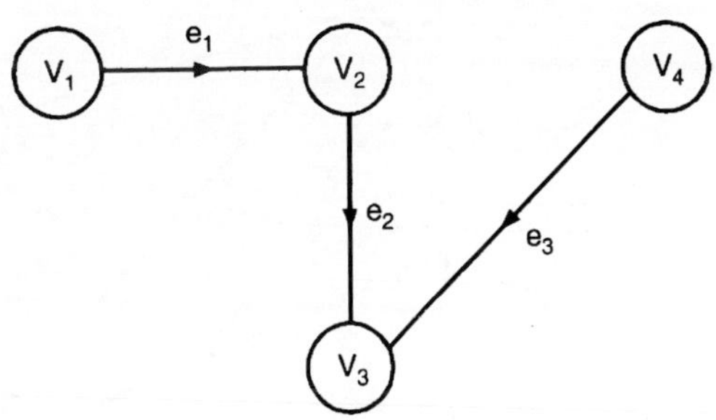

Fig. 2.10. Example of Directed Graph

V_1, V_2, V_3 are the vertices thus represent the neurons and e_1, e_2, e_3 are the edges thus represent weights.

There are 3 fundamentally classes of networks described below:

2.3.1. Single Layer Feedforward Network

Single layer network means, that there are two types of layers present :

(1) **Input layer:** It receives the input signals.

(2) **Output layer:** The neurons in this layer receive the output signals.

Despite the two layers, the network is termed single layer since it is the output layer, which performs computation. And the network here is called feedforard because the links connect the input neurons to output neurons but not vice-versa.

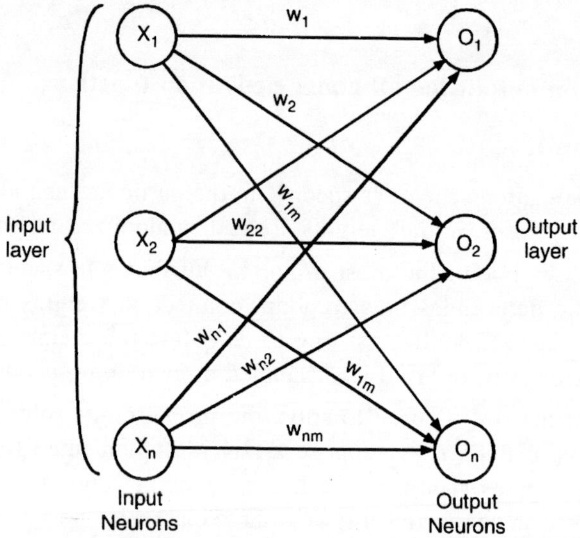

Fig. 2.11. Single Layer Feedforward Network

2.3.2. Multilayer Feedforward Network

Multilayer means there are multiple layers in the network. Alongwith input layer, output layer, one more layer is present in between them that is called hidden layer. The neurons in this layer are called hidden neurons or hidden units. The input neurons are linked to hidden enurons and the hidden neurons are further connected to output neurons.

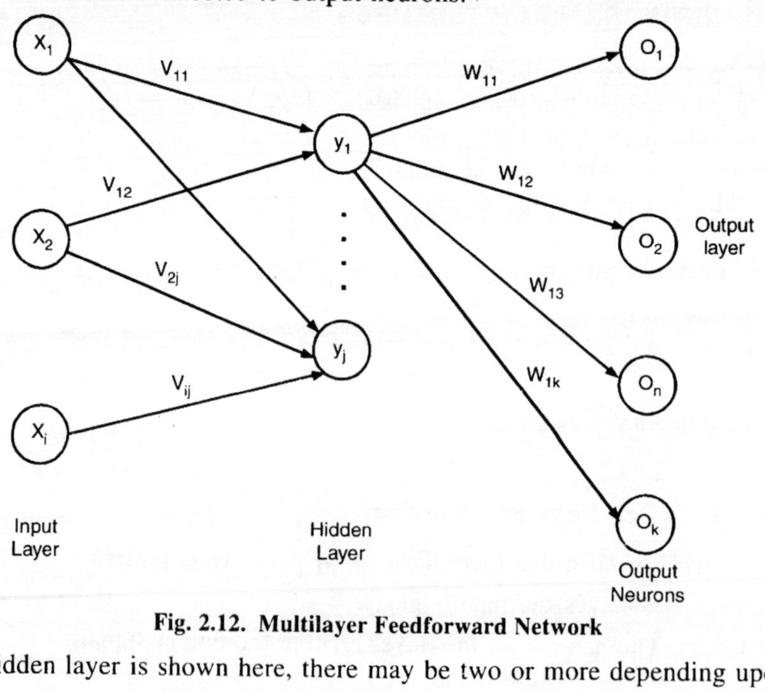

Fig. 2.12. Multilayer Feedforward Network

Only one hidden layer is shown here, there may be two or more depending upon *application.*

2.3.3. Recurrent Networks

In the above describe network architectures, there is no feedback loop. But in recurrent networks there is at least one feedback loop.

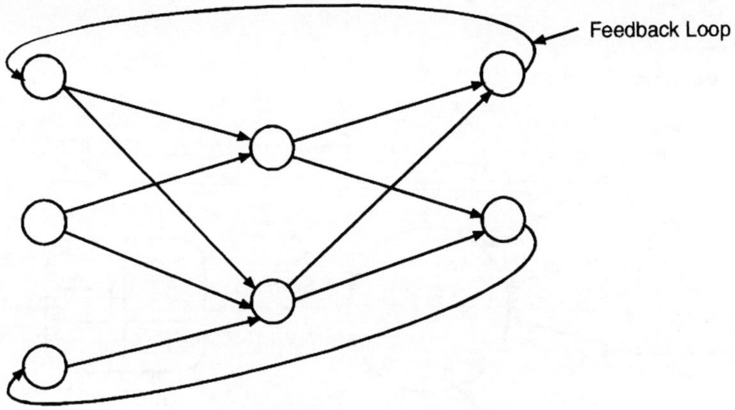

Fig. 2.13. Recurrent Networks

2.4. MODELS OF ARTIFICIAL NEURAL NETWORKS

ANN is defined as an interconnection of neurons such that the neuron outputs are connected , through the weights to all other neurons including themselves ; both lag-free and delay connections are allowed.

2.4.1. McCulloch-Pitts (MCP) Neuron Model **[MDU, May 2010; Dec. 2011]**

The early model of an artificial neuron is introduced by Warren McCulloch and Walter Pitts in 1943. The MCP neuron

(1) Has a number of inputs $I_1, ..., I_n$. These inputs represent the incoming signals received from the neuron's synapses. So I_i can either be 1, which corresponds to the presence of an incoming signal from the ith connection, or I_i is 0, which corresponds to the absence of a signal from the ith connection.

(2) The MCP neuron produces an output y, which can either be 1, corresponding to the neuron sending a signal, or it can be 0, corresponding to the neuron remaining at rest. Mathematically an MCP neuron is a function which takes an n-tuple of 1's and 0's and produces a 1 or a 0. In order to quantify the influence each synapse has on the MCP neuron, we assign a weight w_i to each input I_i.

(3) The weights are decimal numbers, and the size of the weight corresponds to the amount of influence the associated connection has on the MCP neuron. Positive weights correspond to excitatory synapses, and negative weights correspond to inhibitory synapses. Each input is multiplied by the corresponding weight, and these weighted inputs are then added together to produce

$$\text{Sum} = \sum_{i=1}^{N} I_i W_i, \quad \text{...(2.2)}$$

$$y = f(\text{sum}) \quad \text{...(2.3)}$$

where w_i*I_i means w_i multiplied by I_i. This corresponds to the summing of the incoming signals that is assumed to be performed by a neuron. Finally, there is a threshold T, which is a decimal number that is compared to the weighted sum of the signals. This corresponds to the apparent threshold value that the sum of the received signals must exceed in order to cause a neuron to "fire."

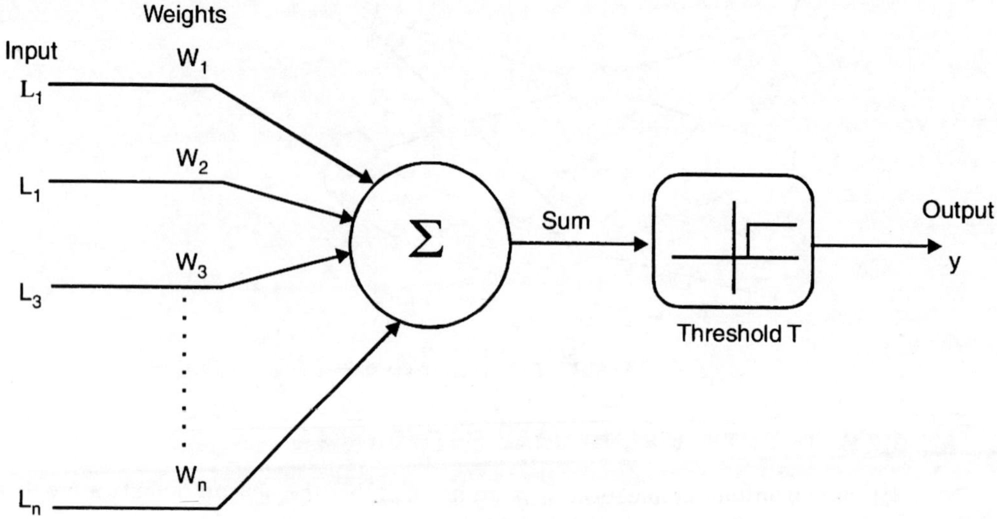

Fig. 2.14. Symbolic Illustration of Linear threshold Gate

Examples:

An MCP neuron is determined by its weights and threshold.

So when we want to describe a particular MCP neuron, we will need to give a list of weights and a threshold. For example, let's consider the MCP neuron with weights $w_1 = 0.2$, $w_2 = -0.3$, $w_3 = 0.5$, $w_4 = -0.4$, and threshold $T = -0.1$. (We could have chosen other weights and threshold.)

So the MCP neuron has four inputs, because there are four weights. Now we'll compute the output of this MCP neuron for several inputs. Recall that each of the four inputs I_1, I_2, I_3, I_4 can be either 0 or 1.

Suppose the inputs are $I_1 = 1$, $I_2 = 0$, $I_3 = 0$, $I_4 = 1$. Then

$$w_1*I_1 + w_2*I_2 + w_3*I_3 + w_4*I_4 = 0.2*1 + (-0.3)*0 + 0.5*0 + (-0.4)*1 = -0.2$$

Now we compare the weighted sum –0.2 to the threshold –0.1. Because $-0.2 < -0.1$, *i.e.*, the weighted sum is less than the threshold, the output will be 0.

Suppose the inputs are $I_1 = 1$, $I_2 = 0$, $I_3 = 1$, $I_4 = 1$. Then

$$w_1*I_1 + w_2*I_2 + w_3*I_3 + w_4*I_4 = 0.2*1 + (-0.3)*0 + 0.5*1 + (-0.4)*1 = 0.3$$

Now when we compare the weighted sum 0.3 to the threshold –0.1, we get 0.3 >= –0.1, *i.e.*, the weighted sum is greater than or equal to the threshold, so the output will be 1.

Exercise: Find the output of the above MCP neuron for each of the following lists of inputs.

1. $I_1 = 0, I_2 = 0, I_3 = 0, I_4 = 0$
2. $I_1 = 1, I_2 = 0, I_3 = 1, I_4 = 0$
3. $I_1 = 0, I_2 = 1, I_3 = 1, I_4 = 0$

Now consider an MCP neuron with $w_1 = 0.7$, $w_2 = 0.3$, and $T = 0.5$. So this MCP neuron has two inputs I_1 and I_2, one input for each weight.

Since 2 weights are given w_1 and w_2.

Consider 2 inputs only I_1 and I_2.

Because each input can only be 1 or 0, there are four possible combinations of inputs:

1. $I_1 = 1$ and $I_2 = 1$
2. $I_1 = 1$ and $I_2 = 0$
3. $I_1 = 0$ and $I_2 = 1$
4. $I_1 = 0$ and $I_2 = 0$

For each possible pair of inputs, we will compute the output of the MCP neuron determined by the chosen weights and threshold.

If $I_1 = 1$ and $I_2 = 1$, then the weighted sum is $w_1*I_1 + w_2*I_2 = 0.7*1 + 0.3*1 = 1$. Because $1 >= 0.5$, our MCP neuron will produce a 1.

If $I_1 = 1$ and $I_2 = 0$, then the weighted sum is $w_1*I_1 + w_2*I_2 = 0.7*1 + 0.3*0 = 0.7$. Because $0.7 >= 0.5$, our MCP neuron will produce a 1.

If $I_1 = 0$ and $I_2 = 1$, then the weighted sum is $w_1*I_1 + w_2*I_2 = 0.7*0 + 0.3*1 = 0.3$. Because $0.3 < 0.5$, our MCP neuron will produce a 0.

If $I_1 = 0$ and $I_2 = 0$, then the weighted sum is $w_1*I_1 + w_2*I_2 = 0.7*0 + 0.3*0 = 0$. Because $0 < 0.5$, our MCP neuron will produce a 0.

TABLE 2.1. MCP Neuron with $w_1 = 0.7$, $w_2 = 0.3$ and $T = 0.5$.

I_1	I_2	$w*I$	$w*I >= T$?	Output
1	1	1.0	Yes	1
1	0	0.7	Yes	1
0	1	0.3	No	0
0	0	0.0	No	0

Exercise: Complete the following table for the MCP neuron with weights $w_1 = 0.5$, $w_2 = 0.4$, and $T = 0.8$.

TABLE 2.2. MCP Neuron with $w_1 = 0.5$, $w_2 = 0.4$ and $T = 0.8$

I_1	I_2	$w*I$	$w*I >= T$?	Output
1	1	0.9	No	0
1	0	0.5	No	0
0	1	0.4	No	0
0	0	0.0	No	0

Implementation of McCulloch–Pitts Network for logic functions

1. AND: The truth table for AND gate is

I_1	I_2	Y (Output)
0	0	0
0	1	0
1	0	0
1	1	1

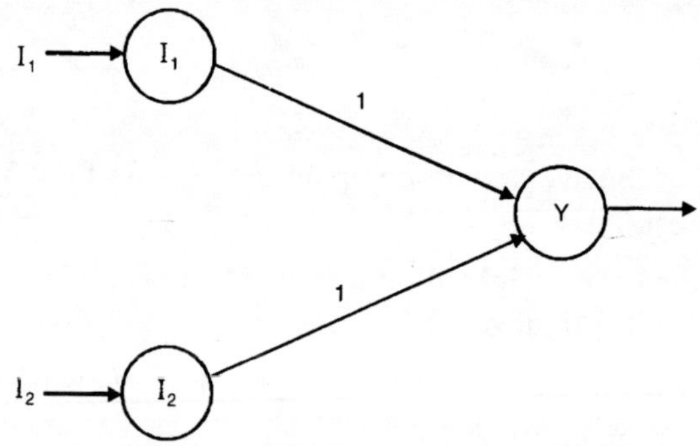

Fig. 2.15

A McCulloch–Pitts neuron to implement AND function is shown in Fig. 2.15. The threshold of the neuron is 2. The output of the neuron Y can be written as $Y = f(y_{in})$

Reason for choosing threshold = 2 is explained below:

The net input is given by $y_{in} = \Sigma$ weight * input

Since inputs are I_1, I_2 and the weights are 1 and 1 respectively, therefore

$$y_{in} = 1 * I_1 + 1 * I_2 = I_1 + I_2$$

from this the activation functions of the output neuron can be formed.

$$Y = f(y_{in}) = 1 \text{ if } y_{in} >= 2$$
$$0 \text{ if } y_{in} < 2$$

where $\qquad Y_{in} = I_1 + I_2$

Now if (*a*) if $I_1 = I_2 = 1$ and $w_1 = w_2 = 1$ then $Y_{in} = 2$ $\therefore$ if Th $\neq$ 2 then answer can't be obtained if both inputs = 1. Threshold value can be chosen according to the application.

and $Y = f(y_{in}) = 1$

(*b*) if $I_1 = 1$ and $I_2 = 0$ then $y_{in} = 1$

and $Y = f(y_{in}) = 0$. Since $y_{in} = 1 < 2$

Similarly for other combinations.

2. OR: The Truth table for OR gate is

I_1	I_2	Y (Output)
0	0	0
0	1	1
1	0	1
1	1	1

The value of OR is 1 if either $I_1 = 1$ or $I_2 = 1$, or both are 1. Otherwise OR is 0. Consider the MCP neuron with $w_1 = 0.5$, $w_2 = 0.5$, and $T = 0.5$. If we use A for the output of this neuron, then we would have the following table.

MCP Neuron with $w_1 = 0.5$, $w_2 = 0.5$, and $T = 0.5$

I_1	I_2	A
1	1	1
1	0	1
0	1	1
0	0	0

Thus this particular MCP neuron is the same as the OR function.

3. Realize NOT gate using MCP neuron: The truth table for NOT function is

I	y
1	0
0	1

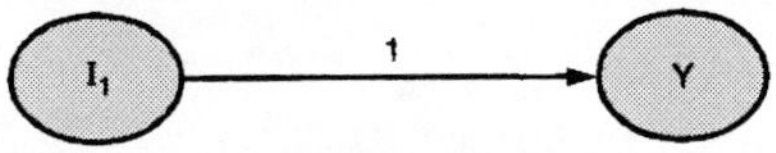

Fig. 2.16 (*a*).

The MCP neuron for this NOT gate is shown in above Fig. 2.16 (*a*). The threshold for this unit y is 1.

The net input is $y_{in} = \Sigma\text{weight} * \text{input}$

Since input is I_1 and the weight is 1, therefore

$$y_{in} = 1 * I_1 = I_1$$

from this the activation functions of the output neuron can be formed.

$$Y = f(y_{in}) = \begin{cases} 0 & \text{if } y_{in} >= 1 \\ 1 & \text{if } y_{in} < 1 \end{cases}$$

Now if (*a*) $I_1 = 1$ and $w_1 = 1$ then $y_{in} = 1$

and $Y = f(y_{in}) = 0$ Since $y_{in} = 1$

(*b*) $I_1 = 0$ and $w_1 = 1$ then $y_{in} = 0$

and $Y = f(y_{in}) = 1$, Since $y_{in} = 0 < 1$

Therefore using the McCulloch-Pitts model logic functions can be modeled. Below the architecture for four logic functions (the truth tables for each function is also shown)

AND

X_1	X_2	Y
1	1	1
1	0	0
0	1	0
0	0	0

AND NOT

X_1	X_2	Y
1	1	0
1	0	1
0	1	0
0	0	0

OR

X_1	X_2	Y
1	1	1
1	0	1
0	1	1
0	0	0

XOR

X_1	X_2	Y
1	1	0
1	0	1
0	1	1
0	0	0

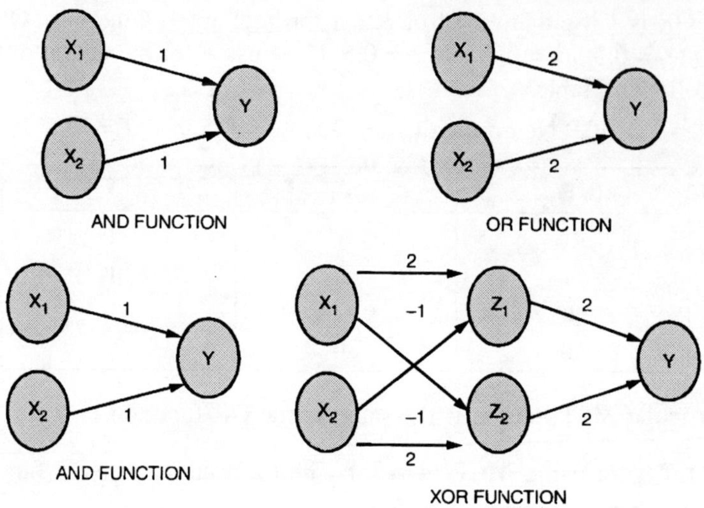

Fig. 2.16(*b*)

AND NOT Function

The truth table for the AND NOT function is shown above.

The function is not symmetric in that an input of 1,0 is treated differently to an input of 0,1. It can be seen from the truth table the that true (value of one) is returned when the first input is true and the second input is false. Again, the threshold on Y is set to 2 and if each of the inputs is applied to the AND NOT network we will find that X1 AND NOT X2 has been modeled.

XOR Function

XOR can be modeled using AND NOT and OR;

```
X1 XOR X2 = (X1 AND NOT X2) OR (X2 AND NOT X1)
```

This explains the network shown above. The first layer performs the two AND NOT's and the second layer performs the OR. Both Z neurons and the Y neuron have a threshold of 2.

Exercise. 1. Construct a table as above for the MCP neuron with $w1 = 0.5$, $w2 = 0.5$, and $T = 0.2$.

2. Construct a table as above for the MCP neuron with $w1 = 0.5$, $w2 = 0.3$, and $T = 0.5$.

2.4.2. Feedforward Network

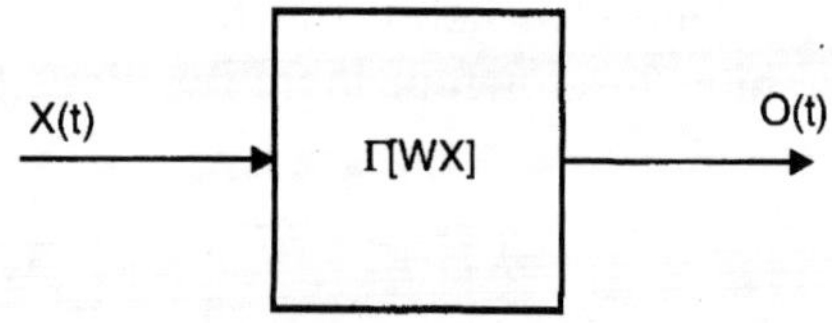

Fig. 2.16 (*c*). Block Diagram

Feedforward neural network is a biologically inspired classification algorithm. It consists of a (possibly large) number of simple neuron-like processing *units*, organized in *layers*. Every unit in a layer is connected with all the units in the previous layer. These connections are not all equal; each connection may have a different strength or *weight*. The weights on these connections encode the knowledge of a network. Often the units in a neural network are also called *nodes*. Data enters at the inputs and passes through the network, layer by layer, until it arrives at the outputs. During normal operation, that is when it acts as a classifier, there is no feedback between layers. This is why they are called *feedforward* neural networks.

Let us consider the architecture of feedforward architecture of m neurons receiving n inputs as shown in Fig. 2.12(*c*). Its inputs and output vectors are respectively

$$O = [O_1 \quad O_2 \ldots O_m]^t$$

$$X = [x_1 \; x_2 \; \ldots x_n]^t \qquad \ldots(2.4)$$

Weight w_{ij} connects the ith neuron with the jth output. Therefore the activation value for the ith neuron is

$$net_i = \sum_{j=1}^{n} w_{ij} x_j \quad \text{for } i = 1, 2, \ldots, n \qquad \ldots(2.5)$$

$$\text{Output is } O_i = f(w_i^t x) \quad \text{for } i = 1, 2, \ldots, m \qquad \ldots(2.6)$$

where weight vector w_i contains the weighting leading towards the ith output node and is defined as

$$w_i = [w_{i1}, w_{i2}, \ldots, w_{in}]^t \qquad \ldots(2.7)$$

Introducing the nonlinear matrix operator Γ the mapping of input space x to output space O implemented by

$$O = \Gamma[wx] \qquad \ldots(2.8)$$

where w is called weight matrix, also called connection matrix.

Since there is no time delay between the input x and the output o, therefore we can write

$$O(t) = \Gamma[wx(t)] \qquad \ldots(2.9)$$

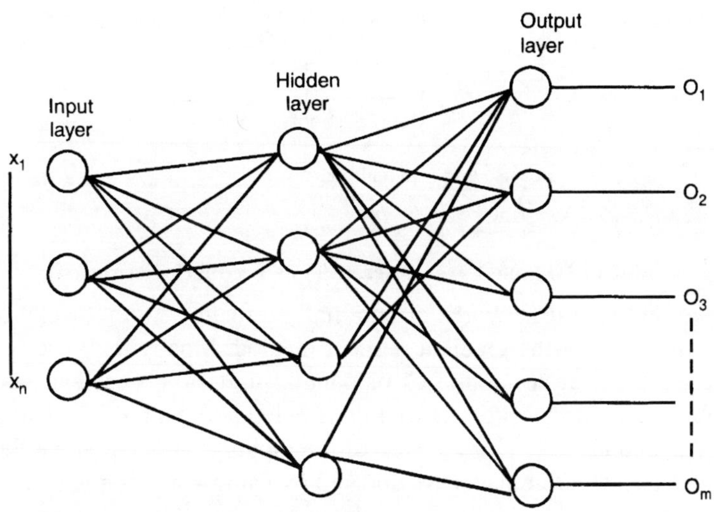

Fig. 2.17. Feedforward Network

TIPS:

- **Information flow unidirectional**
- **Static mapping: y = $f(x)$**
- **Multi-Layer Perceptron (MLP)**
- **Radial Basis Function (RBF)**
- **Kohonen Self-Organising Map (SOM)**

2.4.3. Feedback Network/Recurrent Network [MDU, May 2006]

Recurrent neural networks (RNN) have a closed loop in the network topology. They are developed to deal with the time varying or time-lagged patterns and are usable for the problems where the dynamics of the considered process is complex and the measured data is noisy. Specific groups of the units get the feedback signals from the previous time steps and these units are called *context* unit (Schalkoff, 1997). The RNN can be either fully or partially connected. In a fully connected RNN all the hidden units are connected recurrently, whereas in a partially connected RNN the recurrent connections are omitted partially. Examples of recurrent neural networks are Hopfield networks, Regressive networks, Jordan-Elman networks, and Brain-State-In-A-Box (BSB) networks.

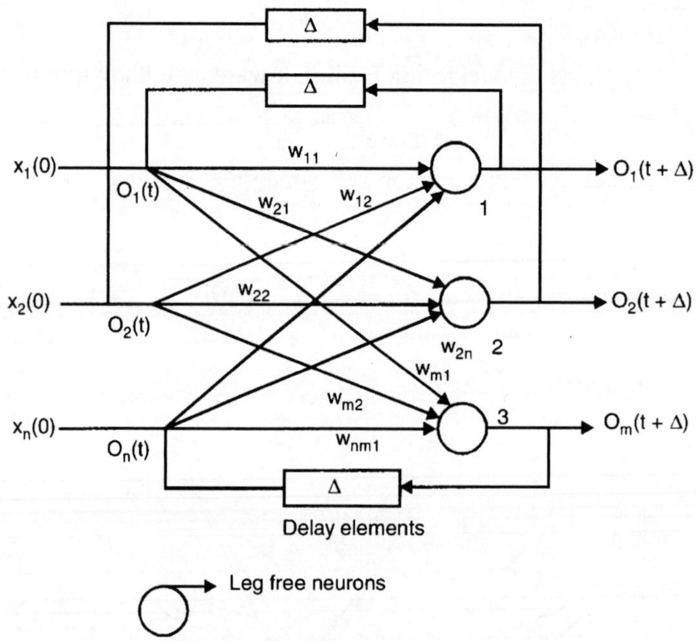

Fig. 2.18. Single Layer Discrete Time Feedback Networks

All types of recurrent neural networks are normally trained with the backpropagation learning rule by minimizing the error by the gradient descent method. Mostly they use some computational units which are called associative memories or context units that can learn associations among dissimilar binary objects, where a set of binary inputs is fed to a matrix of resistors, producing a set of binary outputs. The outputs are '1' if the sum of the inputs is above a given threshold, otherwise it is zero. The weights (which are binary) are updated by using very simple rules based on *Hebbian* learning. These are very simple devices with one layer of linear units that maps *N* inputs (a point in *N* dimensional space) onto *M* outputs (a point in *M* dimensional space). However, they remember the past events.

The time Δ elapsed between t and $t + \Delta$ is introduced by the delay elements in the feedback loop. Using the notation introduced for feedforward networks, the mapping of $o(t)$ into $o(t + \Delta)$ can be written as

$$o(t + \Delta) = \Gamma[Wo(t)] \quad ...(2.10)$$

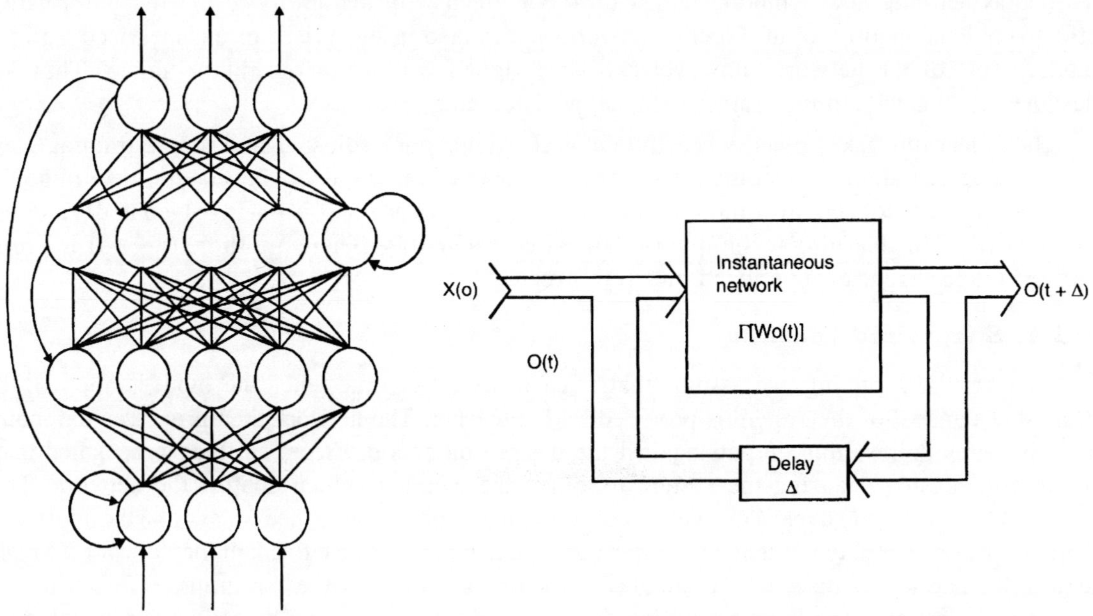

(*a*) Recurrent network (*b*) Block diagram

Fig. 2.19

Note that the input $x(t)$ is only needed to initialize this network so that $O(0) = x(0)$. The input is then removed and the system remains autonomous for $t > 0$.

Recurrent networks typically operate with discrete representation of data; they employ neurons with a hard-limiting activation function. A system with discrete time inputs and a discrete data representation is called an **automaton.**

TIPS:

- ➡ Feedback connections
- ➡ Dynamic memory: $y(t + 1) = f(x(\tau), y(\tau), s(\tau)) \quad \tau \in (t, t - 1, \ldots)$
- ➡ Jordan/Elman ANNs
- ➡ Hopfield
- ➡ Adaptive Resonance Theory (ART)

2.5. LEARNING PARADIGMS [MDU, May 2009; Dec. 2006]

Learning denotes the change in the system that are adaptive in the sense that they enable the system to do the same task or tasks drawn from the same population more efficiently and more effectively the next time.

Learning can refer to either acquiring new knowledge or enhancing or refining skills. Learning new knowledge includes acquisition of significant concepts, understanding of their meanings and relationships to each other and to the domain concerned. The new knowledge should be assimilated and put in a mentally usable form before it can be called "Learned". Thus knowledge acquisition is defined as learning new symbolic information combined with the ability to use that information effectively. The majority of the Neural Networks are trained in supervised or unsupervised learning mode. Some of the networks however can be designed without incremental training. They are designed by "batch learning "rather than stepwise learning.

Batch learning takes place when the network weights are adjusted in a single training step. In this mode of training, the complete set of input, output training data is needed to determine the weights and the feedback information produced by the network itself is not involved in developing the network. This learning technique is also called **recording**. There are three types of learning defined: supervised, unsupervised and reinforced.

2.5.1. Supervised Learning

In supervised training, we assume that at each instant of time when the input is applied, the desired response ***d*** of the system is provided by the teacher. The network then processes the inputs and compares its resulting outputs against the desired outputs ***d***. Errors are then propagated back through the system, causing the system to adjust the weights, which control the network. This process occurs over and over as the weights are continually decreases. Since we assume the adjustable weights, the teacher may implement a reward-and-punishment scheme to adapt the network's weight matrix ***W***. The set of data, which enables the training, is called the "training set". During the training of a network the same set of data is processed many times as the connection weights are ever refined.

Example: Like in the class, a teacher is always present to ask the questions from the students. If the students give the correct answer, then the teacher may give the reward to the students, otherwise if the student is not able to give the right answer, then teacher will again explain that topic to the students until they will well understand the problem.

The current commercial network development packages provide tools to monitor how well an artificial neural network is converging on the ability to predict the right answer. These tools allow the training process to go on for days, stopping only when the system reaches some statistically desired point, or accuracy. However, some networks never learn. This could be because the input data does not contain the specific information from which the desired output is derived. Networks also don't converge if there is not enough data to enable complete learning. Ideally, there should be enough data so that part of the data can be held back as a test. Many layered networks with multiple nodes are capable of memorizing its data in some no significant way, supervised training needs to hold back a set of data to be used to test the system after it has undergone its training (memorization is avoided by not having too many processing elements).

If a network simply can't solve the problem, the designer then has to review the input and outputs, the number of layers, the number of elements per layer, the connections between the layers, summation, transfer, and training functions, and even the initial weights themselves. Those changes required to create a successful network constitutes a process wherein the "art: of neural networking occurs. Another part of designer's creativity governs the rules of training. There are many laws (algorithms) used to implement the adaptive feedback required to adjust the weights during training.

Yet, training is not just a technique. It involves a "feel", and conscious analysis, to insure that the network is not over trained. Initially, an artificially neural network configures itself with the general statistical trends of the data. Later, it continues to "learn" about other aspects of the data, which may be spurious from a general viewpoint.

When finally the system has been correctly trained, and no further learning is needed, the weights can, if desired, be "frozen". In some systems this finalized network is then trained into hardware so that it can be fast. Other systems don't lock themselves in but continue to learn while in production use.

2.5.2. Unsupervised Learning

The other type of training is called unsupervised training. In unsupervised learning, the network is provided with inputs but not with desired outputs. The system itself must then decide what features it will use to group the input data. This is often referred to as self-organization or adaptation.

At the present time, unsupervised learning is not well understood. This adaptation to the environment is a promise, which would enable science fiction types of robots to continually learn on their own as they encounter new situations and new environments. Life is filled with situations where exact training sets do not exist. Some of these situations involve military action where new combat techniques and new weapons might be encountered. Because of this unexpected aspect to life and the human desire to be prepared, there continues to be research into, and hope for, this field. Yet, at present time, the vast bulk of neural network work is in systems with supervised learning. Supervised learning is achieving results.

Example: Learning without supervision corresponds to learning the subject from a videotape lecture covering the material but not including any teacher's involvement. Therefore, in this the students can't get the explanations of the unclear questions.

One of the leading researchers into unsupervised learning is Tuevo Kohonen, an electrical engineer at the Helsinki University of Technology. He has developed a self-organizing network, sometimes called an auto-associator that learns without the benefit of knowing the right answer. It is an unusual looking network in that it contains one single layer with many connections. The weights for those connections have to be initialized and the inputs have to be normalized. The neurons are set up to compete in a winner-take–all fashion.

2.5.3. Reinforcement Learning

This method of learning is mixture of the previous two types. The network is presented with the input, but only told that the answer is wrong, or right. It is expected to organize in some manner to ensure a correct answer is more likely on the next presentation of the input.

Unfortunately, there are many situations where we don't know the correct answers that supervised learning requires. For example, in a flight control system, the question would be the set of all sensor readings at a given time, and the answer would be how the flight control surfaces should move during the next millisecond. Simple neural networks can't learn to fly the plane unless there is a set of known answers, so if we don't know how to build a controller in the first place, simple supervised learning won't help.

For these reasons there has been much interest recently in a different approach known as *reinforcement learning* (*RL*). Reinforcement learning is not a type of neural network, nor is it an alternative to neural networks. Rather, it is an orthogonal approach that addresses a different, more difficult question. Reinforcement learning combines the fields of dynamic programming and supervised learning to yield powerful machine-learning systems. Reinforcement learning appeals to many researchers because of its generality. In *RL*, the computer is simply given a goal to achieve. The computer then learns how to achieve that goal by trial-and-error interactions with its environment.

2.5.4. Learning Algorithms

A neuron is considered to be an adaptive element. Its weights are modifiable depending upon the input signal it receives, its output value and associated teacher response. In some cases the teacher signal is not available and no more error information can be used, thus the neuron will modify its weights based only on input/output. This is the case for unsupervised learning.

General Learning rule: The weight vector $w_i = [w_{i1} \quad w_{i2} \ldots\ldots w_{in}]^t$ increase in proportion to the product of input x and learning signal r. The learning signal r is a general function of w_i, x and teacher's signal $\boldsymbol{d}$ if learning is supervised.

Therefore $$r = r\,(w_i, x, d_i) \qquad \ldots(2.11)$$

The increment of weight vector w_i produced by the learning step at time according to the general learning rule is

$$W_i(t) = cr[W_i(t), x(t), d_i(t)]\; x(t) \qquad \ldots(2.12)$$

c = learning constant (any positive number),

r = learning signal

Determines the rate of learning. The weight adapted at time t becomes at the next instant, or learning step,

$$W_i(t+1) = W_i\,(t) + cr[W_i(t), x(t), d_i(t)]\; x(t) \qquad \ldots(2.13)$$

2.5.4.1. Hebbian Learning Rule [MDU, May 2009; Dec. 2006; Dec. 2009; May 2012]

In 1949, Donald Hebb wrote: "When the axon of cell A is near enough to excite a cell B and repeatedly or persistently takes part in firing it, some growth process or metabolic change takes place in one or both cells such that A's efficiency, as one of the cells firing B, is increased". As a result, a learning method evolved for artificial neural networks, called *Hebbian learning*, which is an instance of an unsupervised learning procedure.

In Hebbian learning, weights between learning nodes are adjusted so that each weight better represents the relationship between the nodes. Nodes which tend to be positive or negative at the same time will have strong positive weights while those which tend to be opposite will have strong negative weights. Nodes that are uncorrelated will have weights near zero. For example, if two nodes A and B are often simultaneously active, Hebbian learning will increase the connection strength between the two so that excitation of either one tends to cause excitation of the other. On the other hand, if nodes A and C were of opposite activations at all times, then Hebbian learning would gradually decrease the connection in between below zero so that an excited A or C would inhibit the other.

For this rule the learning signal is equal simple to the neuron's output.

We have $$r\sim\; = f\,(W_i^t\, x) \qquad \ldots(2.14)$$

The increment ΔW_i of the weight vector becomes

$$\Delta W_i = c\,f\,(W_i^t\, x)\; x \qquad \ldots(2.15)$$

The single weight adjustment using the following increment:

$$\Delta W_{ij} = c\,f\,(W_i^t\, x_i)\; x_j \qquad \ldots(2.16)$$

This can be written as

$$\Delta Wi_j = c\; o_i\, x_j \qquad \text{for } j = 1, 2, 3 \ldots n \qquad \ldots(2.17)$$

Hebbian learning has four features:

(1) First it is unsupervised;

(2) Second it is a local learning rule, meaning that it can be applied to a network in parallel;

(3) Third it is simple and therefore requires very little computation;

(4) Fourth it is biologically plausible.

Example 2.3. This example illustrated Hebbian learning with binary and continuous activation functions of a very simple network.

$$w^1 = \begin{bmatrix} 1 \\ -1 \\ 0 \\ 0.5 \end{bmatrix}$$

Needs to be trained using set of three input vectors as below:

$$x_1 = \begin{bmatrix} 1 \\ -1 \\ 1.5 \\ 0 \end{bmatrix}$$

and $$x_2 = \begin{bmatrix} 1 \\ -0.5 \\ -1 \\ 0 \end{bmatrix}$$

and $$x_3 = \begin{bmatrix} 0 \\ 1 \\ -1 \\ 1.5 \end{bmatrix}$$

Solution.

Step 1: Input x_1 is applied to the networks results in activation net^1

$$net^1 = w^{1t} x_1 = \begin{bmatrix} 1 & -1 & 0 & 0.5 \end{bmatrix} \begin{bmatrix} 1 \\ -1 \\ 1.5 \\ 0 \end{bmatrix} = 2$$

The updated weights are

$$w^2 = w^1 + \text{sgn}(\text{net}^1)\, x_1 = w^1 + (1)\, x_1$$

[sgn(2) = 1 refer activation function]

$$w^2 = \begin{bmatrix} 1 \\ -1 \\ 0 \\ 0.5 \end{bmatrix} + \begin{bmatrix} 1 \\ -1 \\ 1.5 \\ 0 \end{bmatrix} = \begin{bmatrix} 2 \\ -2 \\ 1.5 \\ 0.5 \end{bmatrix}$$

Step 2: Now take this w^2 with 2nd input x_2

$$\text{net}^2 = w^{2t} x_2 = \begin{bmatrix} 2 & -2 & 1.5 & 0.5 \end{bmatrix} \begin{bmatrix} 1 \\ -0.5 \\ -1 \\ 0 \end{bmatrix}$$

$$= [2 + 1 - 1.5 + 0] = 1.5$$

The updated weights are

$$w^3 = w^2 + \text{sgn}\,(net^2)x_2 = w^2 + (1)x_2$$

$$w^3 = \begin{bmatrix} 2 \\ -2 \\ 1.5 \\ 0.5 \end{bmatrix} + \begin{bmatrix} 1 \\ -0.5 \\ -1 \\ 0 \end{bmatrix} = \begin{bmatrix} 3 \\ -2.5 \\ 0.5 \\ 0.5 \end{bmatrix}$$

Step 3: Now take w^3 with 3rd input x_3

$$\text{net}^3 = w^{3t} x_3 = \begin{bmatrix} 3 & -2.5 & 0.5 & 0.5 \end{bmatrix} \begin{bmatrix} 0 \\ 1 \\ -1 \\ 1.5 \end{bmatrix}$$

$$= \begin{bmatrix} 0 - 2.5 - 0.5 + 0.75 \end{bmatrix} = 2.25$$

The updated weights are

$$w^4 = w^3 + \text{sgn}\,(\text{net}^3)\,x_3 = w^2 + (-1)\,x_3$$

$$w^4 = \begin{bmatrix} 3 \\ -2.5 \\ 0.5 \\ 0.5 \end{bmatrix} - \begin{bmatrix} 0 \\ 1 \\ -1 \\ 1.5 \end{bmatrix} = \begin{bmatrix} 3 \\ -3.5 \\ +1.5 \\ -1 \end{bmatrix}$$

It can be seen that learning with discrete f (net) and $c = 1$ results in adding or subtracting the entire input pattern vectors to and from the weight vector, respectively. In the case of a continuous f (net), the weight incrementing and decrementing vector is scaled down to a fractional value of the input pattern.

Now the Hebbian learning example, with continuous bipolar activation function $f(\text{net})$, using input x_1, and initial weights w^1 we obtain neuron output values and the updated weights for $\lambda = 1$ as summarized in Step 1. The only difference compared with the previous case is that instead of $f(\text{net}) = \text{sgn}(\text{net})$, now the neuron's response is computed with the help of

$$f(\text{net}) = \frac{2}{1+\exp(-\lambda\,\text{net})} - 1.$$

$$f(w^t x) = \text{sgn}(w^t x), \quad z = w^t x$$

Exercise: Let $w_0 = \begin{bmatrix} 1.0 \\ -1.0 \\ 0.0 \\ 0.5 \end{bmatrix}$ and $x_0 = \begin{bmatrix} 1.0 \\ -0.5 \\ -2.0 \\ -1.5 \end{bmatrix}, x_1 = \begin{bmatrix} 1.0 \\ 1.1 \\ -1.0 \\ 1.5 \end{bmatrix}, x_2 = \begin{bmatrix} 1.0 \\ 1.0 \\ -1.0 \\ 1.5 \end{bmatrix}$

Hebbian learning with a threshold function results in adding or subtracting the input pattern from the weight vector.

Continuous Neuron with Hebbian Learning

$$f(w^T x) = \frac{2}{1+e^{-z}} - 1, \; z = w^T x$$

Let $w_0 = \begin{bmatrix} 1.0 \\ -1.0 \\ 0.0 \\ 0.5 \end{bmatrix}$ and

$$x_0 = \begin{bmatrix} 1.0 \\ -2.0 \\ 1.5 \\ 0.0 \end{bmatrix}, x_1 = \begin{bmatrix} 1.0 \\ -0.5 \\ -2.0 \\ -1.5 \end{bmatrix}, x_2 = \begin{bmatrix} 0.0 \\ 1.0 \\ -1.0 \\ 1.5 \end{bmatrix}$$

$$z_0 = w_0^T x_0 = [1.0 \;\; -1.0 \;\; 0.0 \;\; 0.5] \begin{bmatrix} 1.0 \\ -2.0 \\ 1.5 \\ 0.0 \end{bmatrix} = 3$$

The update is:

$$w_{k+1} = w_k + f(z_k)x_k, \; y = f(z) = \frac{2}{1+e^{-z}} - 1$$

as $\mu = 1$, thus giving:

$$w_1 = \begin{bmatrix} 1.0 \\ -1.0 \\ 0.0 \\ 0.5 \end{bmatrix} + 0.905 \begin{bmatrix} 1.0 \\ -2.0 \\ 1.5 \\ 0.0 \end{bmatrix} = \begin{bmatrix} 1.905 \\ -2.810 \\ 1.357 \\ 0.500 \end{bmatrix}$$

$$y_1 = -0.077, \; w_2 = w_1 - 0.077x_1 = \begin{bmatrix} 1.828 \\ -2.772 \\ 1.512 \\ 0.616 \end{bmatrix}$$

$$y_2 = -0.932, \; w_3 = w_2 - 0.932x_2 = \begin{bmatrix} 1.828 \\ -3.700 \\ 2.440 \\ -0.783 \end{bmatrix}$$

Hebbian learning with a continuous function results in adding or subtracting a fraction (determined by *beta*) of the input pattern from the weight vector.

Exercise: Using Hebbian learning rule find out the final weight adjustment.

$$w_1 = \begin{bmatrix} 1 \\ 0 \\ -1 \\ 1.5 \end{bmatrix}$$

and 3 input vectors are

$$x_1 = \begin{bmatrix} -1 \\ 1 \\ 0 \\ 1.5 \end{bmatrix}, \quad x_2 = \begin{bmatrix} -1 \\ -0.5 \\ 1 \\ 0 \end{bmatrix}, \quad x_3 = \begin{bmatrix} 1 \\ 1 \\ 0 \\ 1.5 \end{bmatrix}$$

2.5.4.2. Perceptron Learning Rule **[MDU, May 2008; Dec. 2008]**

1. Perceptrons are trained on examples of desired behavior.
2. This rule is for supervised learning mode.
3. The desired behavior can be summarized by a set of input, output pairs where **p** is an input to the network and **d** is the corresponding correct (target) output. The objective is to reduce the error **e**, which is the difference between the neuron responses **a** and the target vector **d**. The perceptron learning rule **learnp** calculates desired changes to the perceptron's weights and biases given an input vector **p**, and the associated error **e**. The target vector t must contain values of either 0 or 1, as perceptrons (With hardlimiting transfer functions) can only output such values.

Each time learnp is executed, the perceptron has a better chance of producing the correct outputs. The perceptron rule is proven to converge on a solution in a finite number of iterations if a solution exists. If a bias is not used, **learnp** works to find a solution by altering only the weight vector w to point toward input vectors to be classified as 1, and away from vectors to be classified as 0. This results in a decision boundary that is perpendicular to w, and which properly classifies the input vectors. There are three conditions that can occur for a single neuron once an input vector *p* is presented and the network's response *a* is calculated:

CASE 1. If an input vector is presented and the output of the neuron is correct ($a = d$, and $e = d - a = 0$), then the weight vector w is not altered.

CASE 2. If the neuron output is 0 and should have been 1 ($a = 0$ and $d = 1$, and $e = d - a = 1$), the input vector p is added to the weight vector w. This makes the weight vector point closer to the input vector, increasing the chance that the input vector will be classified as a 1 in the future.

CASE 3. If the neuron output is 1 and should have been 0 ($a = 1$ and $d = 0$, and $e = d - a = -1$), the input vector p is subtracted from the weight vector w. This makes the weight vector point farther away from the input vector, increasing the chance that the input vector is classified as a 0 in the future. The perceptron learning rule can be written more concisely in terms of the error $e = d - a$, and the change to be made to the weight vector w:

Case 1: If $e = 0$, then make a change w equal to 0.

Case 2: If $e = 1$, then make a change w equal to pD.

Case 3: If $e = -1$, then make a change w equal to $-pD$.

All three cases can then be written with a single expression:

$$w = (d - a)\,p^D = ep^D \qquad \text{...(2.18)}$$

You can get the expression for changes in a neuron's bias by noting that the bias is simply a weight that always has an input of 1:

$$b = (d - a)\,(1) = e \qquad \text{...(2.19)}$$

For the case of a layer of neurons we have

$$w = (d - a)\,(p)^D = e(p)^D \qquad \text{...(2.20)}$$

$$b = (d - a) = e \qquad \text{...(2.21)}$$

The perceptron learning rule can be summarized as follows:

$$W^{\text{new}} = W^{\text{old}} + e(p)^D \qquad \text{...(2.22)}$$

$$b^{\text{new}} = b^{\text{old}} + e \qquad \text{...(2.23)}$$

$$e = (d - a) \qquad \text{...(2.24)}$$

Example 2.4. Let $w_0 = \begin{bmatrix} \mathbf{1.0} \\ \mathbf{-1.0} \\ \mathbf{0.0} \\ \mathbf{0.5} \end{bmatrix}$, $d_0 = -1.0,\ d_1 = -1,\ d_2 = 1$

d_0, d_1, d_2 are teacher's signal.

and $x_0 = \begin{bmatrix} 1.0 \\ -2.0 \\ 0.0 \\ -1.0 \end{bmatrix}, x_1 = \begin{bmatrix} 0.0 \\ 1.5 \\ -0.5 \\ -1.0 \end{bmatrix}, x_2 = \begin{bmatrix} -1.0 \\ 1.0 \\ 0.5 \\ -1.0 \end{bmatrix}$

$$z_0 = w_0^T x_0 = [1.0 \;\; -1.0 \;\; 0.0 \;\; 0.5] \begin{bmatrix} 1.0 \\ -2.0 \\ 0.0 \\ -1.0 \end{bmatrix} = +2.5$$

Note that sgn (2.5) ≠ d_0. (Note $c = 0.1$). The weight update is evaluted as:

$$= d - a$$
$$= d_0 - f(w_0^t x_0)$$
$$= -1 - f(2.5)$$
$$= -1 - 1$$ [because in perceptron learning rule signum function is used and sgn (2.5) = +1]

$$w_1 = w_0 + 0.1(-1-1)x_0$$

$$w_1 = \begin{bmatrix} 1.0 \\ -1.0 \\ 0.0 \\ 0.5 \end{bmatrix} - 0.2 \begin{bmatrix} 1.0 \\ -2.0 \\ 0.0 \\ -1.0 \end{bmatrix} = \begin{bmatrix} 0.8 \\ -0.6 \\ 0.0 \\ 0.7 \end{bmatrix}$$

$$z_1 = w_1^t x_1 = [0.0 \quad 1.5 \quad -0.5 \quad -1.0] \begin{bmatrix} 0.8 \\ -0.6 \\ 0.0 \\ 0.7 \end{bmatrix} = -1.6$$

Since sgn (–1.6) = d_1 no weight update is required. Now

$$w_2 = w_1 + 0.1\ (-1 + 1).x_1$$

$$w_2 = w = \begin{bmatrix} 0.8 \\ -0.6 \\ 0 \\ 0.7 \end{bmatrix}$$

$$x_2 = w_2^T x_2 = [0.8 \quad -0.6 \quad 0 \quad 0.7] \begin{bmatrix} -1.0 \\ 1.0 \\ 0.5 \\ -1.0 \end{bmatrix} = -2.1$$

Note that sgn (–2.1) ≠ d_2. The weight update is evaluated as:

$$w_3 = w_2 + 0.1(1+1)x_2$$

$$w_3 = \begin{bmatrix} 0.8 \\ -0.6 \\ 0.0 \\ 0.7 \end{bmatrix} + 0.2 \begin{bmatrix} -1.0 \\ 1.0 \\ 0.5 \\ -1.0 \end{bmatrix} = \begin{bmatrix} 0.6 \\ -0.4 \\ 0.1 \\ 0.5 \end{bmatrix}$$

Example 2.5. Develop a perceptron network to implement an AND function.

2.5.4.3. Delta Learning Rule [MDU, Dec. 2008; May 2009; Dec. 2009; May 2010; May 2012]

1. Delta learning rule is only valid for continuous activation functions as defined earlier.
2. This rule is used in the supervised training mode.
3. The learning signal for this rule is called delta and is defined as follows:

$$r \cong \left[d_i - f(w_i^t x)\right] f'(w_i^t x) \qquad \text{...(2.26)}$$

The term $f'(w_i^t x)$ is the derivative of the activation function f(net) computed for net = W_{ix}. The explanation of the delta learning rule is shown in diagram. This learning rule can be readily derived from the condition of least squared error between o_i and d_i. Calculating the gradient vector with respect to W_i of the squared error defined as

$$E \cong \frac{1}{2}(d_i - o_i)^2 \qquad \text{...(2.27)}$$

which is equivalent to

$$E \cong \frac{1}{2}\left[d_i - f(w_i^t x)\right]^2 \qquad \text{...(2.28)}$$

Error gradient vector value

$$\nabla E = -[d_i - f(w_i^t x)]\, f'(w_i^t x)\, x \qquad \text{...(2.29)}$$

The components of the gradient vector are

$$\frac{\partial E}{\partial w_{ij}} = -\left[d_i - f(w_i^t x)\right] f'(w_i^t x)\, x_j \qquad \text{...(2.30)}$$

for $j = 1, 2, ..., n$

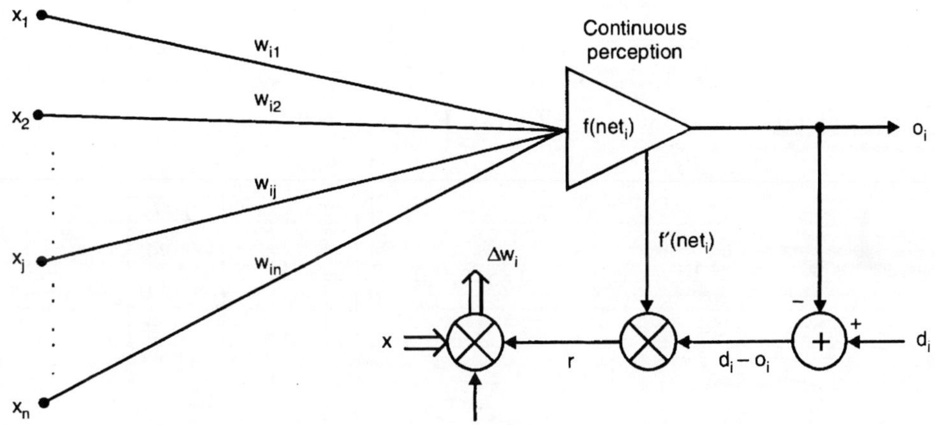

Fig. 2.20

Since the minimization of the error requires the weight changes to be in the negative gradient direction, we take

$$\Delta W_i = -\eta \nabla E \qquad \text{...(2.31)}$$

where η is a +ve equation from (2.30) and (2.31)

$$\Delta W_i = \eta[d_i - f(w_i^t x)]\, f'(w_i^t x)x \qquad \text{...(2.32)}$$

or, for the single weight the adjustment becomes

$$\Delta w_{ij} = \eta(d_i - o_i)\, f'(\text{net}_i)x_j, \qquad \text{...(2.33)}$$

for $j = 1, 2, ..., n$

This delta rule was introduced by McClelland and Rumelhart in 1986. This rule parallels the discrete perceptron training rule.

This rule is also called as continuous perceptron training rule.

Example 2.6. Take the values for W and input vectors from the last example (perceptron learning rule) and $c = 0.1$ and $\lambda = 1$ for the bipolar continuous activation function f (net), use Delta Learning Rule to find out the final weight adjustment. Use $f'(\text{net}) = (1 - O^2)/2$.

Solution. Input is x_1, and initial weight vector given is w^1.

Therefore $\quad \text{net}^1 = w^{1t} x_1$

$$= [1 \ -1 \ 0 \ 0.5] \begin{bmatrix} 1 \\ -2 \\ 0 \\ -1 \end{bmatrix} = [1 + 2 + 0 - 0.5] = 2.5$$

$$O^1 = f(\text{net}^1) = \left[\frac{2}{1 + \exp(-\lambda \text{ net})}\right] - 1$$

Taking value of $\lambda = 1$

$$= \frac{2}{1 + \exp(-1 * 2.5)} - 1 = 0.85$$

$$f'(\text{net}) = \frac{(1 - (O^2))}{2} = \frac{(1 - (0.85)^2)}{2} = 0.140$$

Now new weight is $\quad W^2 = c[d_1 - f(w_1^t x)] f'(w_1^t x) x_1 + W^1 \quad$ from (2.32)

$$w_i^t x = \text{net}$$

Therefore

$$W^2 = 0.1[-1 - 0.85] 0.140 \begin{bmatrix} 1 \\ -2 \\ 0 \\ -1 \end{bmatrix} + \begin{bmatrix} 1 \\ -1 \\ 0 \\ 0.5 \end{bmatrix} = \begin{bmatrix} 0.974 \\ -0.948 \\ 0 \\ 0.526 \end{bmatrix}$$

Step 2: Input vector is x_2

Now

$$\text{net}^2 = w^{2t} x^2 = [0.974 \ -0.948 \ 0 \ 0.526] \begin{bmatrix} 0 \\ 1.5 \\ -1.5 \\ -1 \end{bmatrix} = -1.948$$

$$O^2 = f(\text{net}^2) = \left[\frac{2}{[1 + \exp(-\lambda \text{ net})]}\right] - 1$$

Taking value of

$$\lambda = 1 = \frac{2}{[1 + \exp(-1 * (-1.948)]} - 1 = -0.75$$

$$f'(\text{net}) = \frac{(1 - (O^2))}{2} = \frac{(1 - (0.75^2))}{2} = 0.218$$

Now new weight is $W^3 = c[d_2 - f(w_2{}^t x)]\, f'(w_2{}^t x)x_2 + W^2$ from (2.32)

$w_i^t x = \text{net}$

Therefore $$W^3 = 0.1[-1-(+0.75)]\,0.218\begin{bmatrix}1\\1.5\\-0.5\\-1\end{bmatrix} + \begin{bmatrix}0.974\\-0.948\\0\\0.526\end{bmatrix} = \begin{bmatrix}0.974\\-0.89\\0.02\\0.487\end{bmatrix}$$

Step 3: Input is x_3

Now $$\text{net}^3 = w^{3t}\, x_3 = [0.974 \;\; -0.89 \;\; 0.02 \;\; 0.487]\begin{bmatrix}-1\\1\\0.5\\-1\end{bmatrix} = -2.34$$

$$O^3 = f(\text{net}^3) = \left(\frac{2}{[1+\exp(-\lambda\,\text{net})]}\right)^{-1}$$

Taking value of $\lambda = 1$

$$= \left(\frac{2}{[1+\exp(-1*(-2.46))]}\right)^{-1} = 0.824$$

$$f'(\text{net}) = \frac{(1-(O^2))}{2} = \frac{(1-(-0.842)^2)}{2} = 0.16$$

Now new weight is $W^4 = c[d_3 - f(w_3^t x]\, f'(w_3^t x)\, x_3 + W^3$ from (2.32)

$w/x = \text{net}$

$$W^4 = 0.1[1-(-0.842)]\,0.145\begin{bmatrix}-1\\1\\0.5\\-1\end{bmatrix} + \begin{bmatrix}0.974\\-0.89\\0.02\\0.787\end{bmatrix} = \begin{bmatrix}0.9769\\-0.892\\0.0186\\0.4846\end{bmatrix}$$

2.5.4.4. Widrow-Hoff Learning Rule **[MDU, Dec. 2006; Dec. 2007]**

The Widrow-Hoff learning rule (Widrow 1962) is applicable for

1. The supervised training of neural networks.
2. Weights are initialized at any value in this method.
3. This rule is also called LMS (least mean square).
4. It is independent of the activation function of neurons used since it minimizes the squared error between the desired output value d, and the neuron's activation value $\text{net}_i = \boldsymbol{Wi^t\, x}$. The learning signal r for this rule is

$$r = d_i - w_i^t x \quad ...(2.34)$$

$$\Delta W_i = c(d - w_i^t x)x \quad ...(2.35)$$

For single weight adjustment

$$\Delta W_{ij} = c(d_i - w_i^t x)\, x_j \qquad \text{for } j = 1, 2, ..., n$$

Note: Widrow-Hoff learning rule is a special case of Delta learning rule.

If $f(w_i^t x) = w_i^t x$

Then $f(\text{net}) = \text{net}$ and $f'(\text{net}) = 1$ where net $= wi'^t x$

Then from Delta learning rule, the learning signal is

$$r = [d_i - f(w_i^t x)]\, f'(w_i^t x)$$ becomes

$$r = [d_i - f(w_i^t x)]*1 \quad \text{or} \quad r = [d_i - w_i^t x]$$

Therefore the learning signals of delta learning rule and Widrow Hoff learning rules are equal for this special case.

2.5.4.5. Correlation Learning Rule **[MDU, Dec. 2007; Dec. 2009]**

1. This rule is for supervised learning.
2. By substituting $r = d_i$ in general learning rule eqn. (*xii*) *i.e.*,

$$W_i(t) = cr\big[W_i(t), x(t), d_i(t)\big]\, x(t),$$

We obtain the correlation learning rule. The weight adjustment are

$$\Delta W_i = c(d_i)x \qquad \text{...(2.36)}$$

or for single weight adjustment

$$\Delta W_{ij} = c(d_i)x_j \qquad \text{for } j = 1, 2, \ldots, n$$

This rule states that if d_i is the desired response due to x_j the corresponding weight increase is proportional to their product.

3. The rule typically applies to recording data in memory networks With binary response neurons .
4. This rule is a special case of **Hebbian Learning Rule** with a binary activation function and for $O_i = d_i$. But the main difference is that Hebbian Rule is performed in unsupervised mode and correlation learning rule in supervised learning mode. Similar to Hebbian Learning Rule, this learning rule also requires weight initialization $w = 0$.

2.5.4.6. Winner–Take–All Learning Rule

1. This rule is for unsupervised mode and it differ from all learning rules that were discussed before.
2. In this neurons are arranged in a layer of P units. This rule is an example of competitive learning and is used for unsupervised training mode.
3. Learning is based on the fact that one of the neurons in the layer (suppose m) has maximum response due to input X, then that neuron is declared as Winner. As a result of this Winning event weight vector $w_m = [w_{m1}, w_{m2} \ldots w_{mn}]^t$
4. The increment in the weight vector is computed as

$$\Delta w_m = \alpha(X - w_m) \qquad \text{...(2.37)}$$

for single weight adjustment

$$\Delta w_{mj} = \alpha(X_j - w_{mj}) \qquad \text{for } j = 1, 2, \ldots, n$$

$$\alpha > 0 \Rightarrow \text{ small learning constant.}$$

α decreases as learning constant. increases.

5. The winner selection is based on the criteria of maximum activation among all participating neurons in the competition:

$$\Delta w_{mt} X = \max (w_i^t x) \qquad \text{where } i = 1, 2, ..., P \quad ...(2.38)$$

6. In this rule weights are initialized at random values

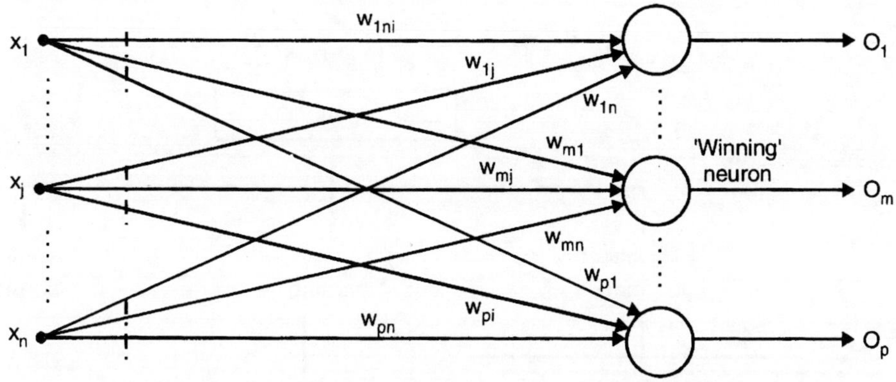

Fig. 2.21. Unsupervised Winner–Take–All Learning Rule

Least Mean Square Error

The LMS rule is actually a gradient descent rule. Error E can be represented as

$$E \cong \frac{1}{2}[d_i - (w_i^t x)]^2$$

Error gradient vector value

$$\nabla E = -[d_i - (w_i^t x)]X$$

The components of the gradient vector are

$$\frac{\partial E}{\partial w_{ij}} = -[d_i - (w_i^t x)]x_j \qquad \text{for } j = 1, 2, ..., n$$

$$\frac{\partial E}{\partial w_{ij}} = -[d_i - (w_j^t x)]x_j \qquad \text{for } j = 1, 2, ..., n$$

Since the minimization of the error requires the weight changes to be in the negative gradient direction, we take

$$\Delta W_i = \eta \nabla E$$

where η is a +ve constt. and called learing rate

$$\Delta W_i = \eta[d_i - (w_i^t x)]X$$

which is LMS rule.

Figure 2.22 illustrates the classification of learning algorithms.

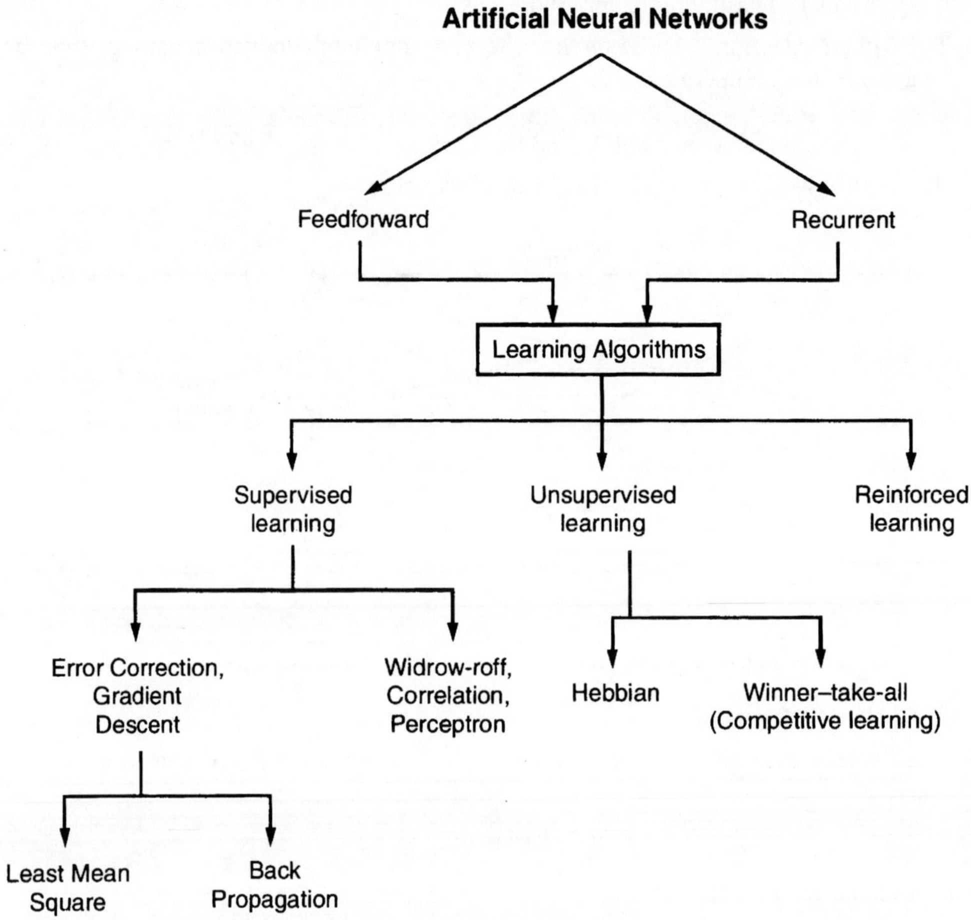

Fig. 2.22

2.6. PERCEPTRON CONVERGENCE THEOREM

Theorem: If there is a set of weights that correctly classify the (linearly separable) training patterns, then the learning algorithm will find one such weight set, w^* in a finite number of iterations (Rosenblatt).

Assumptions:

- At least one such set of weights, w^*, exists, and
- There are a finite number of training patterns.
- The threshold function is uni-polar (output is 0 or 1).

Proof. At the kth iteration, we have:

$$w_{k+1} = w_k + \mu e_k x_k, \quad e_k = d_k - y_k \qquad \text{...(2.39)}$$

or $w_{k+1} = w_k + ce_k x_k$

where c is learning constant

e_k is learning signal
d_k is desired output
y_k is actual output.

where: $e_k = d_k - y_k$

Subtracting w^* from both sides yields:

$$w_{k+1} - w^* = w_k - w^* + \mu e_k x_k \quad ...(2.40)$$

If y_k is correctly classified, there is no update.
Taking the norm of both sides of Eq. (2.40) and noting that $\| e \|^2 = 1$, since $e = \pm 1$:

$$\| w_{k+1} - w^* \|^2 = \| w_k - w^* \|^2 + \mu^2 \| x_k \|^2 + 2\mu e_k (w_k - w^*)^T x_k$$

NORM: The norm (Euclidean) of a vector

$$x = (x_1, x_2, ..., x_n)$$

is given by:

$$(x_1^2, x_2^2, ..., x_n^2)^{1/2} = \| x \|^2$$

$$\| x \|^2 = x^2$$

Now if: $x = a + b \Rightarrow x^2 = a^2 + 2ab + b^2$

$$= \| x \|^2 = \| a \|^2 + 2ab + \| b \|^2$$

where $x, a, b \in R^n$

It can be shown, for each misclassified vector x_k, that

$$e_k w^{*T} x_k = |w^{*T} x_k| \geq 0 \quad ...(2.41)$$

$$e_k w_k^T x_k = -|w_k^T x_k| \leq 0 \quad ...(2.42)$$

Therefore,

$$\| w_{k+1} - w^* \|^2 = \| w_k - w^* \|^2 + \mu^2 \| x_k \|^2 - 2\mu(|w^{*T} x_k| + |w_k^T x_k|) \quad ...(2.43)$$

If μ is chosen sufficiently small,

$$\mu^2 << \mu \quad ...(2.44)$$

$\Rightarrow \mu^2 \| x_k \|^2$ goes to 0.

Therefore, Eq. (2.44) becomes

$$\| w_{k+1} - w^* \| = \| w_k - w^* \|^2 - 2\mu(|w^{*T} x_k| + |w_k^T x_k|)$$

Since $\mu >= 0$ and $| w^{*T} x_k | + | w_k^T x_k | >= 0$, $\| w_{k+1} - w^* \|$ must decrease with each iteration. However, $\|.\|$ cannot go negative, so it must converge in a finite number of steps. Note that this may not necessarily converge to 0 (*i.e.*, $w = w^*$)

Ideal Learning Rule

Rewriting Eq. (2.29) in terms of the weight error, $e = w - w^*$,

$$||e_{k+1}||^2 = ||e_k||^2 + \mu^2 ||x_k||^2 - 2\mu(|w^{*T} x_k| + |w_k^T x_k|)$$

Minimising e_k with respect to μ gives:

$$\mu_{\text{opt}} = \frac{|w^{*T} x_k| + |w_k^T x_k|}{||x_k||^2} = \frac{|(w^* - w_k)^T x_k|}{||x_k||^2} \quad \text{...(2.25)}$$

In practice, it is not possible to evaluate this, since it contains w^* (the optimal weight vector) μ_{opt} can be interpreted as the optimal distance to move w_k along the direction of x_k

2.7. ADALINE AND MADALINE

The adaline (adaptive linear element) developed by Widrow is similar to the perceptron. The adaline has a single output (a single processing element) which receives multiple inputs. The processing element takes the weighted linear sum of the inputs and passes it to a bipolar function, which produces either +1 or –1, depending upon the polarity of the sum. A system with multiple outputs is implemented by using multiple adalines in parallel.

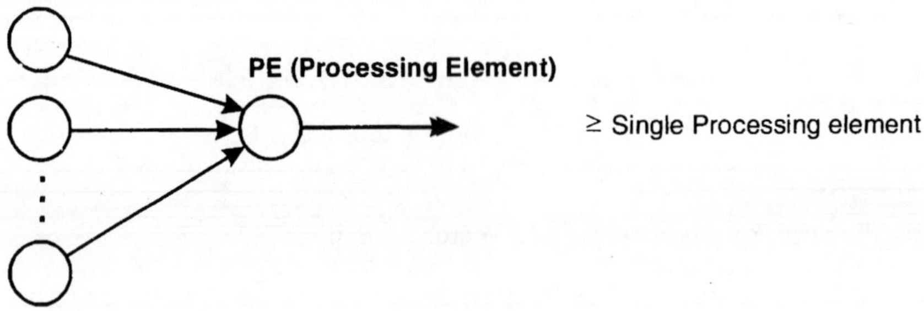

Fig. 2.23

Method: (1) The value in each unit must +1 or –1.

$$\text{net} = \sum_{i=1}^{n} X_i W_i$$

$$X_0 = 1$$

$$\therefore \quad \text{net} = w_0 + w_1 x_1 + w_2 x_2 + \ldots + w_n x_n$$

$$Y = \begin{cases} 1 & \text{if net} \geq 0 \\ -1 & \text{if net} < 0 \end{cases}$$

Different from perception's transfer function.

$$\Delta W_i \geq \eta \geq d \geq Y \geq X_i$$

$$d = \text{expected}$$

$$W_i = W_i + \Delta W_i$$

Limitation : ADALINE can solve only linear problem.

In single-element neural networks, the weights are often trained to classify binary patternsusing binary desired responses. Once the training is complete, the responses of the trained element can be tested by applying various input patterns. If the adaline responds correctly with high probability to input patterns that were not included in the training set, it is said that the generalization has taken place. Learning and generalization are among the most useful attributes of adalines and neural networks.

With n binary inputs and one binary output, a single adaline is capable of implementing certain logic functions. There are 2^n possible input patterns.

A general logic implementation would be capable of classifying each pattern as either +1 or –1, in accordance with the desired responses. Thus, there are 2^{2n} possible logic functions connecting n inputs to a single binary output.

A single adaline is capable of realizing only a small subset of these functions, known as linearly separable logic functions. These are the set of logic functions that can be obtained with all possible weight variations. With two inputs, a single adaline can realize 14 of the 16 possible binary logic functions. The two logical functions that it cannot learn are EX-OR and EX-NOR functions. With many inputs, only a small fraction of all possible logic functions are realizable *i.e.*, linearly separable. Combinations of the elements or networks of elements can be used to realize functions which are not linearly separable.

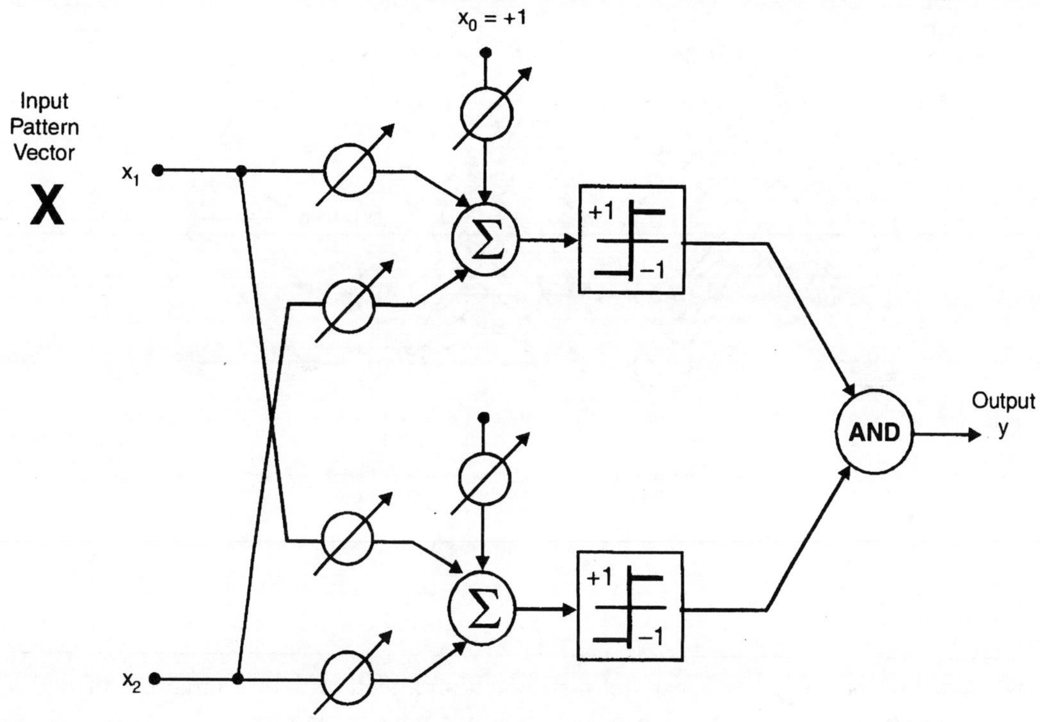

Fig. 2.24. A 2 ADALINE form of MADALINE

The MADALINE refers to the multiple adalines which can be organized into multiple layers.

The ADALINE and the MADALINE have found many useful applications in adaptive signal processing.

The problem dealt with by ADALINE is to determine the weights in such a way that the output response is correct or the error is minimized. The training rule of the ADALINE is the LMS (least mean square) or Widrow-Hoff rule.

The use of multiple ADALINES helps to counter the problem of non-linear separability.

MADALINE: It is composed of many ADALINE (Multilayer Adaline.)

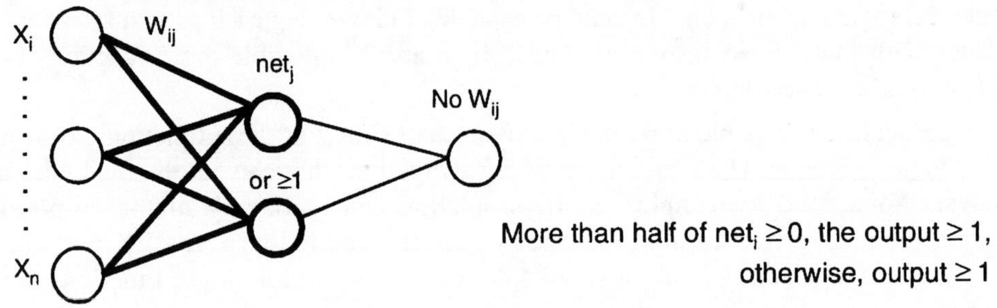

Fig. 2.25.

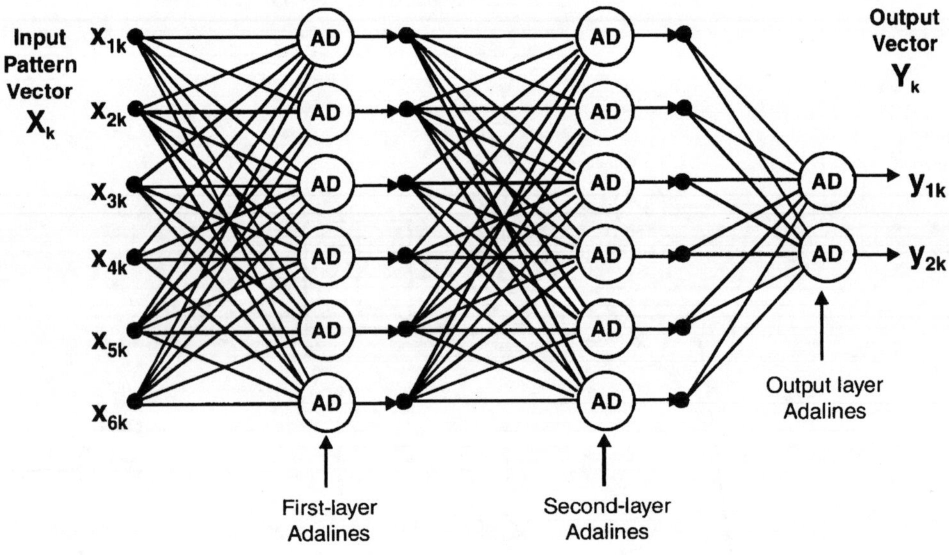

Fig. 2.26. A 3 Layer Adaptive Neural Network

Summary

- A neural network has a parallel-distributed architecture with a large no. of nodes and connections. Each connection points from one node to another and is associated with a weight. The nodes are arranged in 3 layers: Input layer, Hidden layer and Output layer.
- Various activation function used in ANN to get the output from the input. Activation functions act as filter. Various activation function used are: sigmoid function, signum function, **Hyperbolic Tangential, threshold function etc. shown below:**

Function	Definition	Range
Step/Threshold	$f(v) = \begin{cases} 1 & \text{if } v \geq 0 \\ 0 & \text{if } v < 0 \end{cases}$	[0, +1] 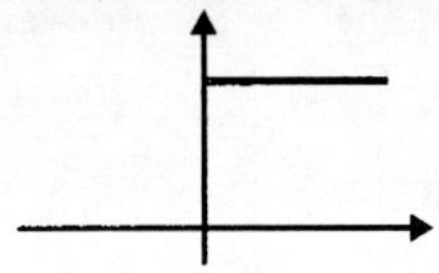
Signum	$f(v) = \begin{cases} 1 & \text{if } v > 0 \\ 0 & \text{if } v = 0 \\ -1 & \text{if } v < 0 \end{cases}$	[0, +1] 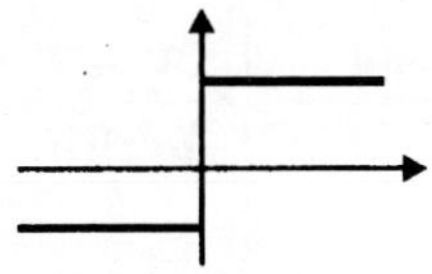
Linear Piecewise	$f(v) = \begin{cases} 1 & \text{if } v \geq 0 \\ v & \text{if } 0.5 > v > -0.5 \\ 0 & \text{if } v < -0.5 \end{cases}$	[0, +1] 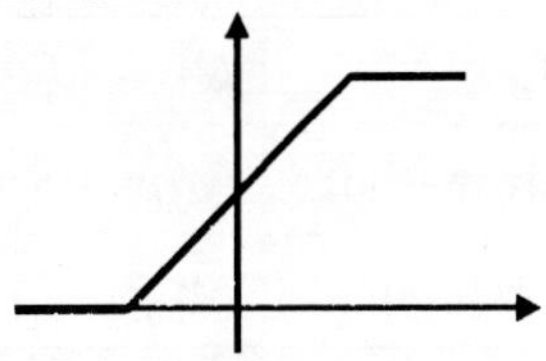
Identity	v	[–inf, +int] 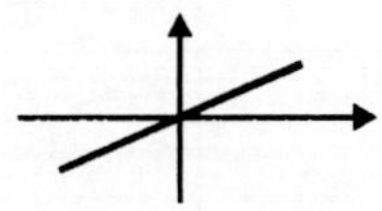
Sigmoidal Logistic	$f(v) = \dfrac{1}{1 + e^{-av}}$	[0, +1] 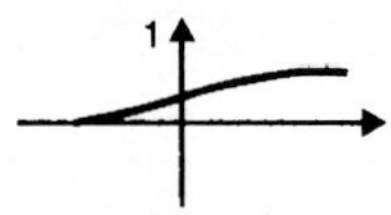
Sigmoidal Hyperbolic	$f(v) = \tanh(v) = \dfrac{e^{v} - e^{-v}}{e^{v} + e^{-v}}$	[–1, +1] 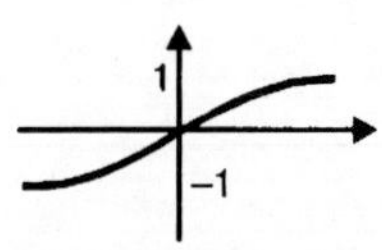
Gaussian	$f(v) = e^{-v}$	[–1, +1] 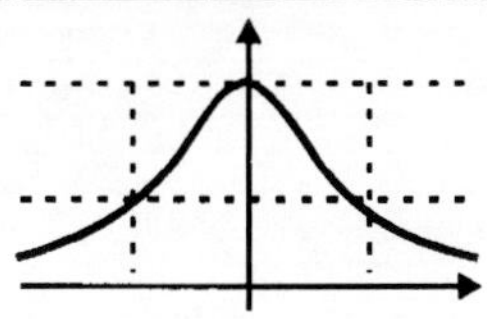

- Three fundamental classes of NN architectures are single layer feedforward, multilayer feedforward and recurrent network architectures.

- The learning rule is one of the most important attributes to specify for a neural network. The learning rule determines how to adapt connection weights in order to optimize the network performance. It indicates how to calculate the weight adjustment during each training cycle. However, the rule is suspended after training is completed. Three types of learning is used in ANN: Supervised learning, Unsupervised learning and reinforcement learning. Example of supervised learning is "TEACHER TEACHING IN A CLASS". And of unsupervised is "VIDEOCONFERENCING". AND reinforcement learning is a mixture of these 2 learning.

➡ Learning algorithms are used in ANN. The summary is shown in the table below:

Name of the Learning Rule	Single weight adjustment	Initial weights	Learning Mode	Neuron Characteristics	Neuron per layer
Hebbian	$C\,O_i\,X_j$	Zero	Unsupervised	Any	Neuron
Perceptron	$\Delta Wi_j = c\,[d_i - \text{sgn}\,(Wi^t x)]\,x_j$	Any	Supervised	Binary Unipolar, Binary Bipolar	Neuron
Delta	$\Delta W_{ij} = c\,(d_i - O_i)\,f'(\,Wi^t x)\,x_j$	Any	Supervised	Continuous	Neuron
Widrow-Hoff	$\Delta W_{ij} = c(d_i - W_i^t x)\,x_j$	Any	Supervised	Any	Neuron
Correlation	$\Delta W_{ij} = c\,(d_i)\,x_j$	Zero	Supervised	Any	Neuron
Winner-Take-All	$\Delta w_{mj} = \alpha(X_j - w_{mj})$ where m is the Winning neuron number	Random Normalized	Unsupervised	Continuous	Layer of P Neurons

➡ *ADALINE Network:* In this, the inputs x_i could be binary, bipolar or real valued but there is only one output neuron and the output values are bipolar (–1 or +1).

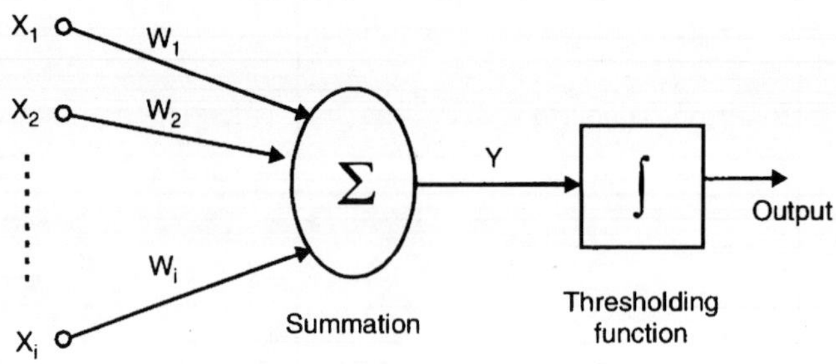

Fig. 2.27. Simple Adaline Network

Review Questions

1. Define activations functions.
2. Define bias and threshold in context of ANN.
3. Define learning in context of ANN.
4. How does threshold help to obtain the output.
5. Differentiate between excitatory and inhibitory connections.
6. Derive the Hebbian and perceptron learning rule. **[MDU May 2010]**
7. Explain Delta learning rule. **[MDU May 2010]**
8. How is "winner-takes-all" process executed by competitive learning?
9. Why Delta learning rule is called as Least mean square rule.

10. Explain three types of layers used in ANN. Explain the significance of hidden layer?
11. Explain the model of ANN.
12. Make an arbitrary feedforward network with 3 neurons in input layer, 2 in hidden and 2 in output layer. Explain the concept of input vector, connection matrix and activation functions. **[MDU May 2009]**

Ans.

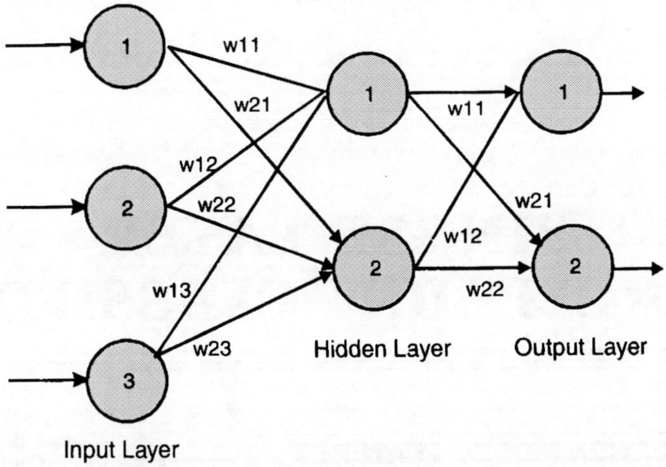

The input vector consists of input signals applied to the neurons. The input element given to a neural network shown by an input vector.
The output vector contains the output signals. The output vector shows all the output functions which is obtained by applying a suitable input.
Connection matrix shows the connection between different layer neurons.

13. What are feedforward and feedback network ? Explain with suitable example. Find the updated weight vectors according to perception leaving rule. The set of input vectors and weight vectors are:

$$x_1 = \begin{bmatrix} 1 \\ -2 \\ 0 \\ -1 \end{bmatrix}, \quad x_2 = \begin{bmatrix} 0 \\ 1.4 \\ 0.5 \\ -1 \end{bmatrix}, \quad x_3 = \begin{bmatrix} -1 \\ 1 \\ 0.5 \\ -1 \end{bmatrix}, \quad w_1 = \begin{bmatrix} 1 \\ -1 \\ 0 \\ 0.5 \end{bmatrix}$$

Desired responses are given as $d_1 = -1$, $d_2 = -1$, $d_3 = 1$ with learining constant $c = 0.1$

14. Draw and explain MCP Model and realize the circuit of NAND using MCP.
15. The network shown in the figure has two inputs x_1, x_2 and one output z. The target output T is as shown in the table. Show learning steps and updated weights of the network using Perceptron rule for single pass. Assume Learning constant $c = 1$, consider $w_1 = 0.4$, $w_2 = -0.2$ and bias $= -1$

x_1	x_2	T
0.8	0.3	1
0.4	0.1	0

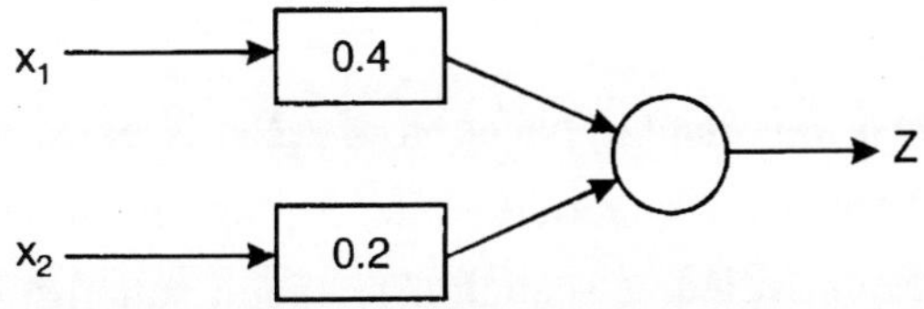

16. What are adaline networks ? Explain the difference between adaline network and the perceptron model.

□□□

CHAPTE

3

SINGLE LAYER PERCEPTION CLASSIFIER

INSIDE THIS CHAPTER

3.1. CLASSIFICATION MODEL, FEATURES AND DECISION REGIONS

3.1.1. Basic Meaning of Classifier

Assume that a set of 8 points a_0, ..., a_7 in three dimensional space is available, where

$$\therefore \quad a_0 = (-1, -1, -1) \qquad a_4 = (1, -1, -1)$$
$$a_1 = (-1, -1, 1) \qquad a_5 = (1, -1, 1)$$
$$a_2 = (-1, 1, -1) \qquad a_6 = (1, 1, -1)$$
$$a_3 = (-1, 1, 1) \qquad a_7 = (1, 1, 1)$$

Elements of this set needs to be classified into two categories.

(1) First category can be set of points with two or more +ve ones.

(2) Second category contains all the remaining points that do not belong to first category.

∴ First category consists of a_3, a_5, a_6, a_7 and remaining points lie in second category.

This is the function of classifier in laymen terms that is to classify or to divide the patterns in different categories.

One of the most important application of ANN is **pattern classification.** **[MDU Dec. 2007]**

Pattern : Pattern is the quantitative description of an object, event or phenomenon.

Patterns can be classified as spatial or temporal patterns.

(*i*) **Spatial :** Examples of spatial patterns are pictures, whether maps video images etc.

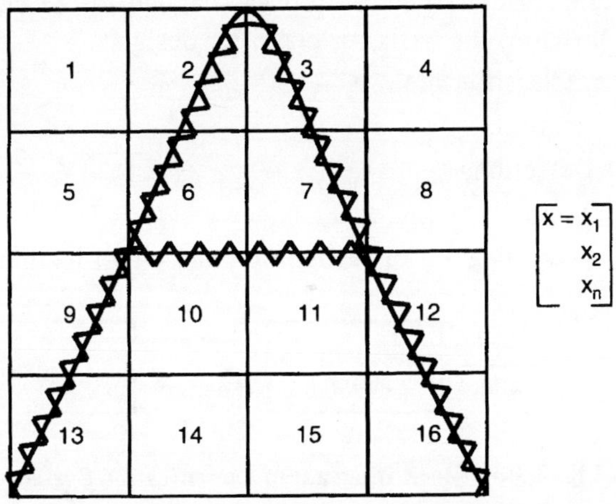

Fig. 3.1. Spatial Pattern

(*ii*) **Temporal :** Examples of temporal patterns are general signals *e.g.* speech signals, ECG. etc. These patterns generally involve the ordered sequence of data appearing in time.

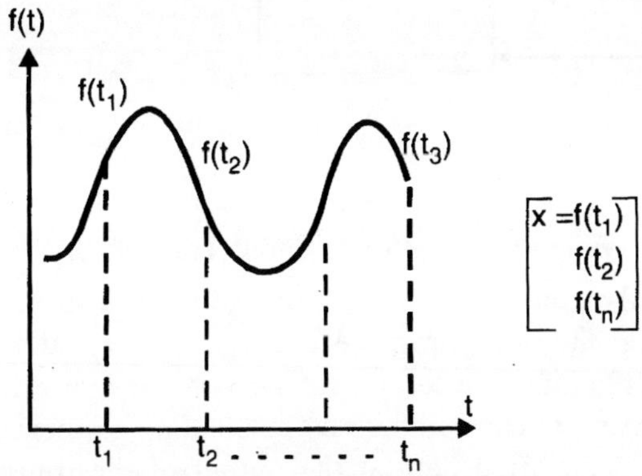

Fig. 3.2. Temporal Pattern

3.2. PATTERN CLASSIFICATION **[MDU, Dec. 2007]**

The main goal of pattern classification is to assign a physical object, event or phenomenon to one of prespecified classes (categories).

For example : We (human beings) classify various objects into different categories like living, non-living things, plants, weather, voice etc. and as discussed in chapter 1 that ANN always learn from the history. The interpretation of data has been learned as a result of repetitive inspection and classification of examples.

When a person perceives a pattern, an inductive inference is made and the perception is associated with some general concepts derived from persons past experience. The problem of pattern classification may be regarded as discriminating the input data within object population via search for the different attributes among members of the population.

In some of the applications, the various classifying aids frame helpful because of the growing complexity of the human environment.

Now the extensive study of classification process has led to the development of an abstract mathematical model that provides the basis for classifier design.

Applications of Pattern Classifier are

(*i*) Fingerprint Identification,

(*ii*) Radar and Signal Detection,

(*iii*) Speech Recognition etc.

Figure 3.3 shows the block diagram of recognition and classification system.

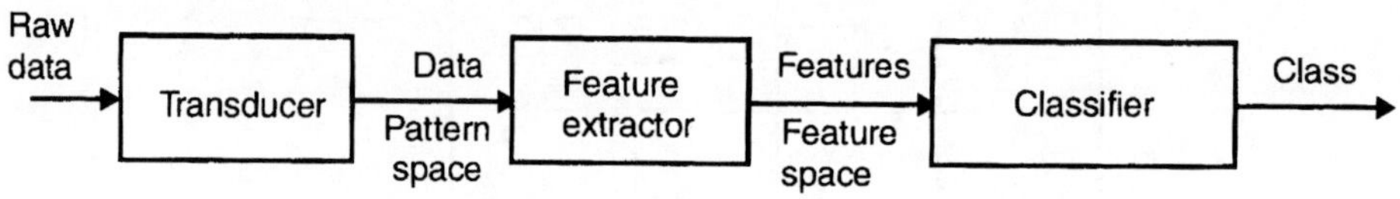

Fig. 3.3(*a*). Block Diagram of Classification System

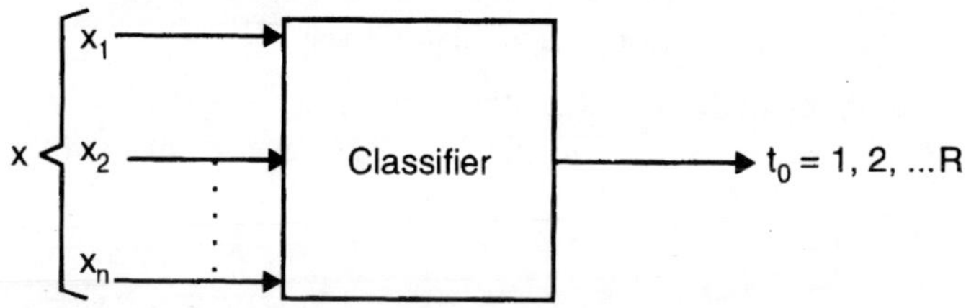

Fig. 3.3(*b*). General Diagram of Pattern Classifier

Diagram Explanation

1. The Classification System Consists of an Input Transducer

   Transducer generally converts one form of energy to another form of energy. Here the input to transducer is the raw data space $x = [x_1, x_2, ..., x_n]$ and output of transducer is data in pattern space that belong to a certain category that is $t_0 = 1, 2, ..., R$. Each member in the pattern space consists of real numbers corresponding to measurement results for a given physical situation. Converted data at the output of the transducer can be compressed. Compressed data is called features.

2. Feature extractor at the input of the classifier performs the reduction of dimensionality. The feature space dimensionality is much smaller than dimensionality of pattern space. Feature vectors retain the minimum number of data dimensions while maintaining the probability of correct classification and hence handling of data become easier.

   For example : To reduce the dimension, the method is the projection of planar data on a single line, thus reducing the feature vector size to a single dimension.

3. Classifier at the end, convert or map the features to the particular class.

Difference between Classification and Recognition

Classification

Classification is a form of neural computation. Assume that a set of input patterns is divided into a number of classes or categories. In response to an

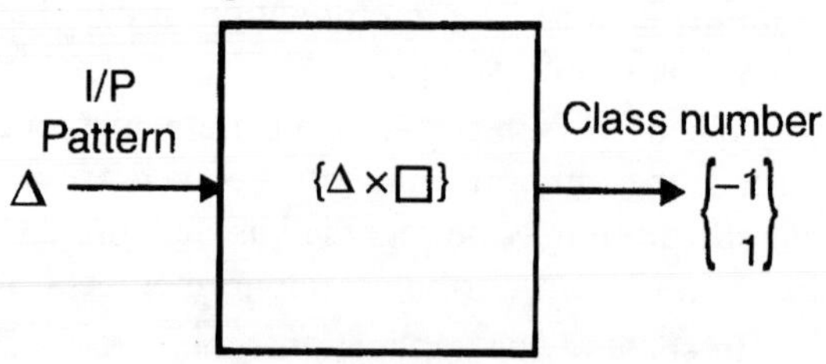

Fig. 3.4(*a*). Classification

input pattern from the set, classifier is supposed to recall the information regarding class membership of input pattern. Classes are generally expressed by discrete valued output vectors and thus output neurons of classifiers would employ binary activation functions.

Recognition

If the network's desired response the class number but the input pattern does not exactly correspond to any of patterns in the set, the processing is called recognition.

When a class membership for one of the patterns in the set is recalled, recognition becomes identical to classification.

TIPS : ***Recognition is of significance when an amount of noise is superimposed on input patterns.***

Figure 3.4(*a*) and 3.4(*b*) shows the block diagram of recognition and classification system.

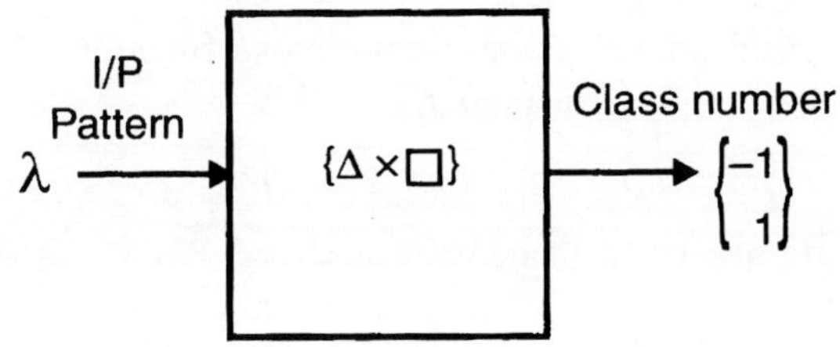

Fig. 3.4(*b*). Recognition

Decision Functions

1. The classification at the system's output is obtained by the classifier, implementing the decision function $i_0(x)$ where the classifier's input component is represented as a vector x.
2. **Responses** represent the categories into which the patterns should be placed.
3. Classification (decision) function $i_0 = i_0(x)$ is provided to do the mapping of n component vector X into one of the category number as shown in Fig. 3.3 (*b*).

$$i_0 = i_0(x)$$

where $$X = \begin{bmatrix} x_1 \\ x_2 \\ \vdots \\ x_n \end{bmatrix}$$

4. In spatial object pattern vector X is represented as

$$X = [x_1, x_2, \ldots\ldots, x_n]^t$$

For spatial patterns, the cell is assigned a value +1 if cell contains a portion of a spatial object, otherwise a value of 0 or –1 is assigned ∴ here

$$X = [\,0\ 1\ 1\ 0\ 0\ 1\ 1\ 0\ 1\ 0\ 1\ 0\ 1\ 0\ 0\ 1]^t$$

for the letter specified in the figure. (for example cell 1 has no position of "A" thus has value 0)

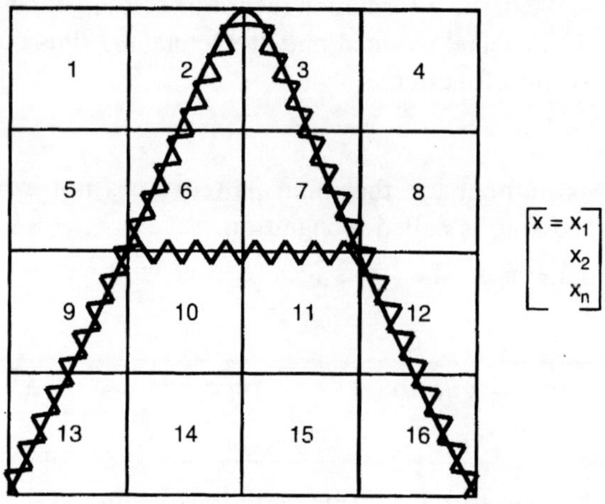

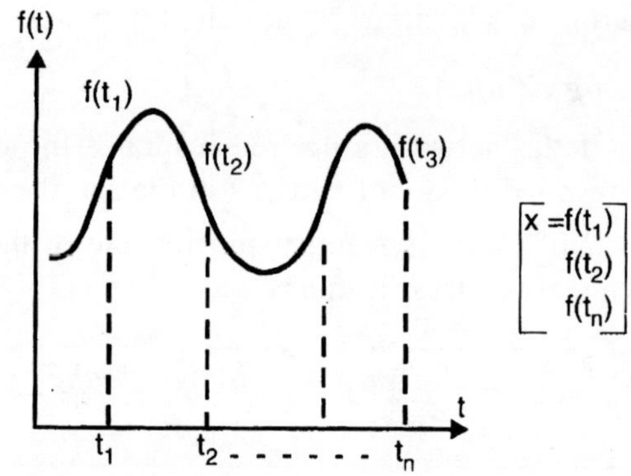

Fig. 3.5. Spatial Object **Fig. 3.6. Temporal Object**

and in case of temporal object, being a continuous function of time t, the pattern vector may be formed at discrete time instants t_i

$$\therefore \qquad x_i = f(t_i) \quad \text{for } i = 1, 2, \ldots, n$$

5. Classification can be described in geometric terms. We all know that any pattern can be represented by a point in n-dimension space. This n-dimension space generally called "EUCLIDEAN SPACE E^n or Pattern Space" classifier then maps sets of points in E^n space into one of the numbers $i_0 = 1, 2, \ldots, R$ according to mapping equation $i_0 = i_0(x)$.
6. Regions that separates the different classes called decision regions.
7. During the classification step, the membership in a category needs to be determined by the classifier based on comparison of R discriminant functions $g_1(x)$, $g_2(x)$ .... $g_R(x)$ computed for input pattern under consideration. It is assumed that discriminant functions $g_i(x)$ are scalar values and pattern X belong to ith category if and only if

$$g_i(x) > g_i(x) \qquad \text{for } i, j = 1, 2, \ldots, R \text{ but } i \neq j$$

3.3. DISCRIMINANT FUNCTION

Discriminant function analysis is the classification of things into groups, classes or the act of distributing categories of the same type.

Discriminant function analysis is a statistical analysis to predict a categorical dependent variable by one or more continuous or binary independent variable. Discriminant function analysis is used when groups are known in advance. Discriminant function analysis finds a set of linear combinations of the variables, whose values are as close as possible within groups and as far apart as possible between the groups. This linear combination are called discriminant functions. It has the following mathematical form:

$$f_{\text{km}} = u_0 + u_1 x_{1\,\text{km}} + u_2 x_{2\,\text{km}} + \ldots\ u_p x_{p\,\text{km}}$$

where f_{km} = the value on the canonical discriminant function for case m in the group k.

$x_{i\,\text{km}}$ = the value on discriminant variable x_i for case m in group k and

u_i = coefficients which produce the desired characteristics in the function

The coefficients for the first discriminant function are derived so as to maximize the difference between the group means. The 2nd discriminant function is orthogonal to the first and 3rd discriminant function is orthogonal to the 2nd and so on. The maximum number of unique functions that can be derived is equal to number of groups minus –1 or equal to number of discriminating variables, whichever is less.

Discriminants

Neural networks can also be used to classify data. Unlike regression problems, where the goal is to produce a particular output value for a given input, classification problems require us to label each data point as belonging to one of *n* classes. Neural networks can do this by learning a **discriminant** function which **separates** the classes. For example, a network with a single linear output can solve a two-class problem by learning a discriminant function which is greater than zero for one class, and less than zero for the other. Figure 3.7 shows two such two-class problems, with filled dots belonging to one class; and unfilled dots to the other. In each case, a line is drawn where a discriminant function that separates the two classes is zero.

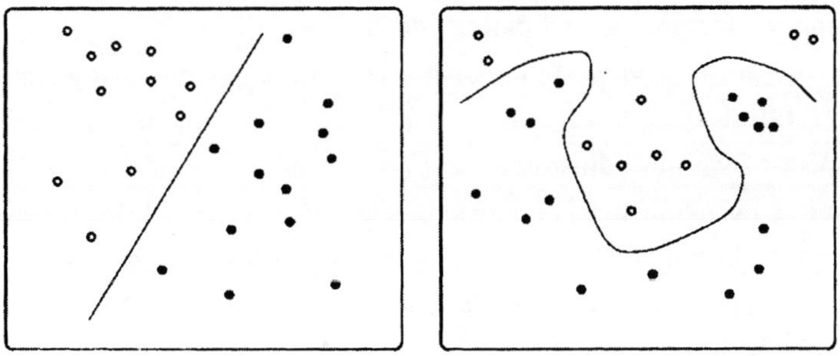

Fig. 3.7.

On the left side, a straight line can serve as a discriminant: we can place the line such that all filled dots lie on one side, and all unfilled ones lie on the other. The classes are said to be **linearly separable**. Such problems can be learned by neural networks without any hidden units. On the right side, a highly non-linear function is required to ensure class separation. This problem can be solved only by a neural network with hidden units.

3.4. TRAINING AND CLASSIFICATION: USING THE DISCRETE PERCEPTRON : ALGORITHM AND EXAMPLE

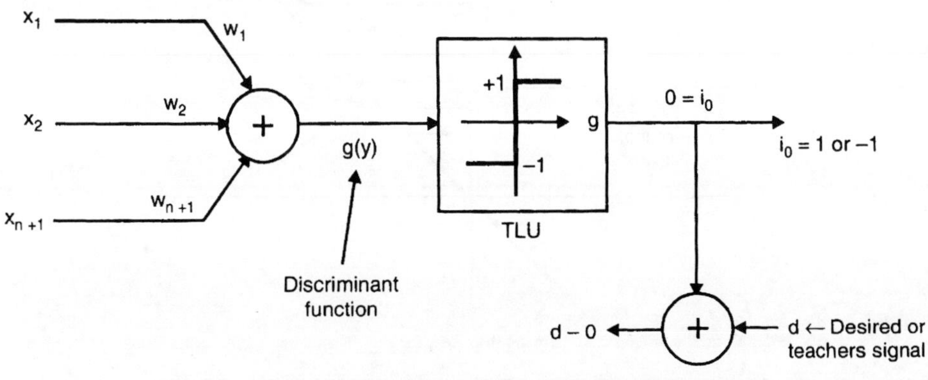

Fig. 3.8. Linear Dichotomizer Using TLU Based Perceptron

Dichotomizer : Classifier that divides the pattern space into 2 classes. (D_i means 2).

i.e., 2 classes *e.g.*, x_1 and x_2 only.

Training Sequence

Sample pattern vectors x_1, x_2, ..., x_p called training sequence *i.e.*, sequence in which ANN is trained. Response is provided by the teacher and it specifies the classification information for each input vector.

The network learns from the experience by comparing the targeted correct response with the actual response to classify structure, and the weight is adjusted after each incorrect response based on error value generated.

In this **discrete perceptron concept** the formula for adjusted weights is **[MDU, May 2009]**

$$w' = w \pm Cy$$

where C is constant, and it is greater than 0 (*i.e.*, $C > 0$) and called correction increment.

here +ve sign is for undetected pattern of class 1

and –ve sign applies for undetected pattern of class 2.

If a correct classification takes place under this rule, no adjustment of weight is made, then new weight (w') = old weight (w).

This is one aspect of weight adjustment.

Another aspect is by using the geometrical relationship correction increment C is chosen in such a way that the weight adjustment setup size is meaningfully controlled.

Let p is distance of a point w' from the plane $w^t y = 0$ in $(n + 1)$ dimensional Euclidean space and it is calculated according to formula

$$p = \pm \frac{w^{1t} y}{\|y\|} \quad \text{...(3.1)}$$

Since p is the distance and by definition of the distance p is always a non-negative scalar.

∴ can be written as

$$p = \frac{w^{1t} y}{\|y\|} \quad \text{...(3.2)}$$

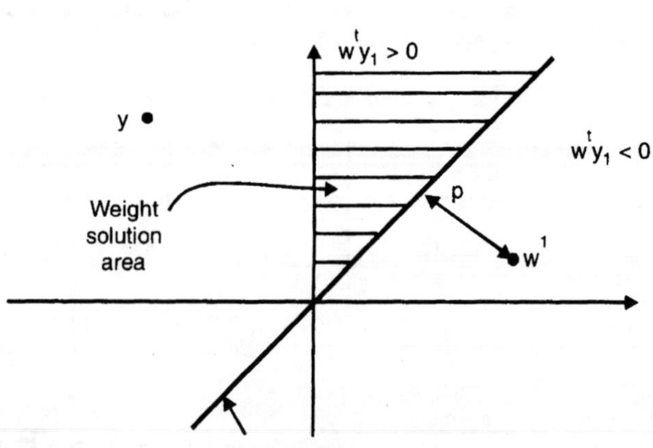

Fig. 3.9.

Now C must be selected such that corrected weight vector w^2 based on $w^1 = w \pm cy$ dislocated on the decision hyperplane $w^{1t}y = 0$ which is the decision hyperplane used for particular correction step.

$$\Rightarrow \qquad w^{2t}y = 0$$

$$(w^1 \pm cy)^t y = 0$$

$$\therefore \qquad \left(w^{1t} + cy^t\right)y = 0$$

$$w^{1t}y \pm cy^t y = 0 \qquad \text{...(3.3)}$$

$$\text{or} \qquad cy^t y = \mp w^{1t} y$$

$$c = \mp \frac{w^{1t} y}{y^t y} \qquad \text{...(3.4)}$$

Since c is correction constant and is positive

$$\therefore \qquad c = \frac{\left|w^{1t} y\right|}{y^t y} \qquad \text{...(3.5)}$$

or length of weight adjustment vector cy can be expressed as

$$\| cy \| = \frac{\left|w^{1t} y\right|}{y^t y} \| y \| \qquad \text{...(3.6)}$$

$$\text{But} \qquad y^t y = \| y \|^2$$

$\Rightarrow$ required distance p from the point w^1 to decision plane and expressed by (3.2) is identical to length of the weight increment vector (3.6).

$\therefore$ For this the correction increment c is therefore not constant and depends on current tracking pattern as expressed of (3.5).

Example 3.1. Given the trained classifier should provide the following classification of four patterns x with known class membership d.

$$x_1 = 1, x_3 = 3, d_1 = d_3 = 1 : \text{Class 1}$$

$$x_2 = -0.5, x_4 = -2, d_2 = d_4 = -1 : \text{Class 2}$$

Input vectors are $$y_1 = \begin{bmatrix} 1 \\ 1 \end{bmatrix}, y_2 = \begin{bmatrix} -0.5 \\ 1 \end{bmatrix}, y_3 = \begin{bmatrix} 3 \\ 1 \end{bmatrix}, y_4 = \begin{bmatrix} -2 \\ 1 \end{bmatrix}$$

Review the training with an arbitrary selection of $C = 1$ and initial weights chosen as $w_1 = [-2.5 \quad 1.75]^t$

Use $$\Delta w^k = \frac{c}{2}\left[d_k - \text{sgn}\,(w^{kt}\, y_k)\right] y^k$$

Total step = 1 to 10. Find out the total outputs and updated weights during training.

Solution. Step 1. y_1 is input

$$O_1 = \text{sgn}(w^{1t}y_1) = \text{sgn}\left[(-2.5 \quad 1.75)\begin{bmatrix}1\\1\end{bmatrix}\right]$$

$$= \text{sgn}(-2.5 + 1.75) = \text{sgn}(-.75)$$

$$= -1$$

$$\therefore \qquad d_1 - O_1 = 1 - (-1) = +2$$

New weight is $\qquad w^2 = w^1 + cy^1$

but $c = 1$ $\left[w^{\text{new}} = w^{\text{old}} \pm cy\right] \Rightarrow w^2 = w^1 + y^1$

$$w^2 = \begin{bmatrix}-2.5\\1.75\end{bmatrix} + \begin{bmatrix}1\\1\end{bmatrix} = \begin{bmatrix}-1.5\\2.75\end{bmatrix}$$

Step 2. y_2 is the input

$$O_2 = \text{sgn}(w^{2t}y_2) = \text{sgn}\left([-1.5 \quad 2.75]\begin{bmatrix}-0.5\\1\end{bmatrix}\right)$$

$$= +1$$

$$\therefore \qquad d_2 - O_2 = -1 - 1 = -2$$

Now $\qquad$ new weight $w^3 = w^2 - cy^2 \qquad c = 1$

$$= \begin{bmatrix}-1.5\\+2.75\end{bmatrix} - \begin{bmatrix}-2.5\\1\end{bmatrix} = \begin{bmatrix}-1\\1.75\end{bmatrix}$$

Here –ve sign taken because $d_2 - O_2$ is –ve.

Step 3. y_3 is the input

$$O_3 = \text{sgn}\left([-1 \quad 2.75]\begin{bmatrix}3\\1\end{bmatrix}\right) = -1$$

$$d_3 - O_3 = 2$$

new weight $w^4 = w^3 + cy^3$ because $d_3 - O_3$ is +ve

$$= \begin{bmatrix}-1\\1.75\end{bmatrix} + \begin{bmatrix}3\\1\end{bmatrix} = \begin{bmatrix}2\\2.75\end{bmatrix}$$

Step 4. y_4 is the input

$$O_4 = \text{sgn}\left([2 \quad 2.75]\begin{bmatrix}-2\\1\end{bmatrix}\right) = \text{sgn}(-4 + 2.75) = -1$$

$$\therefore \qquad d_4 - O_4 = -1 - (-1) = 0$$

Since there is no evidence of correct classification for weights w_4.

$\therefore$ There is an ordered sequence of patterns y_1, y_2 and y_3 is recycled.

$\therefore y_4 = y_1$, $y_5 = y_2$.

Here $w_6 = w_5 = w_4 \Rightarrow$ weight adjustment

$$w^7 = [2.5 \quad 1.75]^t$$

$$w^{10} = w^9 = w^8 = w^7$$

$$w^{11} = [3 \quad 0.75]^6$$

3.5. PERCEPTION CONVERGENCE THEOREM

The classification of linear separable patterns belonging to 2 classes only. The training task for the classifier was to find the weight vector w^* such that

$$w^t y > 0 \text{ for each } x \text{ of } x_1$$

$$w^t y < 0 \text{ for each } x \text{ of } x_2 \qquad ...(3.7)$$

Upon completion of the training with the fixed correction training rule for any initial weight vector and any correction increment constant leads to the following weights

$$w^* = w^{k_0} = w^{k_0+1} = w^{k_0+2} \ldots$$

where w^* is a solution vector for (i)

k_0 is the training step number starting at which no more misclassification occurs and thus no weight adjustments take place for $k_0 \geq 0$.

This theorem is called perceptron convergence theorem.

If the training set is linearly separable, *i.e.*, if there exists a 'w' such that $w.y > 0$ for all adjusted augmented training patterns, then perceptron learning using the fixed increment rule will find a solution w^* in finite time such that $w^*.y > 0$ for all training patterns.

or there exists a k_0 such that

$$w(k_0) = w(k_0 + 1) = w(k_0 + 2) \ldots\ldots\ldots$$

and $$w^* = w(k_0)$$

Proof. Let y be set of all training patterns.

For notational convenience let

$$y(i) \text{ be } y_i$$

and $$w(k) \text{ be } w_k$$

Then training patterns are $y_1, y_2, \ldots. y_k$

and their weights are $w_1, w_2, \ldots., w_k$

Assumption : If there is no change to previous weight vector *i.e.*, $w_{k+1} = w_k$ then that patterns and weight vector are removed from the sequence.

Let $$c = 1$$

$$\Rightarrow \qquad w_{k+1} = w_1 + y_1 + y_2 + \ldots. y_k \qquad ...(3.8) \qquad \left[\begin{array}{l} \because\ w_{k+1} = w_k + cy_k \\ w_k = w_{k+1} - cy_k \end{array}\right]$$

Pattern are linearly separable, then there exist a solution region W such that for any $w \in W$, $w.\ y > 0$ for all training vectors.

Let $$w^* \in W$$

and minimum $y \in Y\ (y\ .\ w^*) = \alpha$ $\qquad \alpha > 0$

and $$w_1 \,.\, w^* = b \qquad ...(3.9)$$

Multiply (3.8) with w^* on both the sides

$$w_{k+1}.w^* = w_1.w^* + y_{1.w^*} + y_2.w^* ..... + y_k.w^* \qquad ...(3.10)$$

Substitutes values from (3.9) to (3.10)

$$w_{k+1}.w^* = \alpha + \beta$$

or $$w_{k+1}.w^* \geq k\alpha + \beta \qquad ...(3.11)$$

$$\left[(w_{k+1}).w^*\right]^2 \leq |w_{k+1}|^2 |w^*|^2$$

or $$|w_{k+1}|^2 \geq \frac{(w_{k+1}.w^*)^2}{|w^*|^2} \geq \frac{(k\alpha+\beta)^2}{|w^*|^2}$$

As $$w_{j+1} = w_j + y_j$$

Taking square on both sides

$$|w_{j+1}|^2 = |w_j|^2 + |y_j|^2 + 2w_j.y_j$$

$$|w_{j+1}|^2 = |w_j|^2 \leq |y_j|^2 \qquad \left[\text{Assume } w_j.y_j \leq 0\right]$$

$$|w_{k+1}|^2 \leq \sum_{j=1}^{k} |y_j|^2 + |w_1|^2 \leq KA|w_1|^2 \qquad ...(3.12)$$

where $$A = \max y \in Y |y|^2$$

for large k, (*iv*) and (*v*) are contradictory.

$\therefore$ K can't be larger than k_m, w_k is the solution.

Hence correction steps must be less than k_m if the patterns are this early separable.

Example 3.2. 8 binary patterns representing one of the 2 classes are submitted to the input of the classifier in an ascending binary sequence. 4 patterns belong to class 2 and 4 patterns belong to class 1, as mentioned in '*d*' columns of the table. Using the table.

Input Data				Class
x_1	x_2	x_3	x_4	d
0	0	0	1	2
0	0	1	1	1
0	1	0	1	2
0	1	1	1	1
1	0	0	1	2
1	0	1	1	1
1	1	0	1	2
1	1	1	1	1

Find final weights.

For class 2 $\quad w^1 = w - cy$

For class 1 $\quad w^1 = w + cy$

Solution. As there are 8 rows given. Therefore, these 8 binary patterns are spanned in 3 dimensional space and representing one of the 2 classes are submitted to the input of classifier.

Starting from initial weight vector $w = [0\ \ 0\ \ 0\ \ 0]$

Now, as first row in the given problem belongs to class 2 and for this class $w' = w - cy$ where c is constant and its value is 1.

$$\therefore \qquad w' = [0\ \ 0\ \ 0\ \ 0] - 1[0\ \ 0\ \ 0\ \ 1]$$
$$= [0\ \ 0\ \ 0\ \ -1]$$

net weight is
$$w^t \,.\, y = \begin{bmatrix} 0 \\ 0 \\ 0 \\ 0 \end{bmatrix} [0\ \ 0\ \ 0\ \ 1] = 0.00$$

for inputs 0 0 1 1, as clear from the table that pattern [0011] belongs to class 1 therefore $w' = w + cy$

$$w^1 = [0\ \ 0\ \ 0\ \ -1] + [0\ \ 0\ \ 1\ \ 1]$$
$$= [0\ \ 0\ \ 1\ \ 0]$$

$$\therefore \qquad w^t y = \begin{bmatrix} 0 \\ 0 \\ 0 \\ -1 \end{bmatrix} .[0\ \ 0\ \ 1\ \ 1]$$
$$= -1$$

Similarly for next inputs, the complete steps for iteration = 1 are shown below.

Training steps Iteration = 1	w_1	w_2	w_3	w_4	Net = $w^t y$
0 0 0 1	0	0	0	0	0
0 0 1 1	0	0	0	–1	–1
0 1 0 1	0	0	1	0	0
0 1 1 1	0	–1	1	–1	–1
1 0 0 1	0	0	2	0	0
1 0 1 1	–1	0	2	–1	0
1 1 0 1	0	0	3	0	0
1 1 1 1	–1	–1	3	–1	0

Iteration = 2 (Consider the values of weights from iteration 1)

Training steps	w_1	w_2	w_3	w_4	Net = $w^t y$
0 0 0 1	0	0	4	0	0
0 0 1 1	0	0	4	–1	3
0 1 0 1	0	0	4	–1	–1
0 1 1 1	0	0	4	–1	3
1 0 0 1	0	0	4	–1	–1
1 0 1 1	0	0	4	–1	3
1 1 0 1	0	0	4	–1	–1
1 1 1 1	0	0	4	–1	3

Iteration = 3 (Consider the values of weights from iteration 2)

Training steps	w_1	w_2	w_3	w_4	Net = $w^t y$
0 0 0 1	0	0	4	–1	–1
0 0 1 1	0	0	4	–1	3
0 1 0 1	0	0	4	–1	–1
0 1 1 1	0	0	4	–1	3
1 0 0 1	0	0	4	–1	–1
1 0 1 1	0	0	4	–1	3
1 1 0 1	0	0	4	–1	–1
1 1 1 1	0	0	4	–1	3

∴ Final weight are : 0 0 4 –1

$$x = \begin{bmatrix} x_1 \\ x_2 \\ x_3 \\ x_4 \end{bmatrix} \text{ is a pattern vector}$$

but $x_4 = 1$ is augmented pattern component

∴ $4x_3 - 1 = 0$ is the decision plane.

Example 3.3. Let us consider pattern classes C1 and C2, where C1: {(0, 2), (0, 1)} and C2: {(1, 0), (1, 1)}. The objective is to obtain a decision surface based on perceptron learning.

Solution. Since, the input vectors consist of two elements, the perceptron structure is simply as follows :

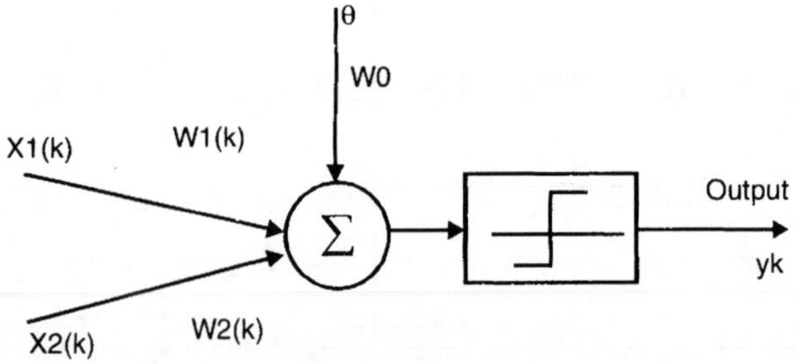

Fig. 3.10

For simplicity, let us assume $\eta = 1$ and initial weight vector $W(1) = [0\ \ 0]$. The iteration weights are as follow:

Iteration 1: $W^T(1)x(1) = [0\ \ 0]\begin{bmatrix}0\\2\end{bmatrix} = 0$ (output ≤ 0 and $x(k)\ \varepsilon\ C1$)

Weight Update:

$$W(2) = W(1) + \eta x(1) = [0\ \ 0] + [0\ \ 2] = [0\ \ 2]$$

Iteration 2: $W^T(2)x(2) = [0\ \ 2]\begin{bmatrix}-0\\-1\end{bmatrix} = 2$ (desired output as output > 0 and $x(k)\varepsilon\ C1$)

Weight Update:

$$W(3) = W(2)$$

Iteration 3: $W^T(3)x(3) = [0\ \ 2]\begin{bmatrix}-1\\-0\end{bmatrix} = 0$ (output ≥ 0 and $x(k)\varepsilon\ C2$)

Weight Update:

$$W(4) = W(3) - \eta\ x(3) = [0\ \ 2] - 0 = [-1\ \ 2]$$

Iteration 4: $W^T(4)x(4) = [-1\ \ 2]\begin{bmatrix}-1\\-1\end{bmatrix} = 1$ (output ≥ 0 and $x(k)\varepsilon\ C2$)

Weight Update:

$$W(5) = W(4) - \eta\ x(4) = [-1\ \ 2] - 1 = [-2\ \ 1]$$

Iteration 5: $W^T(5)x(1) = [-2\ \ 1]\begin{bmatrix}-0\\-2\end{bmatrix} = 2$ (desired output as output > 0 and $x(k)\varepsilon\ C1$)

Weight Update:

$$W(6) = W(5)$$

Iteration 6: $W^T(6)x(2) = [-2\ \ 1]\begin{bmatrix}-0\\-1\end{bmatrix} = 1$ (desired output as output > 0 and $x(k)\varepsilon\ C1$)

Weight Update:

$$W(7) = W(6)$$

Iteration 7: $W^T(7)x(3) = [-2\ \ 1]\begin{bmatrix}-1\\-0\end{bmatrix} = -2$ (desired output as output < 0 and $x(k)\varepsilon\ C2$)

Weight Update:

$$W(8) = W(7)$$

Iteration 8: $W^T(8)x(4) = [-2\ \ 1]\begin{bmatrix}-1\\-1\end{bmatrix} = -1$ (desired output as output < 0 and $x(k)\varepsilon\ C2$)

Weight Update:

$$W(9) = W(8)$$

Therefore, the algorithm converges and the decision surface for the above perceptron is as follows:

$$d(x) = -2x_1 + x_2 = 0$$

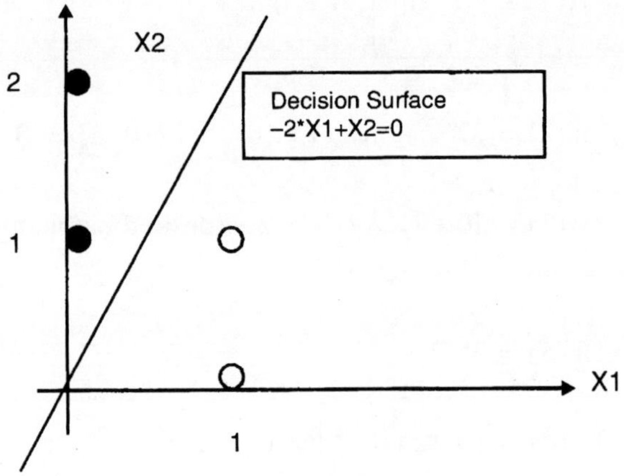

Fig. 3.11

Now, let us consider the input data {1, 2}, which is not in the training set. If we calculate the output :

$$y = W^T x = [-2 \quad 1] \begin{bmatrix} -2 \\ -1 \end{bmatrix} = -3$$

The output is < 0, thus input data belongs to the class C2 as is expected.

3.6. SINGLE DISCRETE PERCEPTRON TRAINING ALGORITHM

[MDU, May 2009; May 2012]

Given there are P training pairs

$\{x_1, d_1, x_2, d_2, ..., x_p, d_p\}$ where

x_i is $(n \times 1)$ and d_i is (1×1), $i = 1, 2, ..., P$

$$\text{Input vector is } y_i = \begin{bmatrix} x_i \\ 1 \end{bmatrix} \quad \text{for } i = 1, 2, ..., P$$

Step 1. $C > 0$ is chosen.

Step 2. Weights are initialized at w at small random values and $k \leftarrow 1, p \leftarrow 1, E \leftarrow 0$ where k is training step.

p is step counter within the training step or cycle.

Step 3. Training cycle starts.

Input is presented and output is computed

$$y \leftarrow y_p,\ d \leftarrow dp,\ o \leftarrow sgn\ (w^t y)$$

Step 4. Weights are updated :

$$w \leftarrow w + \frac{1}{2} c(d - o) y$$

Step 5. Cycle error is computed

$$E \leftarrow \frac{1}{2}(d-o)^2 + E \rightarrow \text{Previous cycle error}$$

Step 6. If $p < P$ then $p \leftarrow p + 1$, $k \leftarrow k + 1$ and go to step 3; otherwise go to step 7.

Step 7. Training is completed. If $E = 0$, terminate the traning, output weight and K.

If $E > 0$ then $E \leftarrow 0$, $P \leftarrow 1$, new traning step or iteration starts, starting from step 3.

3.7. SINGLE LAYER CONTINUOUS PERCEPTRON NETWORKS FOR LINEAR SEPARABLE CLASSIFICATIONS

Linearly separable and non-separable classifications explained in chapter 4.

As the main difference between continuous and discrete perceptron is that in discrete perceptron, TLU (Threshold logic units) is used. ∴ In this topic the TLU elements with weights are replaced by the continuous perceptron. The Aim of replacement of TLU by continuous perceptron are

1. To enable the computation of error gradient.
2. To gain finer control over the training procedure.

It is already known that weight modification problem could be better solved by minimizing the scalar criterion function and such a task can possibly be attempted by using gradient or steepest Descent procedure and the procedure is:

1. Starting from an arbitrary weight vector w, the gradient $\nabla E(w)$ of current error function is computed.
2. The next value of w is obtained by moving in the direction of –ve gradient along the multidimensional error surface. The direction of negative gradient is one of steepest descent.

$$\therefore \qquad w^{k+1} = w^k - \eta \nabla E(w^k) \qquad ...(3.13)$$

where η is the +ve constant called learning constant and the superscript (k) denotes the step number. [derivation of w^{k+1} is in chapter 2 under gradient descent rule]

Error E_K is defined as

$$E_K = \frac{1}{2}\left[(d^k - o^k)^2\right] \qquad ...(3.14)$$

[For d^k = Desired value, O^k = Actual output value]

$$\text{or} \qquad E_K = \frac{1}{2}\left[(d^k - f(w^{kt} y^k)\right]^2 \qquad ...(3.15)$$

The aim is E^K to be minimized.

The error minimization algorithm (3.13) requires the computation of gradient of error as follows :

Taking gradient of (3.14) or (3.15)

$$\nabla E_K(w) = \frac{1}{2}\nabla\left[d_k - f(net_k)\right]^2 \qquad ...(3.16)$$

[$O = f(\text{net})$ and net $= \prod_{k=1}^{n} w^{tk} y^k$]

or in generalized form

$$\nabla E(w) = \frac{1}{2}\nabla\left[d - f(net)\right]^2 \quad ...(3.17)$$

$$\nabla E(w) = \frac{1}{2}\left[d - f(net)\, f'(net)\begin{bmatrix} \partial(net)/\partial w_1 \\ \partial(net)/\partial w_2 \\ \vdots \\ \partial(net)/\partial w_{n+1} \end{bmatrix}\right] \quad ...(3.18)$$

$n + 1$ represent that gradient vector is chosen for $n + 1$ dimensional spare.

or let $\frac{\partial(net)}{\partial w_i} = y_i$ for $i = 1, 2, ..., n + 1$

then (3.18) becomes

$$\nabla E(w) = -(d - f(net))\, f'(net).y_i \quad ...(3.19)$$

which is traning rule for continuous perceptron.

For Bipolar continuous functions

$$f(\text{net}) = \frac{2}{1 + \exp(-\lambda net)} - 1 \quad ...(3.20)$$

$$\therefore \quad f'(net) = \frac{2\,(\exp(-net)}{\left[1 + \exp(-net)\right]^2} \quad ...(3.21)$$

Consider the identity

$$\frac{2\exp(-net)}{\left[1 + \exp(-net)\right]^2} = \frac{1}{2}(1 - O^2) \quad ...(3.22)$$

Let $O = f(\text{net})$, then above equation will become

or
$$\frac{2\exp(-net)}{\left[1 + \exp(-net)\right]^2} = \frac{1}{2}\left[1 - f(net)^2\right] \quad ...(3.23)$$

or
$$\frac{1}{2}\left[1 - f(net)^2\right] = \frac{1}{2}\left[1 - \left(\frac{1 - \exp(-net)}{1 + \exp(-net)}\right)^2\right] \quad ...(3.24)$$

or Taking R.H.S. only

$$\frac{1}{2}\left[1 - \left(\frac{1 - \exp(-net)}{1 + \exp(-net)}\right)^2\right] = \frac{2\exp(-net)}{\left[1 + \exp(-net)\right]^2} \quad ...(3.25)$$

$\therefore$ Eq. (3.22) is proved.

$\therefore$ Gradient $\nabla E(w)$ or (3.19) becomes

$$\nabla E(w) = -(d - O)(1 - O^2)y$$

and complete delta training rule for bipolar continuous activation function

$$w^{k+1} = w^k - \eta \nabla E(w^k) \text{ becomes}$$

$$w^{k+1} = w^k + \frac{1}{2}\eta\left(d^k - O^k\right)\left[(1 - O^k)^2\right]y^k \qquad \text{...(3.26)}$$

k is number of training step.

For discrete perceptron

$$w^{k+1} = w^k + \frac{c}{2}(d^k - O^k)y^k \qquad \text{...(3.27)}$$

1. Main difference between (3.26) [for continuous perceptron] and (3.27) [for discrete perceptron] is the size of weight correction vector. The essential difference is the presence of moderating factor $(1 - O^k)^2$. This factor is always +ve and hence smaller than 1. For weekly committed perceptron, the correction scaling factor will be larger than for those responses generated by a net of large magnitude.
2. Another difference between discrete and contniuous perceptron training is that discrete perceptron training algorithm always leads to a solution for linearly separable problems.

But –ve gradient-based training does not guaranteed solutions for linearly separable patterns.

Algorithm of single continuous perceptron training (SCPTA)

Given, P training pairs

$\{x_1, d_1, x_2, d_2, ..., x_p, d_p\}$

Augmented input vectors are used

$$y_i = \begin{bmatrix} x_i \\ 1 \end{bmatrix} \text{ for } i = 1, 2, ..., P$$

where x_i is $(n \times 1)$, d_i is (1×1), $i = 1, 2, ..., P$

p denotes step counter within training cycle and k denotes the training step.

1. Choose $\eta > 0$, λ (steepress constant) $= 1$, $E_{max} > 0$.
2. Weight are initialized at w at small random values, counters and error are initialized.

   $$k \leftarrow 1, p \leftarrow 1, E \leftarrow 0.$$

3. Training starts $\rightarrow$ input y is presented and output is computed *i.e.*,

   $$y \leftarrow y_p, d \leftarrow dp, o \leftarrow f(w^t y)$$

4. Weights are updated : $w^{k+1} \leftarrow w^k + \frac{1}{2}\eta(d^k - o^k)(1 - o^{k^2})y$
5. Error E is computed $E^{k+1} \leftarrow \frac{1}{2}(d^k - o^k)^2 + E^k$.
6. If $p < P$, then $p \leftarrow p + 1$, $k \rightarrow k + 1$ and repeat step 3, otherwise go to step 7.

7. Training cycle is completed. If $E < E_{max}$ terminate the session.
   If $E \geq E_{max}$ then $E \leftarrow 0$, $p \leftarrow 1$, start new training cycle and go to step 3.

Summary

This chapter provides a good overall base for the discussion of basic ANN concepts:

- List of concepts includes classification, recognition, perceptron convergence theorem, continuous and discrete perceptrons.
- Pattern classification is the most important application of ANN. Patterns may be spatial or temporal.
- Although classification and recognition are different but when a class membership for one of the patterns in the set is recalled, recognition becomes identical to classification.
- The main difference between discrete and continuous perceptron training is that discrete perceptron training algorithm always leads to a solution for linearly separable problem.
- Single layer perceptrons are used for linearly separable problems like OR, AND, etc. But single layer perceptrons are not suitable for linearly non-separable problems *e.g.*, to solve EX-OR type problems. Therefore, Multilayer perceptrons are used.

Review Questions

1. What is classification and how various patterns are classified ?
2. Define Pattern and its types.
3. Define Euclidean Distance.
4. Is there any difference between classification and recognition ?
5. Explain the meaning of Training, Testing and Validation Datasets.
6. Differentiate between the weight modifications scheme when continuous and discrete perceptrons are used.

❑❑❑

CHAPTER

4

MULTI-LAYER FEED FORWARD NETWORKS

INSIDE THIS CHAPTER

4.1. CLASSIFICATION MODELS

A Neural Network classifies a given object presented to it according to the output activation. For binary outputs, 1 corresponds to one class and 0 corresponds to the other. For continuous outputs between 0 and 1, 0.5 can be used as threshold to make a decision. In the case of multiple outputs, the object is assigned to the class corresponding to the output node with maximum activation. 2 types of classification models are discussed below:

1. Single Layer Perceptron
2. Multilayer Layer Perceptron

The Perceptron

The perceptron is a program that learns **concepts**, *i.e.*, it can learn to respond with **True** (1) or **False** (0) for inputs we present to it, by repeatedly "studying" examples presented to it.

Example : For example, how do you teach a child to recognize what a chair is? You show him examples telling him "This is a chair; That one is not a chair" until the child learns the concept of what a chair is. In this stage, the child can look at the examples we have shown him and answer correctly to the question "Is this object a chair?" Furthermore, if we show to the child new objects, that he didn't see before, we could expect him to recognize correctly whether the new object is a chair or not, providing that we've given him enough positive and negative examples. This is exactly the idea behind the perceptron.

Single Layer Perceptron

A Single Layer Perceptron consists of an input layer and an output layer. The activation function applied is hard-limiting function as discussed in chapter 2. An output unit will assume the value 1 if the sum of weighted inputs is greater than its threshold. *i.e.,*

$\Sigma W_{ji} X_i > \theta_j$ where W_{ji} is weight from unit i to unit j.

X_i is the input from unit i and θ_j is the threshold on unit j

Let there are 2 classes A and B. If $\Sigma W_{ji} X_i > \theta_j$ then object will be classified as Class A otherwise Class B.

Suppose there are n inputs then the equation $\Sigma W_{ji} X_i = \theta_j$ where $i = 1, 2 \ldots n$ forms a hyperplane, dividing the space in 2 halves.

4.1.1. Linear Separability **[MDU May 2006, May 2008, Dec. 2008]**

When a linear hyperplane exists to place the instances of one class on one side and those of other class on the other side of plane. Then this is called Linear Separability. For example in case of OR gate we can draw a linear hyperplane to separate the inputs whose corresponding outputs are 1 on one side of the plane and the inputs whose corresponding outputs are 0 on other side of the plane.

TABLE 4.1. OR Gate.

Input *A*	Input *B*	Output
0	0	0
0	1	1
1	0	1
1	1	1

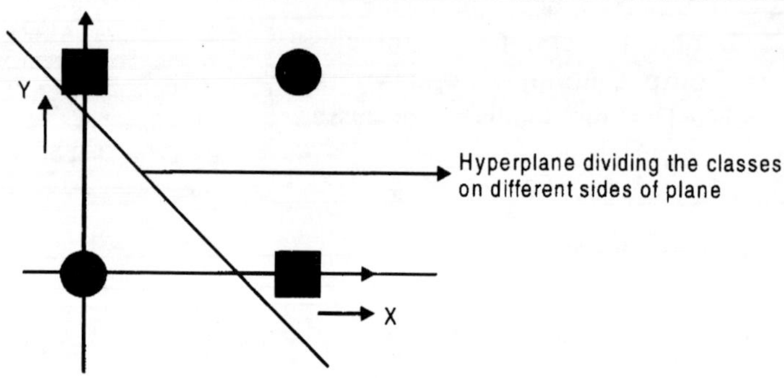

Fig. 4.1. The OR Gate

But all the classification problems are not linear separable. For example the EX-OR problem is not linear separable because EX-OR is non equality gate *i.e.,*

TABLE 4.2. The EX-OR Gate.

Input *A*	Input *B*	Output
0	0	0
0	1	1
1	0	1
1	1	0

In this case we can't draw any hyperplane that divides the space into classes so that different classes are on different sides. To cope up with a problem which is not linear separable, a Multilayer Perceptron is required.

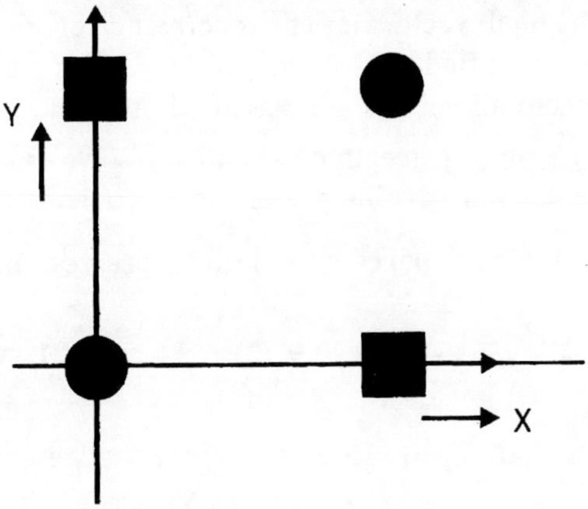

Fig. 4.2. The EX-OR Function

The Perceptron Algorithm

1. **Weight Initialization :**

   Set all the weights and node thresholds to small random numbers.

2. **Calculation of Activation :**

(*a*) The activation level of an input unit is determined by the instance presented to the network.

(*b*) The activation level O_j of an output unit is determined by

$$O_j = F_h(W_{ji}X_i - \theta_j) \quad ...(4.1)$$

where F_h is hard-limiting function:

$$F_h(a) = \begin{cases} 1 & \text{if } a > 0 \\ 0 & \text{Else} \end{cases}$$

3. **Weight Training :**

(*a*) Adjust weights by

$$W_{ji}(t + 1) = W_{ji}(t) + \Delta W_{ji} \quad ...(4.2)$$

where $W_{ji}(t)$ is the weight from unit i to unit j at time t and ΔW_{ji} is the weight adjustment.

(*b*) The weight change may be computed by Delta Learning Rule:

$$\Delta W_{ji} = \eta\delta_j X_i \quad ...(4.3)$$

where Δ is trial –independent learning rate $(0 < \eta < 1)$

and δ_j is the error at unit j and $\delta_j = T_j - O_j$

where T_j is the desired output activation and

O_j is the actual output activation at output unit j

(*c*) Repeat the iterations till convergence.

Limitations

Perceptron networks have several limitations.

First, the output values of a perceptron can take on only one of two values (True or False). Second, perceptrons can only classify linearly separable sets of vectors. If a straight line or plane can be drawn to separate the input vectors into their correct categories, the input vectors are linearly separable and the perceptron will find the solution. If the vectors are not linearly separable learning will never reach a point where all vectors are classified properly.

The most famous example of the perceptron's inability to solve problems with linearly nonseparable vectors is the **boolean exclusive-or problem.** For these problems multilayer perceptron is used.

Example 4.1. In a single layer perceptron unit 1 receives inputs from unit 2 and 3. Given that

$$W_{1,2} = (-3),\ W_{1,3} = (2),\quad X_2,\ X_3 = 1,\ \theta_1 = 1$$

Calculate O_1.

Suppose the desired output $T_1 = 1$. How we adjust the weights? Consider $\eta = 0.3$.

Solution.

$$\begin{aligned} O_1 &= F_h(-3*1 + 2*1 - 1) \\ &= F_h(-2) = 0 && \text{from (4.1)} \\ \delta_j &= 1 - 0 = 1 && \text{where } j = 1 \\ \Delta W_{1,2} &= \eta\delta_1 X_2 = 0.3 * 1 * 1 = 0.3 \\ \Delta W_{1,3} &= \eta\delta_1 X_3 = 0.3*1*1 = 0.3 \\ W_{1,2} &= W_{1,2(old)} + \Delta W_{1,2} \\ &= (-3 + 0.3) = 2.7 \\ W_{1,3} &= W_{1,3(old)} + \Delta W_{1,3} \\ &= (0.3 + 2) = 2.3 \end{aligned}$$

The threshold is always negative of weight from bias unit let $W_{I,b}$ is the bias unit. That is

$$\begin{aligned} W_{I,b} &= (-\theta_1) = (-1) \\ \Delta W_{1,b} &= 0.3*1*1 = 0.3 \\ W_{1,b} &= (-1 + 0.3) = 0.7 \end{aligned}$$

Thus the threshold is changed to 0.7.

4.2. MULTILAYER PERCEPTRON [MDU Dec. 2008]

An MLP is a network of simple neurons called perceptrons. The basic concept of a single perceptron was introduced by Rosenblatt in 1958. A Multilayer Perceptron is a feedforward neural network with at least one hidden layer. It can deal with nonlinear classification problems.

Example : Suppose an output node receives 2 inputs, its threshold is set to 1.5 and its input weights are both set to 1. Therefore when both the input units are active (*i.e.*, 1) the output node will be active. In this case the output node performs a logical AND operation. On the other hand if threshold is set to 0.5, then any active input can activate the node. In this case the output node performs a logical OR operation. Thus by choosing a different set of weights and threshold, a node can implement a different logical operation.

Multilayer-Perceptron Characteristics

Sample Structure	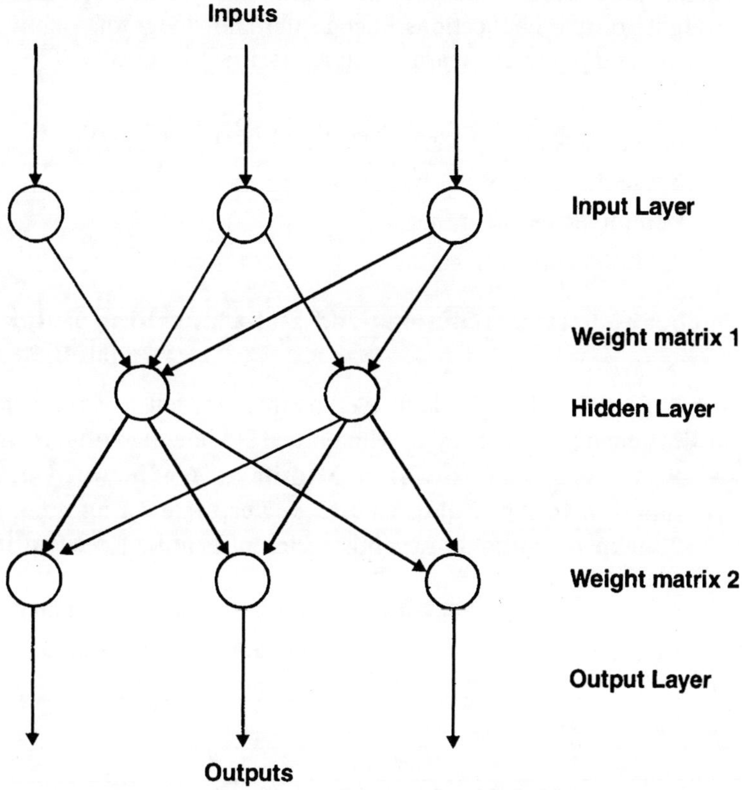
Type	Feed-forward
Neuron Layers	1 input layer 1 or more hidden layers 1 output layer
Input value types	Binary
Activation Functions	Hard limiter/sigmoid
Learning Method	Supervised
Learning algorithm	Delta learning rule Back-propagation (mostly used)

4.3. NEURAL NETWORK LEARNING

The neural network has also been called the "connectionist". It contains a large number of neuron like processing elements and a large number of weighted connections between the elements. The weights on the connections encode the knowledge of a network. It uses a highly parallel, distributed control and can learn to adjust itself automatically. The various learning algorithms are described below:

1. Backpropagation (Generalized Delta Learning Rule)
2. Radial Basis Function Networks
3. Reinforcement Learning
4. ART Networks

1. Backpropagation (Generalized Delta Learning Rule)

[MDU May 2008, Dec. 2008, May 2012]

It was first described by Paul Werbos in 1974, but it wasn't until 1986, through the work of David E. Rumelhart, Geoffrey E. Hinton and Ronald J. Williams, that it gained recognition.

- Backpropagation network is **Multilayer feedforward network** with different transfer function in ANN and a more powerful learning rule. The learning rule is called **"Backpropagation"**, which is a kind of **gradient descent technique** with backward error propagation.
- It is a **Supervised Learning** method, and is an implementation of the **Delta rule**. It requires a teacher that knows, or can calculate, the desired output for any given input.

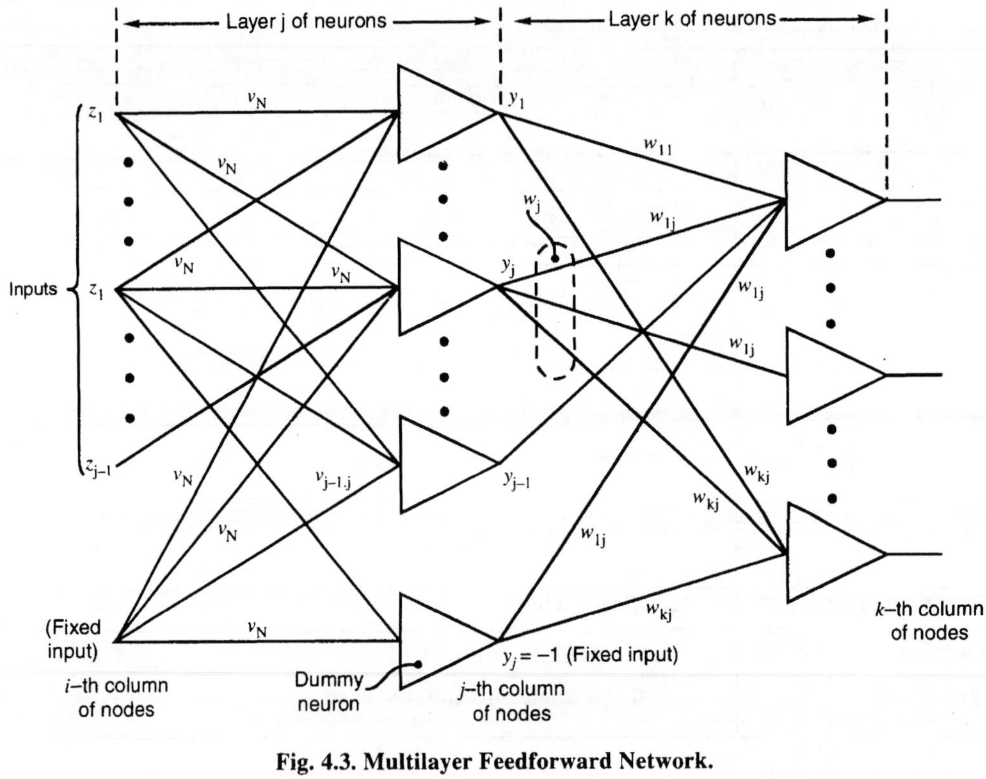

Fig. 4.3. Multilayer Feedforward Network.

When to use (or not!) a BP Neural Network Solution

A back propagation neural network is only practical in certain situations. Following are some guidelines on when another approach should be followed:

- If we write down a flow chart or a formula that accurately describes the problem? If so, then traditional programming method must be used.
- Is there a simple piece of hardware or software that is already doing what the programmer want? If so, then the development time for a NN might not be worth it.
- If we have an easy way to generate a significant number of input/output examples of the desired behavior? If not, then NN won't be able to train anything.
- If the correct answer of the problem can be found in a look-up table of reasonable size then a look-up table is much simpler and more accurate.
- Are precise numeric output values required? NN's are not good at giving precise numeric answers.

Conversely, here are some situations where a BP NN might be a good idea:

- A large amount of input/output data is available, but you're not sure how to relate it to the output.
- The problem appears to have overwhelming complexity, but there is clearly a solution.
- It is easy to create a number of examples of the correct behavior.
- The solution to the problem may change over time, within the bounds of the given input and output parameters (*i.e.*, today 2+2=4, but in the future we may find that 2+2=3.8).
- Outputs can be "fuzzy", or non-numeric.

One of the most common applications of NN's is in image processing. Some examples would be: identifying hand-written characters; matching a photograph of a person's face with a different photo in a database; performing data compression on an image with minimal loss of content. Other applications could be: voice recognition; RADAR signature analysis; stock market prediction. All of these problems involve large amounts of data, and complex relationships between the different parameters.

Limitation of Backpropagation

(1) The training set for the network must be presented many times in order for the interconnection weights between the neurons to settle into a state for correct classification of the input patterns. While the network can recognize patterns to those they have already learned, they do not have the capability to recognize new patterns. This is true for all **Supervised Learning Networks.** In order to recognize new patterns, the networks need to be trained with these patterns along with previously known patterns. If only new patterns are provided for retraining, then old patterns may be forgotten. In this way learning is not incremental over time. This is a **major limitation for supervised learning networks.**

(2) Backpropagation is prone to local minima just like gradient descent algorithm.

The backpropagation network learns a mapping from a set of input patterns to a set of output patterns. This network can be designed and trained to accomplish a wide variety of mappings. This ability comes from the nodes in the **hidden layer** which learns to respond to the features found in the input patterns. As the network is trained with different examples the network has the ability to generalize over similar features found in different patterns. Therefore the hidden units must be trained to extract a sufficient set of general features applicable to all types of instances. To achieve this goal the network should not be overtrained. Overtraining the network will make it memorize the individual input-output training pairs rather than settling in mapping for all cases.

➡ The convergence obtained from backpropagation learning is very slow.

➡ The convergence in backpropagation learning is not guaranteed.

Back Propagation Net Characteristics

Sample Structure	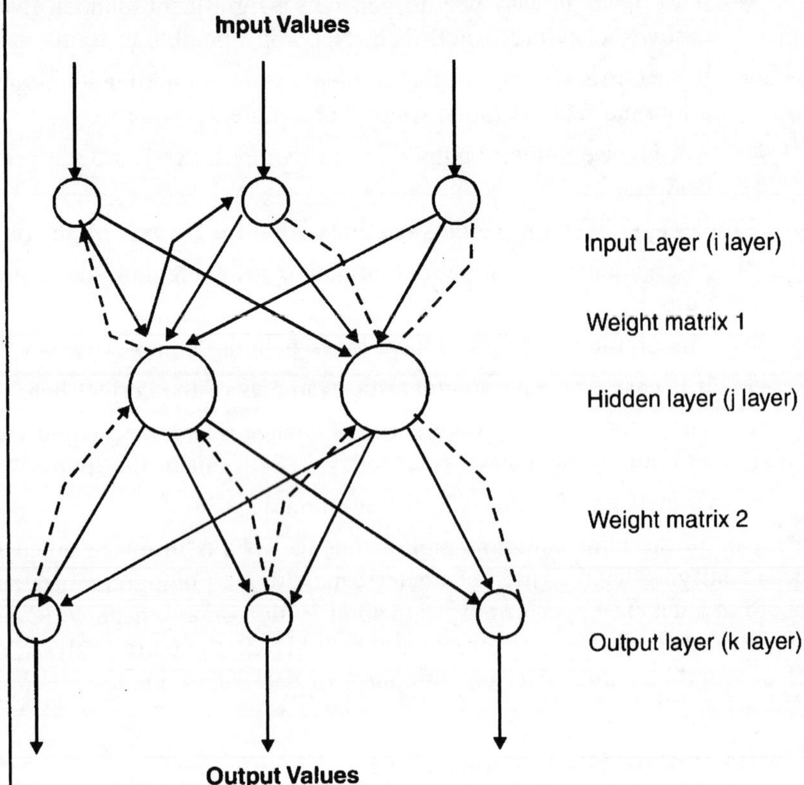
Type	Feed-forward
Neuron Layers	1 input layer 1 or more hidden layers
Input value types	Binary
Activation Functions	Sigmoid
Learning Method	Supervised
Learning algorithm	Back-propagation
Mainly used in	Complex logical operations Pattern classification Speech analysis

The Backpropagation Algorithm

Weight Initialization

Set all the weights and node thresholds to small random numbers. But node threshold is the negative of the weight from the bias unit.

Calculation of Activation

1. The activation level of an input unit is determined by the instances presented to the network.
2. The activation level O_j of hidden and output unit is determined by

$$O_j = F\,(\Sigma W_{ji} O_i - \theta_j)$$

Where W_{ji} is the weight from an input O_i , θ_j is the node threshold, and F is sigmoid function.

$$F(a) = \frac{1}{(1 + e^{-a})} \qquad \text{where } a = \text{weight} * \text{input}$$

Weight Training

1. Start at the output units and work backward to the hidden layers recursively. Adjust weights by

   $$\mathbf{W_{ji}(t + 1) = W_{ji}\,(t) + \Delta W_{ji}}$$

   where $W_{ji}\,(t)$ is the weight from unit i to unit j at time t and ΔW_{ji} is the weight adjustment.
2. The weight change is computed by $\Delta \mathbf{W}_{ji} = \eta \delta_j \mathbf{O}_i$

   where Δ is trial –independent learning rate $(0 < \Delta < 1)$

   and δ_j is the error gradient at unit j. Convergence is sometimes faster by adding a momentum term:

   $$W_{ji}(t + 1) = W_{ji}\,(t) + \eta \delta_j O_i + \alpha[W_{ji}\,(t) - W_{ji}(t - 1)]$$
3. The error gradient is given by :

   For the output units:

   $$\delta_j = O_j(1 - O_j)(T_j - O_j)$$

   Where T_j is the desired output activation and O_j is actual output activation at output unit j.

   For the hidden units:

   $$\delta_j = O_j(1 - O_j)\Sigma\delta_k\, W_{kj}$$

   where δ_k is error gradient at unit k to which a connection points from hidden unit j.
4. Repeat iterations until convergence in terms of the selected error criterion. An iteration include presenting an instance, calculating activations and modifying weights.

NOTE : The name "Backpropagation" comes from the fact that the error of hidden units are derived from propagating backward the errors associated with output units since the target values for the hidden units are not given. In this activation function used is sigmoid function, which compresses the output value into the range between 0 and 1.

Why use Sigmoid Function : The sigmoid function is advantageous because it can accommodate large signals without saturation while allowing the passing of small signals without excessive attenuation. Also, it is a smooth function so that gradients can be calculated which are required for a gradient descent search.

Derivation of Backpropagation Algorithm **[MDU Dec. 2007]**

Consider Fig. 4.3. The negative gradient descent formula for the hidden layer is

$$\Delta v_{ji} = -\eta \frac{\partial E}{\partial v_{ji}}, \quad \text{for } j = 1, 2, ..., J \text{ and } i = 1, 2, ..., I \qquad ...(4.4)$$

and

$$\frac{\partial E}{\partial v_{ji}} = \frac{\partial E}{\partial (net_j)} \cdot \frac{\partial (net_j)}{\partial v_{ji}} \qquad ...(4.5)$$

Let us notice that the inputs to the layer are z_i, for $i = 1, 2, ..., I$. Based on relation

$$\text{net} = \text{weight} * \text{input}$$

$$\text{net}_j = v_{ji} * Z_i$$

The second term in the product (*ii*) is equal to z_i, and the weight adjustment can be expressed as

$$\Delta v_{ji} = \eta \delta_{yj} z_i \qquad ...(4.6)$$

Where δ_{yj} is the error signal term of the hidden layer having output y. This error signal term is produced by the jth neuron of the hidden layer, where $j = 1, 2, ..., J$. The error signal term is equal to

$$\delta_{yj} \triangleq -\frac{\partial E}{\partial (net_j)}, \quad \text{for } j = 1, 2, ..., J \qquad ...(4.7)$$

In contrast to the output layer neurons' excitation net_k, which affected the kth neuron output only, the net_j contributes now to every error component in the error sum containing ***K*** terms. The error signal terms at the node j can be computed as

$$\delta_{yj} = -\frac{\partial E}{\partial_{yj}} \frac{\partial_{yj}}{\partial (net_j)} \qquad ...(4.8)$$

where

$$\frac{\partial E}{\partial_{yj}} = \frac{\partial}{\partial y_j}\left(\frac{1}{2}\sum_{k=1}^{K}\{d_k - f[net_k(y)]\}^2\right) \qquad ...(4.9)$$

The 2nd term of above Eq. (4.9) is

$$\frac{\partial y_j}{\partial (net_j)} = f_j'(net_j) \qquad ...(4.10)$$

Equation (4.9) will become (4.10) after taking the derivative

$$\frac{\partial E}{\partial y_j} = -\sum_{k=1}^{K}(d_k - o_k)\frac{\partial}{\partial y_j}\{f(net_k(y)]\} \qquad ...(4.11)$$

Calculation of the derivative in braces of expression (4.11) becomes

$$\frac{\partial E}{\partial y_j} = -\sum_{k=1}^{K}(d_k - o_k)f'(net_k)\frac{\partial (net_k)}{\partial y_j} \qquad ...(4.12)$$

Now after simplification of above equation, we get

$$\frac{\partial E}{\partial y_j} = -\sum_{k=1}^{K} \delta_{ok} w_{kj} \qquad ...(4.13)$$

Combining the result from (4.13) and (4.9)

$$\delta_{ji} = f_j'(net_j) \sum_{k=1}^{K} \delta_{ok} w_{kj}, \quad \text{for } j = 1, 2, ..., J \qquad ...(4.14)$$

The weight adjustment in the hidden layer now becomes

$$\Delta v_{ji} = \eta f_j'(net_j) z_i \sum_{k=1}^{K} \delta_{ok} w_{kj}, \quad \text{for } j = 1,2,...,J \text{ and } i = 1, 2, ..., I \qquad ...(4.15)$$

The Eq. (4.15) is called generalized Delta Rule.

The adjustment of weights leading to neuron j in the hidden layer is proportional to the weighted sum of all values at the adjacent following layer of nodes connecting neuron j with the output. The weights that fan out from node j are themselves the weighting factors. The weights affecting hidden neuron have been highlighted in the output layer in figure. The modified weights of the hidden layer can be expressed now as

$$v'_{ji} = v_{ji} + \eta f_j'(net_j) z_i \sum_{k=1}^{K} \delta_{ok} w_{kj}, \quad \text{for } j = 1, 2, \ldots, J \text{ and } i = 1, 2, \ldots, I \qquad ...(4.16)$$

The hidden layer weight adjustment based on the generalized delta training rule. For the network can be stated in vector notation as

$$V' = V + \eta \delta_y z'$$

where

$$z = \begin{bmatrix} z_1 \\ z_2 \\ \vdots \\ z_I \end{bmatrix}$$

$$V = \begin{bmatrix} v_{11} & v_{12} & \cdots & v_{1I} \\ v_{21} & v_{22} & \cdots & v_{2I} \\ \vdots & \vdots & \vdots & \vdots \\ v_{j1} & v_{j2} & \cdots & v_{jI} \end{bmatrix}$$

Defining now the jth column of matrix W as w_j, vector δy can be expressed compactly as follows :

$$\delta y = w_j' \delta_0 f_y' \qquad ...(4.17)$$

where f'_y is the column vector with entries f'_{yj} expressed for each hidden layer neuron 1, 2, ..., J, for unipolar and bipolar activation functions, respectively, as

$$f'_{yj} = y_j(1 - y_j)$$

$$f'_{yj} = \frac{1}{2}(1 - y_j^2)$$

4.4. FEED FORWARD RECALL AND ERROR BACKPROPAGATION TRAINING

Feedforward Recall

In general, the layer network is mapping the input vector z into the output vector o as follows :

$$o = N(z)$$

Where N denotes a composite nonlinear matrix operator. For the two-layer net the mapping $z \rightarrow o$ can be represented as a mapping within a mapping, or

$$o = G[W\Gamma[Vz]]$$

$$\Gamma[Vz] = y$$

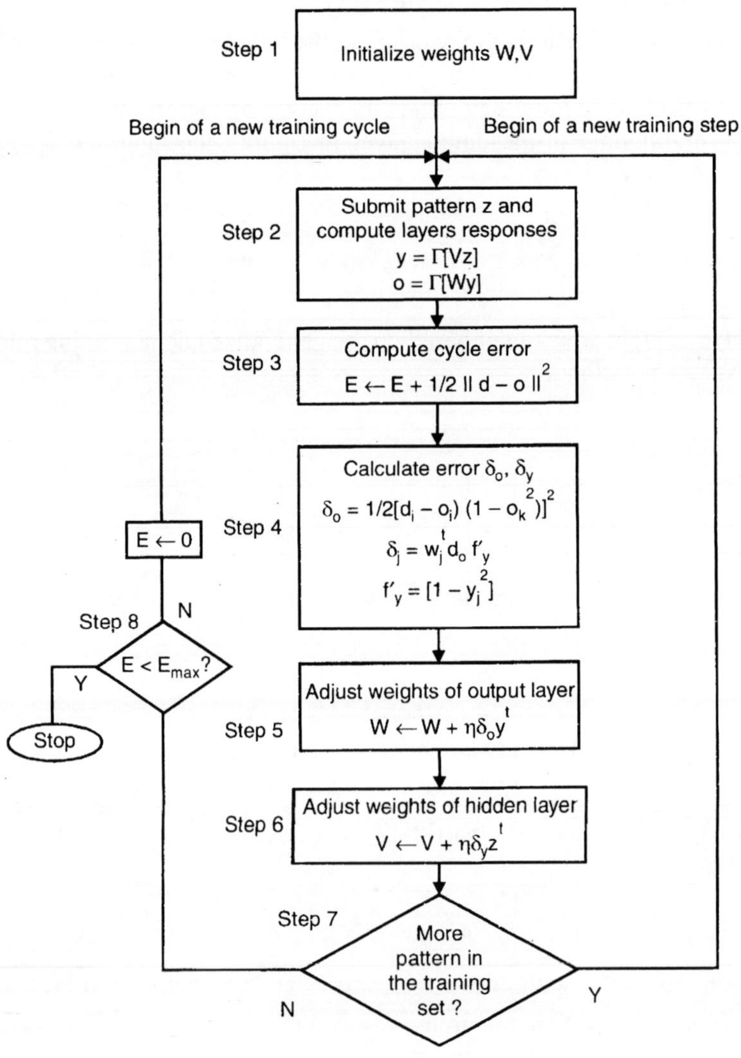

and it related to the hidden layer mapping $z \rightarrow y$. Note that the right arrows denote mapping of one space into another. Each of the mappings is performed by a single-layer of the layered network. The operator Γ is a nonlinear diagonal operator.

Algorithm

Given are P training pairs

$$\{z_1, d_1, z_2, d_2, ..., z_p, d_p\},$$

where z_i is $(I \times 1)$, d_i is $(K \times 1)$, and $i = 1, 2, ..., P$. Note that the Ith component of each x_i is of value -1 since input vectors have been augmented. Size $J - 1$ of the hidden layer having outputs y is selected. Note that the Jth component of y is of value -1, since hidden layer outputs have also been augmented; y is $(J \times 1)$ and o is $(K \times 1)$.

Step 1 : $\eta > 0$, E_{max} chosen

Weights W and Y are initialized at small random values; W is $(K \times J)$, V is $(J \times I)$.

$q \leftarrow 1, p \leftarrow 1, E \leftarrow 0$

Step 2 : Training step starts here (See Note 1 at end of list.)

Input is presented and the layers output computed [f(net) as in (2.3a) is used]

$$z \leftarrow z_p, \; d \leftarrow d_p$$

$$y_j \leftarrow f(v_j^t z), \text{ for } j = 1, 2, ..., J$$

where v_j, a column vector, is the jth row of V and

$$o_k \leftarrow f(w_k^t y), \text{ for } k = 1, 2, ..., K$$

where w_k, a column vector, is the kth row of W.

Step 3 : Error value is computed :

$$E \leftarrow \frac{1}{2}(d_k - o_k)^2 + E, \text{ for } k = 1, 2, ..., K$$

Step 4 : Error signal vectors δ_0 and δ_y of both layers are computed.

Vector δ_0 is $(K \times 1)$, δ_y is $(J \times 1)$. (See Note 2 at end of list.)

The error signal terms of the output layer in this step are

$$\delta_{ok} = \frac{1}{2}(d_k - o_k)(1 - o_k^2), \text{ for } k = 1, 2, ..., K$$

The error signal terms of the hidden layer in this step are

$$\delta_{yj} = \frac{1}{2}(1 - y_j^2)\sum_{k=1}^{K}\delta_{ok} w_{kj}, \quad \text{for } j = 1, 2, ..., J$$

Step 5 : Output layer weights are adjusted :

$$w_{kj} \leftarrow w_{kj} + \eta\delta_{ok} y_j, \quad \text{for } k = 1, 2, ..., K \text{ and } j = 1, 2, ..., J$$

Step 6 : Hidden layer weights are adjusted

$$v_{ji} \leftarrow v_{ji} + \eta\delta_{yj} z_i, \quad \text{for } j = 1, 2, ..., J \text{ and } i = 1, 2, ..., I$$

Step 7 : If $p < P$ then $p \leftarrow p + 1$, $q \leftarrow q + 1$ and go to Step 2; otherwise, go to Step 8.

Step 8 : The training cycle is completed.

For $E < E_{max}$ terminate the training session. Output weights W, V, q and E.

If $E > E_{max}$, then $E \leftarrow 0$, $p \leftarrow 1$, and initiate the new training cycle by going to Step 2.

How to solve the numerical based on Backpropagation

Consider three layers input layer, hidden layer and output layer.

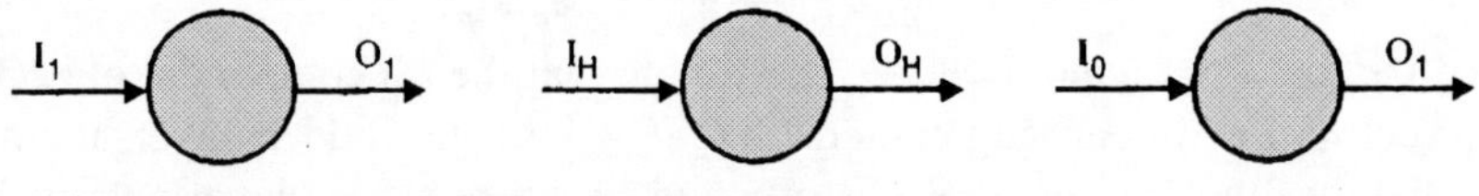

Fig. 4.4.

The input to the input layer is represented by I_I and the output of this layer O_I.

The input to the input layer is represented by I_H and the output of this layer O_H.

The input to the input layer is represented by I_O and the output of this layer O_O.

1. Write the input for input layer from the given question. The output of this layer is same as the input because there is no processing in the input layer.
2. Now calculate I_H.

   $I_H = [V]^t * O_I$. Where $[V]$ is the weight between the input and the hidden layer.
3. Next is to calculate the output of the hidden layer O_H.

   $O_H = 1/[1 + \exp(-\lambda net)$ where λ is a constant and net is the value calculated above. *i.e.*, net = weight*input for that layer.

   $$i.e., \quad O_H = 1/[1 + \exp(-\lambda . I_H)$$
4. Calculate I_O. $I_O = [W]^t * O_H$. where $[W]$ is the weight between the hidden layer and the output layer.
5. Calculate O_O. $O_O = 1/[1 + \exp(-\lambda net)$. where λ is a constant and net is the value calculated above. *i.e.*, net = weight*input for that layer *i.e.*, $O_O = 1/[1 + \exp(-\lambda . I_O)$
6. Calculate the error by $E = [1/2]*(d - O)^2$.
7. Calculate the error signal term for the output layer

   $$\delta_o = \frac{1}{2}[(d_k - o_k)(1 - o_k^2)]^*$$
8. Next calculate error signal term for hidden layer

   $$\delta_y = w_j^t \delta_0 f_y^t.$$

   Where $$f'_y = O_H(1 - O_H)$$
9. Adjust the weights of the output layer

   $$W_{new} = W_{old} + \eta \delta_0 . O_H^t$$
10. Adjust the weight of the hidden layer

    $$V^{new} = V^{old} + \eta \delta_y I_I^t$$

Example 4.1. To solve the EX-OR problem, we build a backpropagation network as shown in Fig 4.3. The network will be trained on the following instances:

The weights are initialized randomly as follows:

$$W_{13} = 0.02,\ W_{14} = 0.03,\ W_{12} = -0.02,\ W_{23} = 0.01,\ W_{24} = 0.02,\ W_{1b} = -0.01,\ W_{2b} = -0.01$$

Calculation of Activation: Consider a training instance with the input vector = (1,1) and the desired output vector = 0.

$$O_3 = O_4 = 1$$
$$O_2 = 1/[1 + e^{-(1*0.01+1*0.02\ -\ 1*0.01)}] = 0.505$$
$$O_1 = 1/[1 + e^{-(0.505*-0.02+1*0.02+1*0.03-1*0.01)}] = 0.508$$

Weight Training : Assume that the learning rate $\eta = 0.3$

$$\delta_1 = 0.508(1 - 0.508)\ (0\text{-}0.508) = -0.127$$

Weight adjustment $\Delta W_{13} = \eta\delta_1\ O_3 = 0.3*(-0.127)*1 = -0.038$

$$\delta_2 = 0.505(1 - 0.505)\ (-0.127* - 0.02) = -0.0006$$
$$\Delta W_{23} = 0.3*0.0006*1 = 0.0002$$

As the weight adjustments are decreasing therefore the rest of the weight adjustments are omitted.

Example 4.2. Consider the simple network below **[MDU May 2009]**

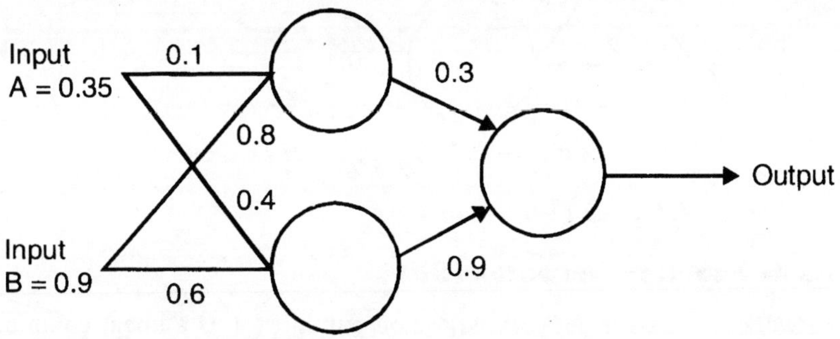

Fig. 4.5.

Assume that the neurons have a Sigmoid activation function and

(*i*) Perform a forward pass on the network.

(*ii*) Perform a reverse pass (training) once (target = 0.5).

(*iii*) Perform a further forward pass and comment on the result.

Solution. (*i*) Input to top neuron = (0.35*0.1) + (0.9*0.8) = 0.755.

$$\text{Out} = 1/1 + \exp(-1*0.755) = 0.68.$$

Input to bottom neuron = (0.9*0.6) + (0.35*0.4) = 0.68. Out = 0.6637.

Input to final neuron = (0.3*0.68) + (0.9*0.6637) = 0.80133. Out = 0.69.

(*ii*) Output error $\delta = (t - 0)(1 - 0)0 = (0.5 - 0.69)(1 - 0.69)0.69 = -0.0406.$

New weights for output layer

$$w1^{new} = w1 + (\delta * \text{input}) = 0.3 + (-0.0406*0.68) = 0.272392.$$
$$w2^{new} = w2 + (\delta * \text{input}) = 0.9 + (-0.0406*0.6637) = 0.87305.$$

Errors for hidden layers:

$$\delta_1 = \delta * w1 = -0.0406 * 0.272392 * (1 - o)o = -2.406*10^{-3}$$

$$\delta_2 = \delta * w2 = -0.0406 * 0.87305 * (1 - o)o = -7.916*10^{-3}$$

New hidden layer weights:

$$w3^{new} = 0.1 + (-2.406 \times 10^{-3} \times 0.35) = 0.09916.$$

$$w4^{new} = 0.8 + (-2.406 \times 10^{-3} \times 0.9) = 0.7978.$$

$$w5^{new} = 0.4 + (-7.916 \times 10^{-3} \times 0.35) = 0.3972.$$

$$w6^{new} = 0.6 + (-7.916 \times 10^{-3} \times 0.9) = 0.5928.$$

(*iii*) Old error was –0.19. New error is –0.18205. Therefore error has reduced.

3. Try a training pass on the following example. Target = 1, Learning rate = 1:

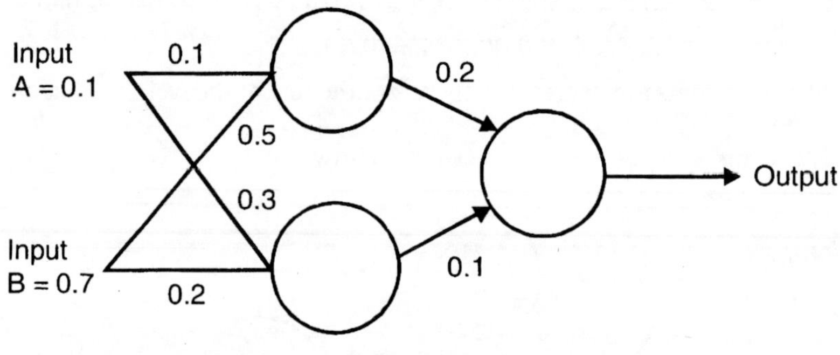

Fig. 4.6.

Output is 0.5, consider value of δ = 1. δ = 0.6.

2. Radial Basis Function Networks (RBF)

A RBF network is a 2 layer network whose output unit forms a linear combination of the basis functions computed by hidden units. And each hidden unit implements a radial activated function. The output unit implements a weighted sum of hidden output units. The input to RBF network is nonlinear while the output is linear. Due to their nonlinear properties, RBF networks are able to model complex mappings. The most common basis function can be viewed as activation function in the hidden layer.

The most common basis function chosen is Gaussian function, in which case the activation level O_j of the hidden unit j is calculated as

$$O_j = \exp\left[-(X - W_j)^* (X - W_j)/2s_j^2\right]$$

Where X is the input vector W_j is the weight vector associated with hidden unit j and s_j^2 is the normalization factor. The output of the hidden unit lie between 0 and 1; the closer the input to center of the Gaussian, the larger the response of the node . Because the node produces an identical output for inputs with equal distance from the centre of the Gaussian, it is called a radial basis. The activation level of an output unit is determined by

$$O_j = \partial W_{ji}\, O_i$$

The output units form a linear combination of nonlinear basis functions, and thus the overall network performs a nonlinear transformation of the input.

Learning is Done in Two Stages in RBF Networks

(1) Learning in the hidden layer is performed using unsupervised methods such as clustering algorithms will be described later. While learning in output layer is performed using supervised methods such as least mean square (LMS) algorithm.

(2) After the initial solution is found by this approach, a supervised learning algorithm *e.g.*, backpropagation can be applied to both the layers to fine tune the parameters of the network.

The normalization vector s represents a measure of the spread of the data in the cluster associated with the hidden unit. The value of s for hidden unit is

$$s_j^2 = [1*\Delta(X - W_j) * (X - W_j)]/M$$

where X is training pattern in the cluster, W_j is the center of the cluster associated with hidden unit j, and M is the number of training instances in that cluster.

RBF network can be used for both classification and functional approximation. For classification, the network can determine how close the given input is to center of a Gaussian basis by the response of the corresponding hidden unit. If only a single hidden unit is employed, the decision region is circular.

Difference between RBF and Backpropagation

The main difference is in their Basis Function. The radial basis function is used in RBF and it can cover only small regions, whereas the sigmoid function assumes large number of nonzero values over an infinitely large region of the input space. Depending upon the problem to be solved either can be used.

RBF Network

1. **Weight Initialization :** Weights in the hidden layer are determined by the clustering algorithm and weights in the output layer are initialized to small random values.
2. **Calculation of activation :** The activation level O_j of the hidden unit j is calculated as

$$O_j = \exp\,[-(X - W_j)^* (X - W_j)/2\sigma_j^2]$$

The activation level of an output unit is determined by

$$O_j = \Sigma W_{ji}\, O_i$$

3. **Weight Learning**

In the hidden layer, use k-means clustering algorithm

In the output layer :

(*a*) adjust the weights by $W_{ji}(t + 1) = W_{ji}(t) + \Delta W_{ji}(t)$

where $W_{ji}(t)$ is the weight from unit I to unit j at time j and $\Delta W_{ji}(t)$ is the weight adjustment.

(*b*) the weight change is computed by $\Delta \boldsymbol{W}_{ji} = \eta\delta_j O_i$

where Δ is a trial independent learning rate and δ_j is error at unit j :

$$\delta_j = T_j - O_j$$

where T_j is desired output activation and O_j is actual output activation at output unit j.

(*c*) repeat the iterations until convergence.

4.5. ART NETWORK

Adaptive resonance theory (ART) networks are most useful for pattern clustering, classification and recognition. They can also perform pattern association with some modifications. These networks work on binary or analog- valued input. Their ability to generalize is limited because ART networks can lack the hidden layer of neurons which perform feature recognition or pattern recognition in the backpropagation network.

The basic ART system is an unsupervised learning model. It typically consists of a comparison field and a recognition field composed of neurons, a vigilance parameter, and a reset module. The vigilance parameter has considerable influence on the system: higher vigilance produces highly detailed memories (many, fine-grained categories), while lower vigilance results in more general memories (fewer, more-general categories). The comparison field takes an input vector (a one-dimensional array of values) and transfers it to its best match in the recognition field. Its best match is the single neuron whose set of weights (weight vector) most closely matches the input vector. Each recognition field neuron outputs a negative signal (proportional to that neuron's quality of match to the input vector) to each of the other recognition field neurons and inhibits their output accordingly. In this way the recognition field exhibits lateral inhibition, allowing each neuron in it to represent a category to which input vectors are classified. After the input vector is classified, the reset module compares the strength of the recognition match to the vigilance parameter. If the vigilance threshold is met, training commences. Otherwise, if the match level does not meet the vigilance parameter, the firing recognition neuron is inhibited until a new input vector is applied; training commences only upon completion of a search procedure. In the search procedure, recognition neurons are disabled one by one by the reset function until the vigilance parameter is satisfied by a recognition match. If no committed recognition neuron's match meets the vigilance threshold, then an uncommitted neuron is committed and adjusted towards matching the input vector.

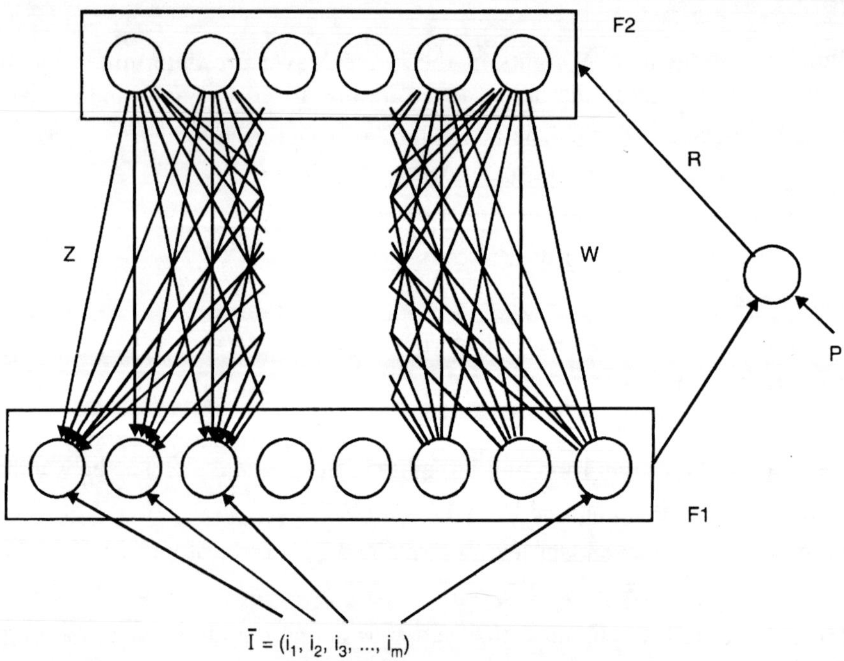

Fig. 4.7.

Training

There are two basic methods of training ART-based neural networks: slow and fast. In the slow learning method, the degree of training of the recognition neuron's weights towards the input vector is calculated to continuous values with differential equations and is thus dependent on the length of time the input vector is presented. With fast learning, algebraic equations are used to calculate degree of weight adjustments to be made, and binary values are used. While fast learning is effective and efficient for a variety of tasks, the slow learning method is more biologically plausible and can be used with continuous-time networks (*i.e.,* when the input vector can vary continuously).

ART 1

ART 1is the simplest variety of ART networks, accepting only binary inputs.

ART 2

ART 2 extends network capabilities to support continuous inputs.

ART 2-A

ART 2-A is a streamlined form of ART-2 with a drastically accelerated runtime, and with qualitative results being only rarely inferior to the full ART-2 implementation.

ART 3

ART 3 builds on ART-2 by simulating rudimentary neurotransmitter regulation of synaptic activity by incorporating simulated sodium (Na+) and calcium (Ca2+) ion concentrations into the system's equations, which results in a more physiologically realistic means of partially inhibiting categories that trigger mismatch resets.

Fuzzy ART

Fuzzy ART implements fuzzy logic into ART's pattern recognition, thus enhancing generalizability. An optional (and very useful) feature of fuzzy ART is complement coding, a means of incorporating the absence of features into pattern classifications, which goes a long way towards preventing inefficient and unnecessary category proliferation.

ARTMAP

ARTMAP, also known as **Predictive ART**, combines two slightly modified ART-1 or ART-2 units into a supervised learning structure where the first unit takes the input data and the second unit takes the correct output data, then used to make the minimum possible adjustment of the vigilance parameter in the first unit in order to make the correct classification.

Fuzzy ARTMAP

Fuzzy ARTMAP is merely ARTMAP using fuzzy ART units, resulting in a corresponding increase in efficacy.

4.6. FACTORS AFFECTING THE BACKPROPAGATION TRAINING [MDU, May 2010]

(1) Initialization of Weights : The weights of the network to be trained are initialized at small random values. The initialization affects the complete solution. If all the weights start out with equal weight values and if the solution requires that unequal weights be developed, the network may not be trained properly. Artificial neural networks typically start out with randomized weights for all their neurons. This means that they don't "know" anything and must be trained to solve the particular problem for which they are intended. Broadly speaking, there are two methods for training an ANN, depending on the problem it must solve.

A self-organizing ANN (often called a Kohonen after its inventor) is exposed to large amounts of data and tends to discover patterns and relationships in that data. Researchers often use this type to analyze experimental data.

A back-propagation ANN, conversely, is trained by humans to perform specific tasks. During the training period, the teacher evaluates whether the ANN's output is correct. If it's correct, the neural weightings that produced that output are reinforced; if the output is incorrect, those weightings responsible are diminished. This type is most often used for cognitive research and for problem-solving applications.

Implemented on a single computer, an artificial neural network is typically slower than a more traditional algorithmic solution. The ANN's parallel nature, however, allows it to be built using multiple processors, giving it a great speed advantage at very little development cost. The parallel architecture also allows ANNs to process very large amounts of data very efficiently. When dealing with large, continuous streams of information, such as speech recognition or machine sensor data, ANNs can operate considerably faster than their linear counterparts.

If training continues beyond a certain low error level, it will result in undesirable drift of weights. This causes error to increase and the quality of mapping by the network decreases. Therefore the choice of initial weights is however only one of several factors affecting training of error towards an acceptable error minimization.

(2) Learning Constant: Careful selection of step size (learning constant η) is often necessary to ensure the smooth convergence. However optimum value of η depends on problem being solved and there is no single learning constant value suitable for different training cases.

$$\eta = \frac{1.5}{N_1^2 + N_2^2 + \ldots + N_n^2}$$

Where N_1 is the number of patterns of type 1 and there are total n different patterns.

The learning coefficient cant be –ve because this would cause the change of weight vector to move from ideal weight vector position. If the value of η is zero then, no learning takes place and hence the value of ç must be +ve. From past experiences it is clear that a large step size may cause network to become paralyzed. When network paralysis occurs, further training does little for convergence. On the other hand, if the size is too small, convergence may be too slow *i.e.,* If η is too small, the algorithm will take a long time to converge (Fig. 4.8).

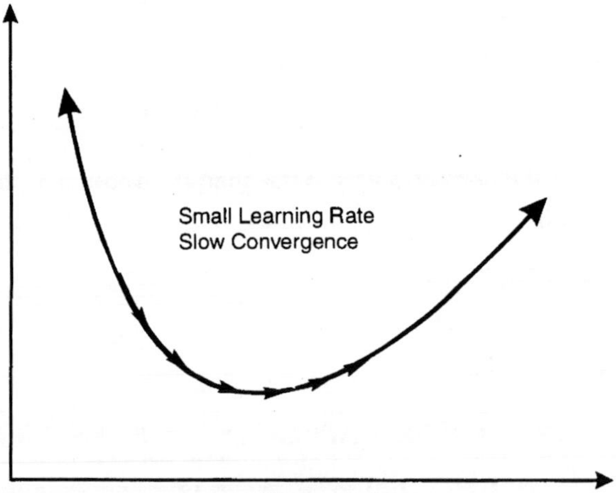

Fig. 4.8.

Therefore this leads to conclusion that η should be chosen experimentally for each problem.

Conversely, if η is too large, we may end up bouncing around the error surface out of control - the algorithm **diverges** (Fig. 4.9). This usually ends with an overflow error in the computer's floating-point arithmetic.

Therefore the value of learning coefficient must be between zero and one.

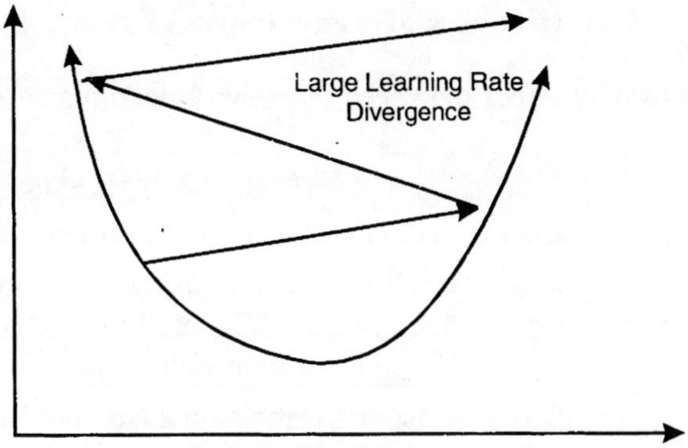

Fig. 4.9.

(3) Momentum Method : The purpose of the momentum method is to accelerate the convergence of the error back-propagation learning algorithm. The method involves supplementing the current weight adjustments with a fraction of the most recent weight adjustment. This is usually done according to the formula

$$\Sigma W(t) = (-)\delta\ \Delta E/\Delta t + \alpha\Sigma W(t-1)$$

where the arguments t and $t-1$ are used to indicate the current and the most recent training step, respectively, and it is a user-selected positive momentum constant. The second term, indicating a scaled most recent adjustment of weights, is called the ***momentum term.*** If momentum is zero, then smoothening is minimum and the entire weight adjustment comes from newly calculated change if momentum is 1, new weight adjustment is ignored and the previous one is repeated. Between 0 and 1 is a region where the weight adjustment is smoothened by an amount proportional to the momentum factor. Typically, α is chosen between 0.1 and 0.8 to increase the speed of learning without leading to oscillations.

When the gradient keeps changing direction, momentum will smooth out the variations. This is particularly useful when the network is not well-conditioned. In such cases the error surface has substantially different curvature along different directions, leading to the formation of long narrow valleys. For most points on the surface, the gradient does not point towards the minimum, and successive steps of gradient descent can oscillate from one side to the other, progressing only very slowly to the minimum (Fig. 4.10). Figure 4.10 shows how the addition of momentum helps to speed up convergence to the minimum by damping these oscillations.

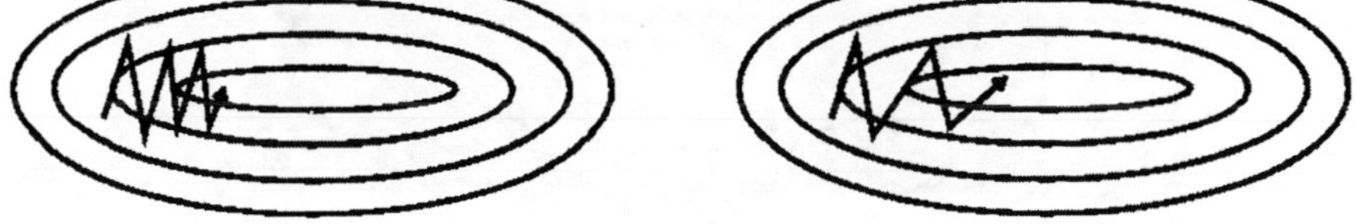

Fig. 4.10.

For the total of N steps using the momentum method, the current weight change can be expressed as :

$$\Delta w(t) = -c\sum_{n=0}^{N}\alpha^n \nabla E(t-n)$$

Typical value of α constant chosen less than unity

(4) Steepness of Activation Function : Both choice and shape of the activation function strongly affect the speed of network learning. $F(\text{net}) = \dfrac{2}{(1 + \exp(-\lambda net))} - 1$

Then $F'(\text{net}) = 2\ \lambda\exp(-\lambda\text{net}) / (1 + \exp(-\lambda\text{net}))^2$

It reaches a maximum value of ½ λ at net = 0

Previous researches show that using activation function with large λ may give result as in the case of large learning constant. Generally λ is kept at standard value of 1 and to control the learning speed using the learning constant rather than controlling both λ and η.

(5) Network Size : Let the Neural Network has three layers input, hidden, output layers represented by *I, J, K*. The number of input nodes is determined by dimensions or the size of input vector to be classified , generalized or associated with certain output quantity. Input vector size usually corresponds to number of distinct features of input patterns. The network should have such a size as to capture the structure of the data and to model the problem. With some prior knowledge about the problem structure, one can form a good estimate of the proper network size.

But a trade-off usually exists between number of training patterns P needed to create the training set and dimension I of the input pattern vector.

Let the characters I and T represented on 3*3 grids. This means total 3*3=9 pixels can be used to depict each of the characters.

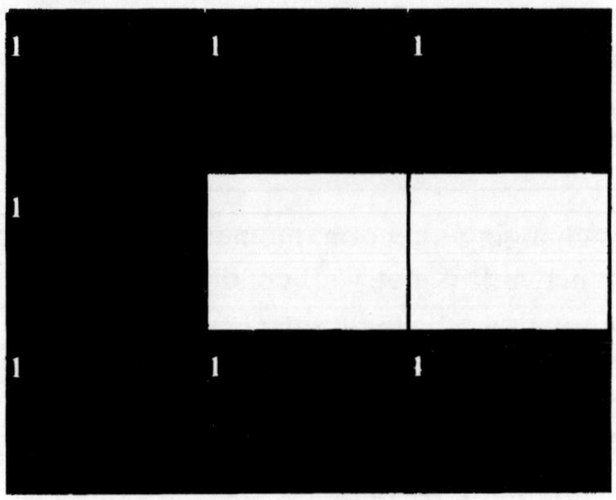

Fig. 4.11. C is represented by 1 .

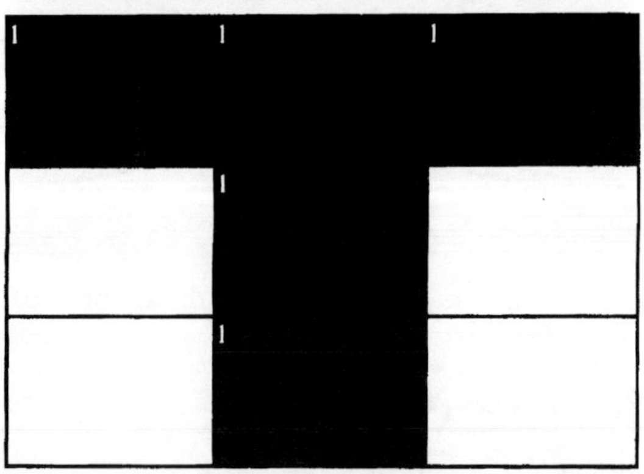

Fig. 4.12. T is represented by 1's in the cells.

On the figure, pixels of character *C*, *I* and *T* are represented using this approach. Assuming that the size of the input pattern vector is nine, we have $P = 3$ and the following three training vectors in the training pattern set $\{x_1, x_2, x_3\}$:

$$X_1 = [1\ 1\ 1\ 1\ 0\ 0\ 1\ 1\ 1]^t \quad : \text{Class C}$$
$$X_2 = [0\ 1\ 0\ 0\ 1\ 0\ 0\ 1\ 0]^t \quad : \text{Class I}$$
$$X_3 = [\ 1\ 1\ 1\ 0\ 1\ 0\ 0\ 1\ 0]^t \quad : \text{Class T}$$

Conditions for selecting the number of output neurons. The number of neurons in the output layer, *K*, can be made equal to the dimension of vectors to be associated if the network is employed as an associator of input to output vectors. If the network works as an auto-associator, which associates the distorted input vector with the undistorted class prototype, then we obviously have $I = K$. In the case of a network functioning as a classifier, *K* can be made equal to the number of classes.

(6) Necessary number of Hidden neurons : The size, of a hidden layer is one of the most important considerations when solving actual problems using multilayer feedforward networks Single-hidden layer networks can form arbitrary decision regions in n dimensional input pattern space. There exist certain useful solutions as to the number J of hidden neurons needed for the network to perform properly. The solutions also determine the lower bound on the number of different patterns P required in the training set. As will be shown, the number of hidden neurons depends on the dimension n of the input vector and on the number of separable regions in *n*-dimensional Euclidean input space.

(7) Sigmoidal gain : If sigmoidal function is selected, the input-output relationship of the neuron can be set as $O = 1/[1 + \exp(-\lambda(I + O))$

Where λ is a scaling factor known as sigmoidal gain.

In some problems, when the weights become large and force the neuron to operate in a region where the sigmoidal function is very flat, a better method of coping with the network paralysis is to adjust the sigmoidal gain.

By decreasing the scvaling factor, this activation function can be spread out on wide range so that training proceeds faster.

(8) Threshold Value : *O* in above equation is called as threshold value of a neuron, or bias or the noise factor. A neuron generates the output if the weighted sum of the input exceeds the threshold value.

(9) Stopping Criteria : The process of adjusting the weights based on the gradients is repeated until a minimum is reached. There are several stopping criteria that can be considered.

(1) **Based on the error to be minimized.** In various applications like pattern recognition the stopping criteria may be when the training data are correctly classified. When this is not the case, a fixed threshold is used so that the procedure is stopped if the error is below it. But this method does not guarantee the generalization to new data.

(2) **Based on gradient.** The algorithm is terminated when the gradient is sufficiently small. The gradient will be zero at a minimum by definition.

(3) **Based on cross-validation performance.** This criterion can be used to monitor generalization performance during learning and to terminate the algorithm when there is no more improvement.

The first 2 criteria are sensitive to the choice of parameters and may lead to poor results if the parameters are improperly chosen.

4.7. VARIATIONS ON THE BACKPROPAGATION ALGORITHM

1. Backpropagation using Genetic Algorithm

In BPN, gradient descent technique is used to determine the weights. Instead of this genetic algorithm can be a more robust search and optimization technique to obtain the solutions of problems, more quickly and accurately.

2. Augmented Backpropagation Networks

This is another variation of conventional BPN in which logarithmic and exponential neurons are added to neural network's input and output layers. The logarithmic neuron in the input layer receives a natural logarithmic transformation of a corresponding input value of the training data.

$$A_i = \ln (1.175X_i + 1.543)$$

where X_i is the ith input value of the training data and A_i is the output of ith logarithm neuron in the input layer.

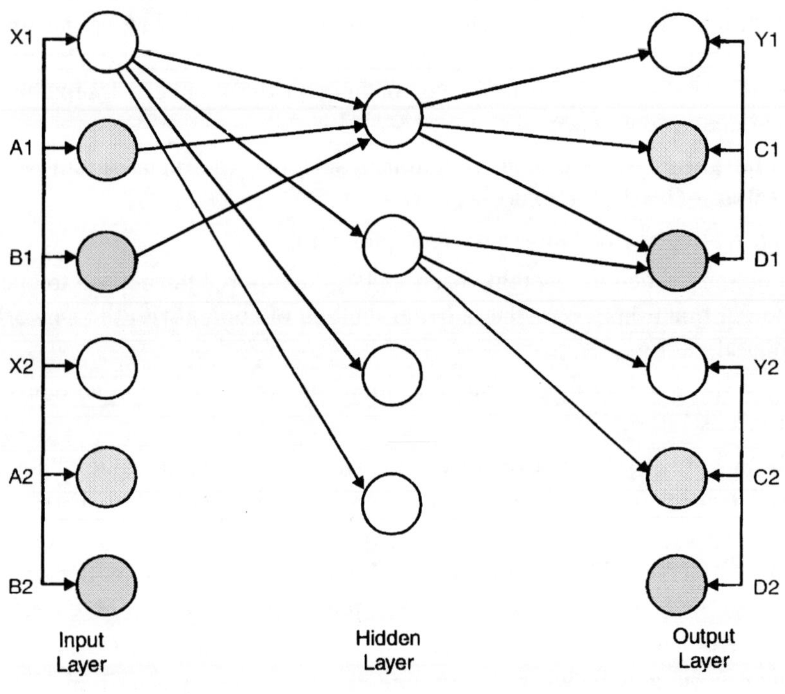

Fig. 4.13

4.8. OTHER IMPORTANT TOPICS

Cumulative Update of Weights

Another technique that has an effect on convergence speed is to only update the weights after a certain number of input training patterns. The number of training patterns presented during the accumulation is referred to as an **epoch**. *This may correspond either to the complete set of training patterns or a subset.*

Accumulate the weight correction term, and average over the entire training set or a given batch size. In this case, we get an increase in convergence of the neural network but perhaps to a local minimum.

$$Dw_{batch} = \frac{1}{P}\sum_{1}^{P} \Delta w$$

where P = number of training patterns in batch = epoch

ADAPTIVE LEARNING RATES: HEURISTIC APPROACH

Let each weight have its own learning rate, which varies as training progresses

If weight change in the same direction (increase or decrease) for several training patterns, the learning rate for the weight is increased.

If the direction of weight change alternates, the learning rate is decreased

DELTA-BAR-DELTA RULE

Notation:

$w_{ij}(t)$ = weight between neuron i and neuron j at time t.

$a_{ij}(t)$ = learning rate between neuron i and neuron j at time t.

The weight update formulas are given by

$$\Delta w_{jk}(t+1) = \alpha_{jk}(t+1)\,\delta_k z_j + \mu \Delta w_{jk}(t)$$

$$\Delta v_{ij}(t+1) = \alpha_{ij}(t+1)\,\delta j x_j + \mu \Delta v_{ij}(t)$$

or equivalently,

$$\Delta w_{jk}(new) = \alpha_{jk}(new)\,\delta_k z_j + \mu \Delta w_{jk}(old)$$

where $\Delta v_{ij}(new) = \alpha_{ij}(new)\,\delta_j x_j + \mu \Delta v_{ij}(old)$

where

$$\alpha_{jk}(t+1) = \begin{cases} \alpha_{jk}(t) + K & \text{if } d > 0 \\ (1-\gamma)\alpha_{jk}(t) & \text{if } d > 0 \\ \alpha_{jk}(t) & \text{otherwise} \end{cases}$$

γ = constant learning rate decrement factor

K = constant learning rate increment

where Delta-bar-delta $\quad d = \bar{\Delta}_{jk}(t-1)\Delta_{jk}(t)$

For each output neuron, we define a "delta"

Batch vs. Online Learning

Above we have accumulated the gradient contributions for all data points in the training set before updating the weights. This method is often referred to as **batch** learning. An alternative approach is **online** learning, where the weights are updated immediately after seeing each data point. Since the gradient for a single data point can be considered a noisy approximation to the overall gradient G (Fig. 4.14), this is also called **stochastic** (noisy) gradient descent.

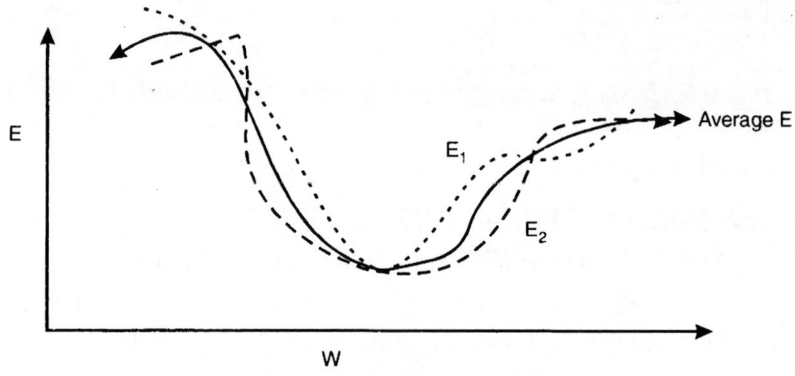

Fig. 4.14

Online learning has a number of advantages:

- it is often much faster, especially when the training set is **redundant** (contains many similar data points),
- it can be used when there is no fixed tr training set (new data keeps coming in),
- it is better at tracking **nonstationary** environments (where the best model gradually changes over time),
- the noise in the gradient can help to escape from **local minima** (which are a problem for gradient descent in nonlinear models).

These advantages are, however, bought at a price: many powerful optimization techniques (such as: conjugate and second-order gradient methods, support vector machines, Bayesian methods, etc.) - which we will not talk about in this course! - are batch methods that cannot be used online

Local Rate Adaptation

If we are willing to be a little more sophisticated, we go a lot further than the above global methods. First let us define an online weight update that uses a local, time-varying learning rate for each weight:

$$w_{ij}(t+1) = w_{ij}(t) + \mu_{ij}(t)\Delta w_{ij}(t)$$

The idea is to adapt these local learning rates by gradient descent, while simultaneously adapting the weights. At time t, we would like to change the learning rate (before changing the weight) such that the loss $E(t+1)$ at the next time step is reduced. The gradient we need is

$$\frac{\partial E(t+1)}{\partial \mu_{ij}(t)} = \frac{\partial E(t+1)}{\partial w_{ij}(t+1)} \frac{\partial w_{ij}(t+1)}{\partial \mu_{ij}(t)} = -\Delta w_{ij}(t)\,\Delta w_{ij}(t-1)$$

Ordinary gradient descent in μ_{ij}, using the meta-learning rate q (a new global parameter), would give

$$\mu_{ij}(t) = \mu_{ij}(t-1) + q\Delta w_{ij}(t)\,\Delta w_{ij}(t-1)$$

We can already see that this would work in a similar fashion to momentum: increase the learning rate as long as the gradient keeps pointing in the same direction, but decrease it when you land on the opposite slope of the loss function.

Problem : μ_{ij} might become negative! Also, the step size should be proportional to μ_{ij} so that it can be adapted over several orders of magnitude. This can be achieved by performing the gradient descent on $\log(\mu_{ij})$ instead:

$$\log(\mu_{ij}(t)) = \log(\mu_{ij}(t-1)) + q\Delta w_{ij}(t)\,\Delta w_{ij}(t-1)$$

Exponentiating this gives

$$\mu_{ij}(t) = \mu_{ij}(t-1)e^{q\Delta w_{ij}(t)\Delta w_{ij}(t-1)}$$
$$\approx \mu_{ij}(t-1)\max(0.5, 1 + q\Delta w_{ij}(t)\,\Delta w_{ij}(t-1))$$

where the approximation serves to avoid an expensive exp function call. The multiplier is limited below by 0.5 to guard against very small (or even negative) factors.

Problem : The gradient is noisy; the product of two of them will be even noisier—the learning rate will bounce around a lot. A popular way to reduce the stochasticity is to replace the gradient at the previous time step $(t-1)$ by an **exponential average** of past gradients. The exponential average of a time series $u(t)$ is defined as

$$\bar{u}(t) = m\bar{u}(t-1) + (1-m)u(t)$$

where $0 < m < 1$ is a new global parameter.

Problem : If the gradient is ill-conditioned, the product of two gradients will be even worse - the condition number is squared. We will need to normalize the step sizes in some way. A radical solution is to throw away the magnitude of the step, and just keep the sign, giving

$$\mu_{ij}(t) = \begin{cases} r\mu_{ij}(t-1) & \text{if } \Delta w_{ij}(t)\,\overline{\Delta w_{ij}}(t-1) > 0 \\ (1/r)\,\mu_{ij}(t-1) & \text{otherwise} \end{cases}$$

where $r = e^q$. This works fine for batch learning, but...

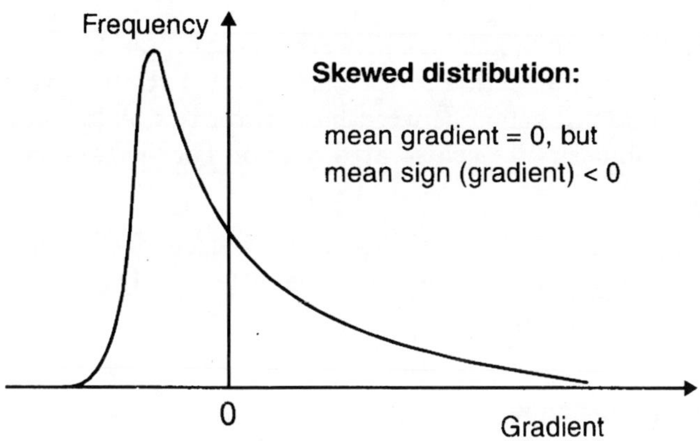

Fig. 4.15.

Problem : Nonlinear normalizers such as the sign function lead to systematic errors in stochastic gradient descent (Fig. 4.14): a skewed but zero-mean gradient distribution (typical for stochastic equilibrium) is mapped to a normalized distribution with non-zero mean. To avoid the problems this is causing, we need a linear normalizer for online learning. A good method is to divide the step by $\overline{\Delta w_{ij}^2}$ an exponential average of the squared gradient. This gives

$$\mu_{ij}(t) = \mu_{ij}(t-1)\max\left(0.5, 1+q\Delta w_{ij}(t)\frac{\overline{\Delta w_{ij}(t-1)}}{\overline{\Delta w_{ij}^2(t)}}\right)$$

Problem : Successive training patterns may be correlated, causing the product of stochastic gradients to behave strangely. The exponential averaging does help to get rid of short-term correlations, but it cannot deal with input that exhibits correlations across long periods of time. If you are iterating over a fixed training set, make sure you **permute** (shuffle) it before each iteration to destroy any correlations. This may not be possible in a true online learning situation, where training data is received one pattern at a time.

Binomial

To use a neural network for classification, we need to construct an equivalent function approximation problem by assigning a target value for each class. For a **binomial** (two-class) problem we can use a network with a single output y, and binary target values: 1 for one class, and 0 for the other. We can thus interpret the network's output as an estimate of the probability that a given pattern belongs to the '1' class. To classify a new pattern after training, we then employ the **maximum likelihood** discriminant, $y > 0.5$.

A network with linear output used in this fashion, however, will expend a lot of its effort on getting the target values *exactly* right for its training points - when all we actually care about is the correct positioning of the discriminant. The solution is to use an activation function at the output that saturates at the two target values: such a function will be close to the target value for any net input that is sufficiently large and has the correct sign. Specifically, we use the logistic sigmoid function

$$f(u) = \frac{1}{1+e^{-u}} \quad f'(u) = f(u)\,(1-f(u))$$

Given the probabilistic interpretation, a network output of, say, 0.01 for a pattern that is actually in the '1' class is a *much* more serious error than, say, 0.1. Unfortunately the sum-squared loss function makes almost no distinction between these two cases. A loss function that is appropriate for dealing with probabilities is the **cross-entropy** error. For the two-class case, it is given by

$$E = -t \ln y - (1-t) \ln (1-y)$$

When logistic output units and cross-entropy error are used together in backpropagation learning, the error signal for the output unit becomes just the difference between target and output:

$$\frac{\partial E}{\partial_{net}} = \ldots = y - t$$

In other words, implementing cross-entropy error for this case amounts to nothing more than omitting the f'(net) factor that the error signal would otherwise get multiplied by. This is not an accident, but indicative of a deeper mathematical connection: cross-entropy error and logistic outputs are the "correct" combination to use for binomial probabilities, just like linear outputs and sum-squared error are for scalar values.

Multinomial

If we have multiple *independent* binary attributes by which to classify the data, we can use a network with multiple logistic outputs and cross-entropy error. For **multinomial** classification problems (1-of-n, where $n > 2$) we use a network with n outputs, one corresponding to each class, and target values of 1 for the correct class, and 0 otherwise. Since these targets are not independent of each other, however, it is no longer appropriate to use logistic output units. The correct generalization of the logistic sigmoid to the multinomial case is the **softmax** activation function:

$$f(net_i) = \frac{e^{net_i}}{\Sigma_0 e^{net_o}}$$

where o ranges over the n output units. The cross-entropy error for such an output layer is given by

$$E = -\sum_o t_o \ln y_o$$

Since all the nodes in a softmax output layer interact (the value of each node depends on the values of all the others), the derivative of the cross-entropy error is difficult to calculate. Fortunately, it again simplifies to

$$\frac{\partial E}{\partial net} = \ldots = y - t$$

Summary

- **Multilayer Perceptron (MLP)** is a network of simple neurons called perceptrons. The basic concept of a single perceptron was introduced by Rosenblatt in 1958. A Multilayer Perceptron is a feedforward neural network with at least one hidden layer. It can deal with nonlinear classification problems.
- The various learning algorithms are described in this chapter:
  (*i*) Backpropagation (Generalized Delta Learning Rule)
  (*ii*) Radial Basis Function Networks
  (*iii*) Reinforcement Learning
  (*iv*) ART Networks
- Backpropagation or generalized delta learning rule has been discussed in this chapter. The algorithm uses least mean square error minimization strategy. Modifications in the weight is taken along the –ve gradient direction to reduce the error.
- The activation function used in the BP is sigmoidal function.
- The limitation of BP is that it is prone to local minima just like gradient descent formula.
- While solving the problems based on BP following formulas are used.

  For computing the error

  $$E \leftarrow E + \frac{1}{2}||d - o||^2$$

  For error signal term at output layer

  $$\delta_0 = \frac{1}{2}[(d_k - o_k)(1 - o_k^2)]$$

  For error signal term at hidden layer

  $$\delta_y = w_j^t \delta_o f_y',$$

where $f'_y = \frac{1}{2}[1 - y_j^2]'$

At the output layer the weights are adjusted according to

$$W \leftarrow W + \eta\delta_o y'$$

where y is the output of the hidden layer node.

At the hidden layer the weights are adjusted according to

$$V \leftarrow V + \eta\delta_y z'$$

1. Where z is the input to the input layer.
2. At the last the various factors affecting Backpropagation like initializing weights, learning constant, momentum, number of hidden layer nodes were discussed.

➡ Many variations of BPN algorithms such as decremental iterative procedure, adaptive BP, GA based BPN, augmented BPN, sequential learning single hidden layer NN are possible. Few of them discussed in this chapter.

Review Questions

1. Explain the term perceptron. What is the difference between single layer and Multilayer perceptron.
2. State whether single layer perceptrons are suitable for linearly non separable problems.
3. Why Backpropagation is called Generalized Delta learning rule?
4. Which activation function is used in Backpropagation algorithm and Why?
5. Explain the various factors affecting Backpropagation algorithm.
6. Derive Backpropagation equation for computing the error signal term at output layer, and hidden layer.

❑❑❑

CHAPTER

5

SINGLE LAYER FEEDBACK NETWORK

INSIDE THIS CHAPTER

Dynamical systems are those systems that processes the initial condition information over time while moving through a sequence of states. Single layer feedback networks are introduced in this chapter. The feedback networks presented in the last chapter have a feedback loop and thus produce time delay. But the feedforward networks have no feedback activity during the information recall phase. The feedback is used for adjustment of weights during training that gradually reduced the overall output error. This reduction of error provides the feedback type interaction in supervised learning. The NN introduced in this chapter represent the dynamical systems evolving in time in either a continuous or discrete output space. This reduction of error provides the feedback type iteraction in supervised learning.

Following steps are used to recall information stored in the network.

1. Input pattern is applied and networks output is initialized accordingly.
2. The initializing pattern is removed and the initial output forces the new, updated input through feedback connections.
3. The first updated input forces the first updated output. This in turn, produces the 2nd updated input and the 2nd updated response.

This process is continued till the further updation is not possible and network has reached the equilibrium.

Single layer feedback networks are used in continuous and discrete-time systems. These types of networks provide associations or classifications, optimization problem solution mapping networks etc.

5.1. BASIC CONCEPTS OF DYNAMICAL SYSTEM

The neural networks considered here are dynamical systems. This means that they process the initial condition information over time while moving through a sequence of states.

Definitions:

Attractor is a state toward which the system evolves in time starting from certain initial conditions. The initial conditions initiates the evolution terminating in that attractor.

The set of conditions for an attractor is called **BASIN of Attraction.**

If an attractor is a unique point in state space, then it is called a **fixed point.**

If attractor is a periodic sequence of states then it is called a **limit cycle.**

For example : Let us assume that dynamical system is designed so that it has 3 vectors representing the letters *A*, *B*, *C* as fixed point.

Then basin of attraction (condition) can be shaped in such a way that it will evolve toward an appropriate character vector even if initializing vector does not exactly resemble the original character. Thus, a certain amount of noise or distortion of character vector not only can be tolerated by the system, but even removed under the appropriate conditions.

Therefore, dynamical neural networks need to have their numerous parameters selected carefully. These are :

1. Learning of parameters of a dynamical system is dependent on the network model adopted.
2. Large number of learning methods available to store the information is single layer feedback networks. For example Correlation, Hebbian learning rules etc.
3. Learning of dynamical systems is accomplished without a teacher *i.e.,* unsupervised learning.
4. Drawback of dynamical systems is their limited explanation capability.

5.2. HOPFIELD MODEL [MDU May 2008, Dec. 2008, Dec. 2011]

The importance of the different Hopfield networks in practical application is limited due to theoretical limitations of the network structure but in certain conditions, they may form interesting models. Hopfield networks are typically used for classification problems with binary pattern vectors.

Hopfield networks are also called associative networks since they associate a class pattern to each input pattern.

The single-layer feedback neural network is shown in Fig. 5.1.

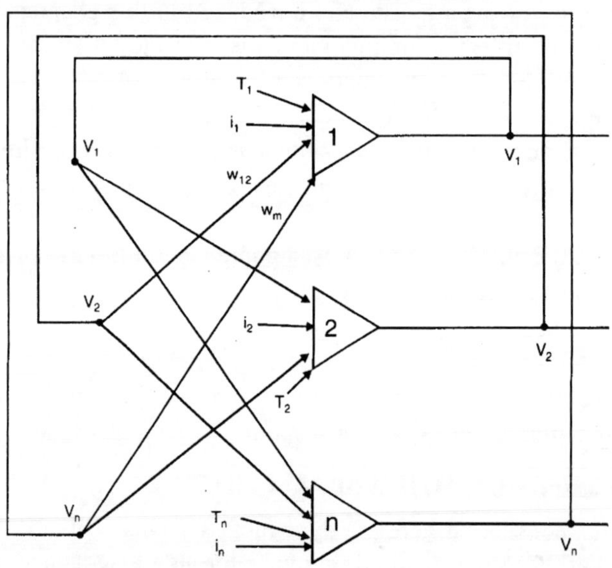

Fig. 5.1. Single Layer Feedback NN

Hopfield networks are constructed from artificial neurons shown in Fig. 5.2.

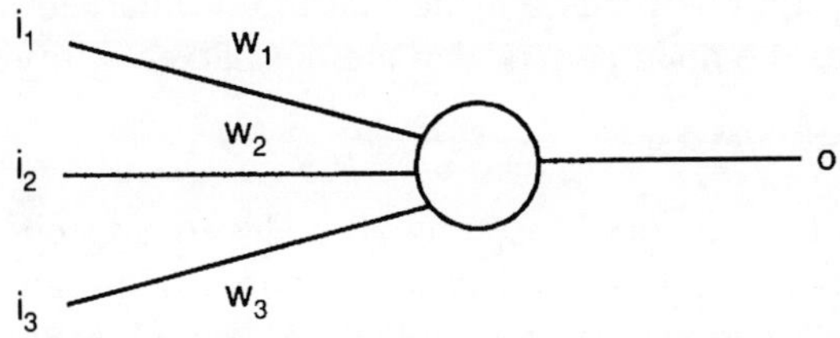

Fig. 5.2. An artificial neuron as used in a Hopfield network.

These artificial neurons have N inputs. With each input i there is a weight w_i associated. They also have an output. The state of the output is maintained, until the neuron is updated. Updating the neuron entails the following operations:

- The value of each input, x_i is determined and the weighted sum of all inputs, $\sum_{i=1}^{n} w_i x_i$ is calculated.
- The output state of the neuron is set to +1 if the weighted input sum is larger or equal to 0. It is set to –1 if the weighted input sum is smaller than 0.
- A neuron retains its output state until it is updated again.

A Hopfield network is a network of N such artificial neurons, which are fully connected. The connection weight from neuron j to neuron i is given by a number w_{ij}. The collection of all such numbers is represented by the weight matrix W, whose components are w_{ij}.

Two ways of updating the neurons are : Synchronous and Asynchronous

In asynchronous updation of the neurons,

- **Asynchronous :** In asynchronous updation of the neurons, one neuron is picked up, and the weighted Input and updation is calculated immediately. This can be done in a fixed order or neurons can be picked at random, that is why this is called asynchronous random updating.
- **Synchronous :** The weighted input sums of all neurons are calculated without updating the neurons. Then all neurons are set to their new value, according to the value of their weighted input sum.

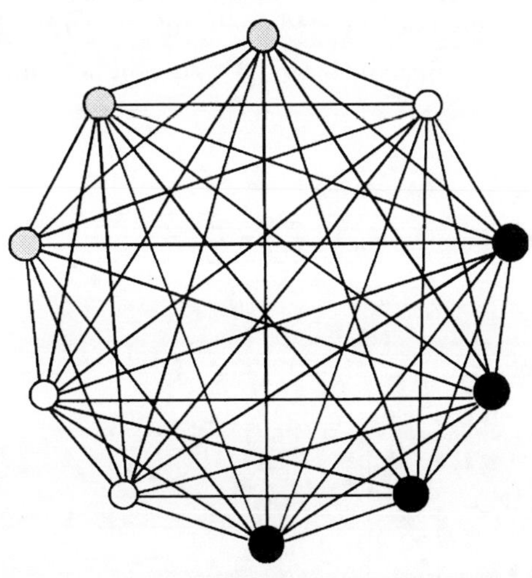

Fig. 5.3.

A Hopfield network works as an autoassociator. One enters a pattern in darken nodes and let the network evolve. After a while one reads out the that are not darken nodes. The memory of the Hopfield network associates the node pattern that are not darken with the darken node pattern.

USE OF HOPFIELD NETWORK

A pattern is entered in the network by setting all nodes to a specific value, or by setting only part of the nodes. The network is then subject to a number of iterations using asynchronous or synchronous updating. This is stopped after a while. The network neurons are then read out to see which pattern is in the network. The idea behind the Hopfield network is that patterns are stored in the weight matrix. The input must contain part of these patterns. The dynamics of the network then retrieve the patterns stored in the weight matrix. This is called Content Addressable Memory (CAM). The network can also be used for auto-association. The patterns that are stored in the network are divided in two parts: cue and association (see Fig. 5.3). By entering the cue into the network, the entire pattern, which is stored in the weight matrix, is retrieved. In this way the network restores the association that belongs to a given cue.

The stage is now almost set for the Hopfield network, we must only decide how we determine the weight matrix. Two conditions on the weight matrix are :

- symmetry: $w_{ij} = w_{ji}$
- No self connections: $w_{ii} = 0$

TIPS :

The energy of a Hopfield network can only decrease or stay the same. If an update cause a neuron to change sign, then the energy will decrease, otherwise it will stay the same.

Hopfield network (model) consists of

(1) A set of neurons and a corresponding set of unit delay forming a multiple-loop feedback system, and n neurons having threshold value T_i.

(2) Number of feedback loops is equal to number of neurons.

(3) The output of each neuron is feedback via a unit delay element to each of the other neurons in the network *i.e.*, there is no self feedback in the network.

(no self feedback as V_1 at output is connected to V_1 that is 0. Let us $V_1 \ldots V_n$. not with 1st neuron.

(4) The feedback input to ith neuron is equal to weighted sum of neuron outputs V_j where $j = 1, 2, \ldots, n$

∴ Total input net_i of ith neuron is

$$\text{net}_i = \sum_{\substack{j=1 \\ j \neq i}}^{n} w_{ij} V_j + \underbrace{i_i - T_i}_{\text{extended input}} \qquad i = 1, 2, \ldots, n$$

or

$$\text{net}_i = w_i^t V + i_i - T_i \qquad i = 1, 2, \ldots, n \quad \ldots(5.1)$$

where

$$w_i = \begin{bmatrix} w_{i1} \\ w_{i2} \\ \vdots \\ w_{in} \end{bmatrix}$$

$$V = \begin{bmatrix} V_1 \\ V_2 \\ \vdots \\ V_n \end{bmatrix}$$

V is neural network output vector.

or in more generalized form

where $$\text{net} = WV + i - t \qquad ...(5.2)$$

$$i = \begin{bmatrix} i_1 \\ i_2 \\ \vdots \\ i_n \end{bmatrix} \text{ and net} = \begin{bmatrix} net_1 \\ net_2 \\ \vdots \\ net_n \end{bmatrix}$$

$$t = \begin{bmatrix} T_1 \\ T_2 \\ \vdots \\ T_n \end{bmatrix}$$

are the vectors denoting input to each neuron 1 activation and threshold vectors.

Matrix W is called connectivity matrix is $n \times n$ matrix, the weight matrix W in this model is considered to be symmetric *i.e.*, $w_{ij} = w_{ji}$ and

$$\therefore \qquad W = \begin{bmatrix} 0 & w_{12} & w_{13} & \cdots & w_n \\ w_{21} & 0 & w_{23} & \cdots & w_{2n} \\ \vdots & \vdots & \vdots & \vdots & \vdots \\ w_{31} & w_{32} & w_{33} & \cdots & 0 \end{bmatrix}$$

With diagonal entries equal explicitly to zero *i.e.*,

$$w_{ij} = w_{ji} \qquad \text{and} \qquad w_{ii} = 0$$

As discrete network evolve TLUs, therefore output v is either -1 or $+1$

i.e., $$v_i \to -1 \text{ if } \text{net}_i < 0$$

$$v_i \to +1 \text{ if } \text{net}_i > 0 \qquad ...(5.3)$$

the rule (5.3) is used in an asynchronous fashion this means that for a given time only a single neuron is allowed to updated its output and only one entry in vector V is allowed to change.

As discrete networks are used. ∴ activation function that is used assumed to be sgn (.).

$$\therefore \qquad V_i^{k+1} = \text{sgn}\,(w_i^t\, V^k + i_i - T_i) \qquad ...(5.4) \quad \text{for} \begin{cases} i = 1, 2, \ldots. n \\ k = 0, 1, \ldots \end{cases}$$

k is index of recursive updation.

In Eq. (5.4) $k + 1$ is used ⇒ there is delay in the discrete system because of the feedback loop used in the system.

The scalar valued energy function is

$$E \triangleq -\frac{1}{2} V^t WV - i^t V + t^t V \qquad ...(5.5)$$

in expanded form

$$E = -\frac{1}{2} \sum_{\substack{i=1 \\ i \neq j}}^{n} \sum_{j=1}^{n} w_{ij} V_i V_j - \sum_{i=1}^{n} i_i . V_i + \sum_{i=1}^{n} t_i . V_i \qquad ...(5.6)$$

and $\qquad V_i^{k+1} - V_i^k = \Delta V_i$

taking gradient of eqn. (5.6) w.r.t. V_i

$$\nabla E = -\frac{1}{2}\left(w_{ij} V_j + w_{ij} V_i^t\right) - i^t + t^t$$

$$= -\frac{1}{2}(W^t + W)V - i^t + t^t \qquad \left[\because V_i^t = V_j\right]$$

For symmetrical matrix $W^t = W$

$$\therefore \qquad \nabla E = -\frac{1}{2}(2W)V - i^t + t^t$$

$$= -WV - i^t + t^t$$

Energy increment $\qquad \Delta E = (\nabla E)^t . \Delta V$

Since only i^{th} output is updated, we have

$$\Delta V = \begin{bmatrix} 0 \\ \vdots \\ \Delta V_i \\ \vdots \\ 0 \end{bmatrix}$$

$\therefore$ Energy increment út reduces to

$$\Delta t = (-w_i^t V - i_i^t + t_i^t)\, \Delta V_i$$

or $\qquad \Delta t = -\text{net}_i\, \Delta V_i$

$\therefore$ when $\qquad \text{net}_i < 0$ then $\Delta V_i \leq 0$

$\qquad \text{net}_i > 0$ then $\Delta V_i \geq 0$

$\Rightarrow$ Product term $\text{net}_i\, \Delta V_i$ is always non-negative.

$\therefore$ Change in ΔE are always non-positive.

$\Rightarrow$ *NN* under going transitions will either decrease or retain its energy E as a result of each individual update.

This is mathematical foundation of discrete time hopfield network.

Example : Let $$v(0) = \begin{bmatrix} 1 \\ 1 \\ -1 \\ -1 \end{bmatrix};\ W = \begin{bmatrix} 0 & 1 & 1 & -1 \\ 1 & 0 & 1 & -1 \\ 1 & 1 & 0 & -1 \\ -1 & -1 & -1 & 0 \end{bmatrix};\ I = T = 0$$

Then $$v(1) = \text{sgn}\ [Wv(0)] = \text{sgn} \begin{bmatrix} 1 \\ 1 \\ 3 \\ -1 \end{bmatrix} = \begin{bmatrix} 1 \\ 1 \\ 1 \\ -1 \end{bmatrix}$$

$$v(2) = \text{sgn}[Wv(1)] = \text{sgn} \begin{bmatrix} 3 \\ 3 \\ 3 \\ -3 \end{bmatrix} = \begin{bmatrix} 1 \\ 1 \\ 1 \\ -1 \end{bmatrix} = v(1)$$

$[1\ \ 1\ \ 1\ \ -1]^T$ and $[-1\ \ -1\ \ -1\ \ 1]^T$ are the two stable attractors.

Note that

$$W = \begin{bmatrix} 1 \\ 1 \\ 1 \\ -1 \end{bmatrix} [1\ 1\ 1\ -1] - \begin{bmatrix} 1 & & & \\ & 1 & & \\ & & 1 & \\ & & & 1 \end{bmatrix} = \begin{bmatrix} 0 & 1 & 1 & -1 \\ 1 & 0 & 1 & -1 \\ 1 & 1 & 0 & -1 \\ -1 & -1 & -1 & 0 \end{bmatrix}$$

5.3. THE NEURO DYNAMICAL MODEL OR MATHEMATICAL FOUNDATIONS OF GRADIENT TYPE HOPFIELD NETWORKS

Neuro dynamical model is based on additive model of a neuron which is described as follows :

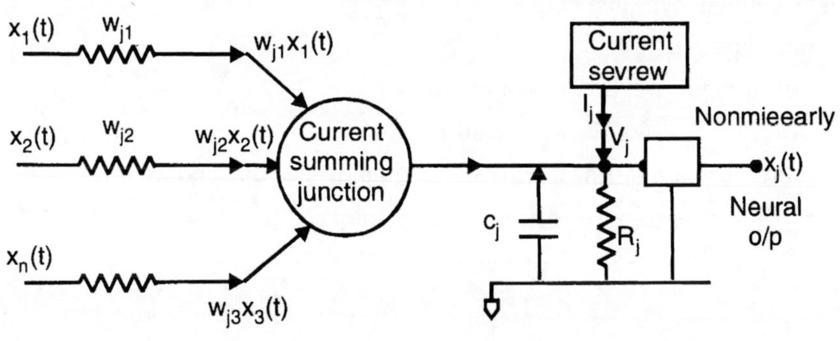

Fig. 5.4. Additive model of a neuron.

Here :

1. Synaptic weights $w_{j1}, w_{j2} ...., w_{nN}$ represents conductances.
2. Respective inputs $x_1(t), x_2(t) ...., x_N(t)$ represents potential and N is number of inputs.

These inputs are applied to a current running function characterized as follows :

- Low input resistance
- Unity current gain
- High output resistance

∴ Total current flowing toward the input node of the non-linear element (activation function) is therefore :

$$\sum_{i=1}^{n} w_{ji}\, x_i(t) + I_j \quad [I_j = \text{Externaily applied bias}]$$

3. Let $V_j(t)$ denotes induced local field at the input of non-linear activation function $\phi(c)$. Therefore total current flowing away from the input node of non-linear element as follows :

$$\underbrace{\frac{V_j(t)}{R_j}}_{\text{due to resistance}} + \underbrace{\frac{C_j \,.\, dV_i(t)}{dt}}_{\text{due to capacitance}}$$

Kirchoff's Current Law

Total current flowing towards any node of an electrical circuit is zero.

∴ Applying this law to the input node non-linearity

$$\frac{C_j \,.\, dV_i(t)}{dt} + \frac{V_j(t)}{R_j} = \sum_{i=1}^{N} w_{ji}\, x_i(t) + I_j$$

This capacitance term is the simplest way to add dynamics to the model of a neuron.

The *RC* model is commonly referred to as additive model.

Dynamics of the network is

$$\frac{C_j\, dV_j(t)}{dt} = \frac{-V_j(t)}{R_j} + \sum_{i=1}^{N} w_{ji}\, x_i(t) + I_j \qquad \text{...(5.7) for } j = 1, 2, ..., N$$

for hopfield gradient type network.

Certain assumption are

1. Matrix of synaptic weights is symmetric

$$w_{ji} = w_{ij} \text{ for all } i \text{ and } j$$

2. Each neuron has a non-linear activation of its own and hence use of

$$\phi_i(s) \qquad \text{...(5.8)}$$

3. Inverse of non-linear activation function exists

$$\therefore \quad v = \phi_i^{-1}(x) \qquad \text{...(5.9)}$$

4. Let activation function is defined by (using tangent function)

$$x = \phi_i(v) = \tanh\left(\frac{a_i v}{2}\right) = \frac{1 - \exp(-1_i\, v)}{1 + \exp(-a_i v)} \qquad \text{...(5.10)}$$

$$\text{or} \quad \frac{a_i}{2} = \left.\frac{d\phi_i}{dv}\right|_{v=0}$$

a_i is referred as gain of neuron i.

Inverse input output relation may be rewritten as

$$v = \phi_i^{-1}(x) = -\frac{1}{a_i}\log\left(\frac{1-x}{1+x}\right) \qquad ...(5.11)$$

The standard form of inverse output-input relation of unity gain is

$$\phi^{-1}(x) = -\log\left(\frac{1-x}{1+x}\right) \qquad ...(5.12)$$

or

$$\phi_i^{-1}(x) = \frac{1}{a_i}\phi^{-1}(x) \qquad ...(5.13)$$

The energy (Lyapunov) function of the hopfield network is defined as

$$E = -\frac{1}{2}\sum_{i=1}^{N}\sum_{j=1}^{N} w_{ji}\, x_i x_j + \sum_{i=1}^{N}\frac{1}{R_j}\int_0^{x_i}\phi_j^{-1}(x)\,dx - \sum_{j=1}^{N} I_i\, x_j \qquad ...(5.14)$$

Differentiating E w.r.t. time

$$\frac{dE}{dt} = \sum_{j=1}^{N}\left(\sum_{i=1}^{N} w_{ji}\, x_i - \frac{V_j}{R_j} + I_j\right)\frac{dx_j}{dt} \qquad ...(5.15)$$

$$\frac{dE}{dt} = -\sum_{j=1}^{N} C_j\left(\frac{dV_j}{dt}\right)\frac{dx_j}{dt} \qquad ...(5.16)$$

because $\sum_{i=1}^{N} w_{ji}\, x_i - \frac{V_j}{R_j} + I_j = C_j \cdot \frac{dV_j}{dt}$

or

$$\frac{dE}{dt} = -\sum_{j=1}^{N} C_j\left[\frac{d}{dt}\left(\phi_i^{-1}(x_j)\right)\right]\frac{dx_j}{dt}$$

putting value of

$$V_j = -\sum_{j=1}^{N} C_j\left(\frac{dx_j}{dt}\right)^2\left[\frac{d}{dx_j}\phi_j^{-1}(x_j)\right] \qquad ...(5.17)$$

$\frac{d}{dx_j}\phi_j^{-1}(x_j) \geq 0$ for all x_j. [$\because$ $\phi_{j-1}(x_j)$ is a increasing function of output x_j]

also $\left(\frac{dx_j}{dt}\right)^2 \geq 0$ for all x_j.

$\therefore$ For energy function defined eqn. (5.14)

$$\frac{dE}{dt} \leq 0$$

$\therefore$ From this it is clear that E is bounded.

$\therefore$ According we may make following 2 statements

1. Energy function E is a Lyapunov function of continuous Hapfield model.
2. Model is stable in accordance with Lyapunov's theorem.

Stability of Hopfield Nets

Selection of a Liapunor function as energy function of Hopfield net ensures stability.

Suppose neuron K changes its state from time t to time $t + 1$.

$\therefore$ Change in energy level

$$\Delta E = E(t+1) - E(t)$$

$$= -\left[\left(\frac{2}{2}\sum_{i} w_{ki} - V_i\right) + I_k - Q_k\right]\Delta V_k \text{ because for neuron k.}$$

$$\Delta V_k = V_k(t+1) - V_k(t)$$

Since a neuron assumes a state of either 0 or 1.

$\therefore$ 3 possible cases for ΔV_k

$$\Delta V_k = \begin{cases} 1 \\ -1 \\ 0 \end{cases}$$

$\therefore$ In the first case, neuron k change its state from 0 to 1, $\therefore$ its total input must be greater than its threshold. It follows that $\Delta E < O$.

In 2nd case, neuron k changes its state from 1 to $\therefore$ total input must be less than its threshold again $\Delta E < O$.

In 3rd case $\Delta E = 0$

$\Rightarrow$ Energy level is always non-increasing (decreasing or remaining constant).

Whenever a neuron changes its state. The network will eventually reach a minimum energy state and step.

The result follows.

5.4. HOPFIELD NETWORK AS ASSOCIATIVE MEMORY

A primary application of the Hopfield network is an associative memory. In this case, the weights of the connections between the neurons have to be set, that the states of the system corresponding with the patterns which are to be stored in the network are stable. These states can be seen as 'dips' in cnergy space. When the network is cued with a noisy or incomplete test pattern, it will render the incorrect or missing data by iterating to a stable state which is in some sense 'near' to the cued pattern.

$$w_{jk} = \begin{cases} \sum_{n=-\infty}^{\infty} x_j^p x_k^p & \text{if } j \neq k \\ 0 & \text{otherwise} \end{cases}$$

i.e., if x_j^p and x_k^p are equal, w_{jk} is increased, otherwise decreased by one (note that, in the original Hebb rule, weights only increase). It appears, however, that the network gets saturated very quickly, and that about 0:15N memories can be stored before recall errors become severe. There are two problems associated with storing too many patterns:

1. The stored patterns become unstable;
2. Spurious (false) stable states appear (*i.e.*, stable states which do not correspond with stored patterns).

5.5. OPTIMIZATION MODELS

The optimization model offers solutions to various combinational optimization problems which back the efficient solution on a digital computer.

Hopfield networks can also be applied to optimization and constraint satisfaction problems in addition to serving as autoassociatives.

The idea is to encode each hypothesis as a unit to encode constraints between the hypothes is by weights.

+ve weights encode mutual supporting relationships

whereas –ve weights encode incompatible relationships.

In a recurrent neural network like hopfield network, the major issue is of stability. Recurrent networks are stable if

$$w_{ij} = w_{ji} \qquad \text{for } i \neq j \text{ and } w_{ji} = 0 \text{ for all } i.$$

This can be proved using a Liapunor function as energy function of the network.

$$E = -\frac{1}{2}\sum_i\sum_j W_{ji}O_iO_j - \sum_j I_jO_j + \sum_j \theta_iO_j$$

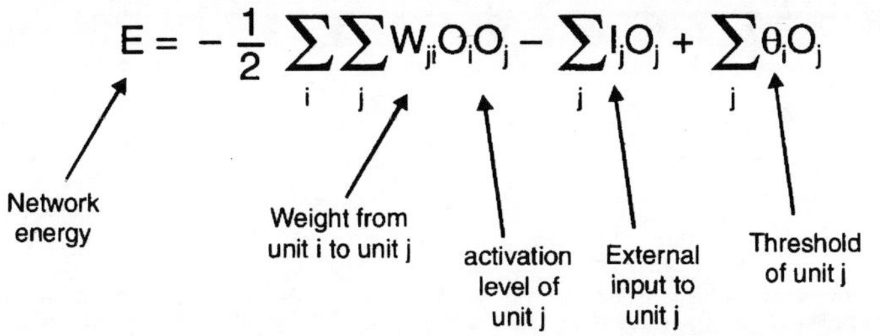

In solving an optimization problem, **energy function** is compared **with another function** that is built from problem constraints in order to determine network weights.

5.5.1. Travelling Salesman Problem

Given a group of n cities and distance between each city, find the shortest tour that visits each city only once and returns to the starting point.

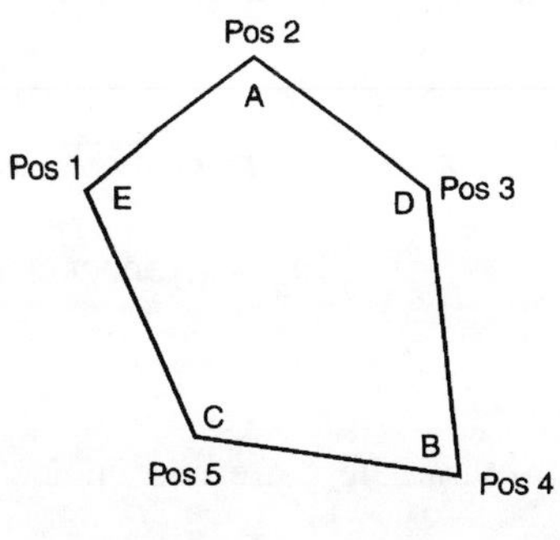

(*a*)

City \ Position	1	2	3	4	5
A	0	1	0	4	0
B	0	0	0	1	0
C	0	0	0	0	1
D	0	0	1	0	0
E	1	0	0	0	0

(*b*)

Fig. 5.5. Travelling salesman problem.

Example 5.1.

(*a*) Goal : To find shortest path

(*b*) Given : City/Position matrix

∴ From Fig. 5.5(*a*) it is clear that Position 2 referred to city *A*

∴ and in the table corresponding to row "A" there is only one "1" at 2nd position and all other are zero.

For *E*, Position 1 ⇒ Column 1 is 1 and rathers are 0.

Solution. Given *n* cities ⇒ there are *n* ! possible solutions. In hapfield network n_2 units are represented as an $n \times n$ array.

➡ Ox_i denotes activation level of unit denoting that city *x* is *i*th city in tour.

➡ *dxy* denotes the distance between cities x and y.

➡ Energy function constructed from the problem constraint is

$$E = E_1 + E_2 + E_3 + E_4$$

where

$$E_1 = \frac{A}{2} \Sigma_x \Sigma_i \Sigma_{i \neq j} O_{xi} O_{xj}$$

$$E_2 = \frac{B}{2} \Sigma_i \Sigma_x \Sigma_{x \neq y} O_{xi} O_{yj}$$

$$E_3 = \frac{C}{2} \left[(\Sigma_x \Sigma_i O_{xi}) - n \right]^2$$

$$E_4 = \frac{D}{2} \Sigma_x \Sigma_{x \neq y} \Sigma_i \, dxy O_{xi} (O_{y,i+1} + O_{y,i-1})$$

where *A*, *B*, *C*, *D* are constants.

> **Important**
>
> E_1 reflects the constraint that each row contains no more than a single 1.
>
> E_2 reflects the constraint that each column contains no more than a single 1.
>
> E_3 reflects that there should be exactly *n* 1's in the array.
>
> E_4 reflects the constrain that shortest tour favoured.

By comparing the energy functions with (1), weights is given by

$$W_{xi,yj} = A\delta_{xy}(1 - \delta_{ij}) - B\delta_{ij}(1 - \delta_{xy}) - C - D\,d_{xy}(\delta_{j,i+1} + \delta_{j,i-1})$$

[d_{ij} = Kronecker delta function]

where $\delta_{ij} = 1$ if $i = j$ otherwise $S_{ij} = 0$

Once the weights are determined and other system parameters are chosen, network uses the **algorithm formulated below to solve the problem.**

1. Write an energy function based on problem constraints.

2. Compare above energy function with following energy function of Hapfield network (a Liapunov function) to determine weights :

$$E = -\frac{1}{2}\sum_i\sum_j W_{ji}\, O_i\, O_j - \sum_j I_j O_j + \sum_j \theta_j\, O_j$$

3. Calculation of activation function

(*a*) At time = 0

$$O_j(0) = \text{a randomized small value}$$

where $O_j(t)$ is activation level of unit j at time t.

(*b*) At time $t > 0$

$$O_j(t+1) = F_h\left(\sum_i W_{ji}\, O_i(t) + I_j\right)$$

[For F_h = Hard limiting or discrete function]

and $$F_h(a) = \begin{cases} +1 \text{ if } a > O_j \\ O_j(t) \quad a = O_j \\ -1 \text{ if } \quad a = O_j \end{cases}$$

(*c*) Repeat step 2 until equilibrium. Then pattern of activation upon equilibrium represents the optimized solution.

Example 5.2. Use the weight matrix $W = \sum_i \left(X_i^T X_i - I_n\right)$ where X is n-dimensional bipolar (–1 and +1) vector to be stored in the net and I_n is $n \times n$ identity matrix.

Solution. Given $X_1 = (1, -1, -1)$

$X_2 = (-1, 1, -1)$

$X_3 = (-1, -1, 1)$

Compute the weight matrix

$$W = \left(X_1^T X_1 - I_3\right) + \left(X_2^T X_2 - I_3\right) + \left(X_3^T X_3 - I_3\right)$$

Thus $$W = \begin{bmatrix} 0 & -1 & -1 \\ -1 & 0 & -1 \\ -1 & -1 & 0 \end{bmatrix}$$

Now we use X_1 as input pattern

$$F_h(X, W) = F_h\begin{bmatrix}2 & 0 & 0\end{bmatrix} = \begin{bmatrix}1 & -1 & -1\end{bmatrix}$$

which is exactly X_1.

Summary

Important points :

Hopfiled Network

- Auto-associative network
- Fully connected with symmetric weights
- State of node = f(inputs)
- Weight values based on Hebbian principle.
- Performance : Much less computation than in backpropagation networks.

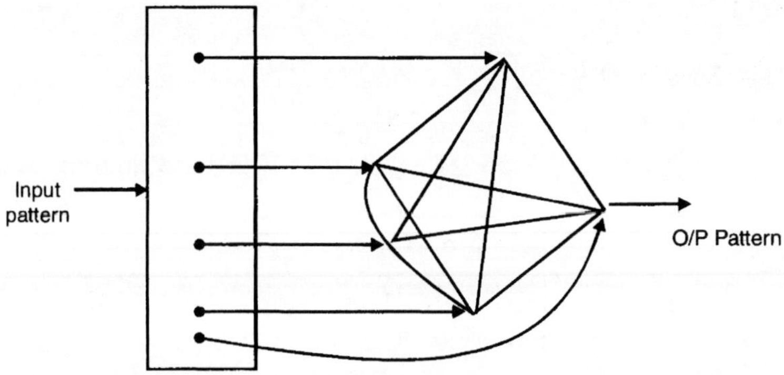

Fig. 5.6.

- Discrete node update rule :

$$x_{pk}(t+1) = \text{sgn}\left(\sum_{i=1}^{n} w_{kj}\, x_{pj}(t) + I_{pk}\right) \qquad \text{[For } I_{pk} = \text{Input value]}$$

Numerical

1. Patterns to remember

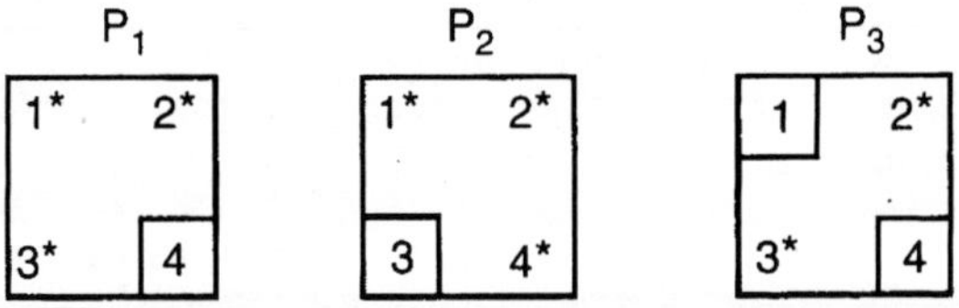

2. Hebbian weight initialization.

Average	P_1	P_2	P_3	Avg.
W_{12}	1	1	–1	+1/3
W_{13}	1	–1	–1	–1/3
W_{14}	–1	+1	+1	1/3
W_{23}	+1	–1	+1	1/3
W_{24}	–1	+1	–1	–1/3
W_{34}	–1	–1	–1	–1

3. Build Network

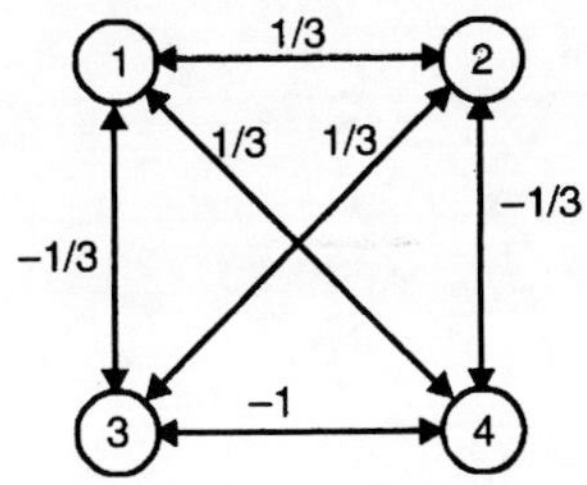

4. Enter test pattern

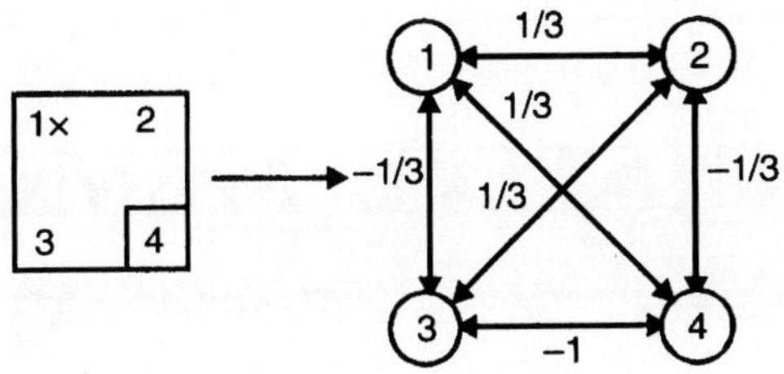

Review Questions

1. Explain : (*a*) Attractor; (*b*) Basin of attraction; (*c*) Fixed point; (*d*) Limit cycle.
2. State the difference between static and dynamical systems.
3. Explain the features of Hopfield Network.
4. State whether Hopfield network and associative memory same.
5. How Hopfield Networks are used to solve Travelling salesman problem ?
6. What are the main applications of Hopfield Network?

❑❑❑

CHAPTER

6

ASSOCIATIVE MEMORIES

INSIDE THIS CHAPTER

This chapter is concerned with associative learning and retrieval of information (vector patterns) in neural networks. These networks are usually referred to as associative neural memories (or associative memories) and they represent one of the most extensively analyzed class of artificial neural networks. Various associative memory architectures are presented with emphasis on dynamic (recurrent) associative memory architectures. These memories are treated as nonlinear dynamical systems where information retrieval is realized as an evolution of the system's state in a high-dimensional state space.

Dynamic associative memories (DAM's) are a class of recurrent artificial neural networks which utilize a learning/recording algorithm to store vector patterns (usually binary patterns) as stable memory states. The retrieval of these stored "memories" is accomplished by first initializing the DAM with a noisy or partial input pattern (key) and then allowing the DAM to perform a collective relaxation search to arrive at the stored memory which is best associated with the input pattern.

Associative memories belong to a class of NN that learns according to a certain recording algorithm. They usually acquire information a priori, and their connectivity matrices also need to be formed in advance. As we know there are 2 types of operations done on memory—Read or Write. Writing into memory produces changes in the neural interconnections while reading of the stored information from memory is considered as the transformation of the input signals by the network.

Associative memory also called *Content Addressable Memory (CAM)* data searches within a stored data file in *parallel.* The searches in this memory are done according to contents rather than the addresses as done in other type of memories like RAM, ROM, etc.

The chapter starts by presenting some simple networks which are capable of functioning as associative memories and derives the necessary conditions for perfect storage and retrieval of a given set of memories. The chapter continues by presenting additional associative memory models.

6.1. BASIC OF ASSOCIATIVE MEMORY [MDU, May 2006, May 2008, Dec. 2011]

Figure 6.1 shows block diagram of an associative memory performing an associative mapping of an input vector X into an output vector V.

$$\therefore \qquad V = M[X] \qquad ...(6.1)$$

where M = denotes a general non-linear diagonal operator and it is different for each type of memory model used. For dynamic memories, M also involves time variable. Thus, V is available at the memory output at a later time than the input has been applied.

i.e., $$V(t) = M[X(t)]$$

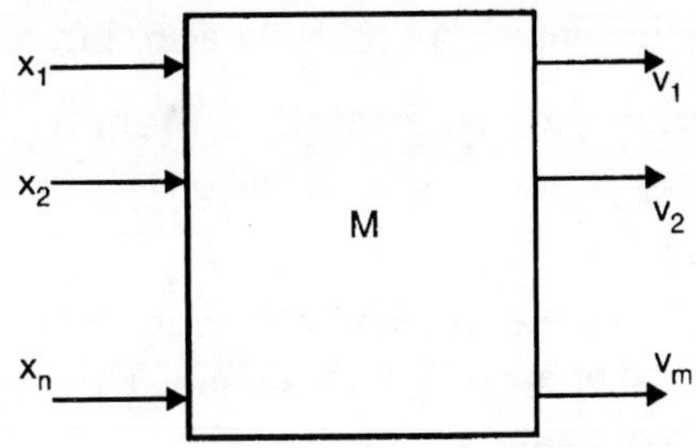

Fig. 6.1. Block Diagram of an Associative Memory

The computation of M is done with the help of "recording or storage algorithm". The mapping of V to X is called retrieval. The storage algorithm depends on whether an auto associative or a hetero-associative type of memory is designed.

The mapping as in Eq. (6.1) performed on a key vector X is called a retrieval. It is not necessary that Recall always provide a desired solution prototype. Sometimes it provides undesired prototype.

The storage algorithm depends on whether an autoassociative or a heteroassociative type of memory is used.

$$X = [x_1, x_2, ..., x_n]^t$$
$$V = [V_1, V_2, ..., V_m]^t$$

Assumption for retrieving the stored pattern is that certain prototype vector are stored in memory in such a way that once a key input has been applied, an output produced by the memory and associated with the key is the memory response.

Associative Memory Classes

As stated before, there are two classes of associative memory

- Auto associative memory
- Hetero associative memory

(1) Auto Associative Memory : An auto-associative memory, also know as auto-associative correlator, is used to retrieve a previously stored pattern that most closely resembles the current pattern.

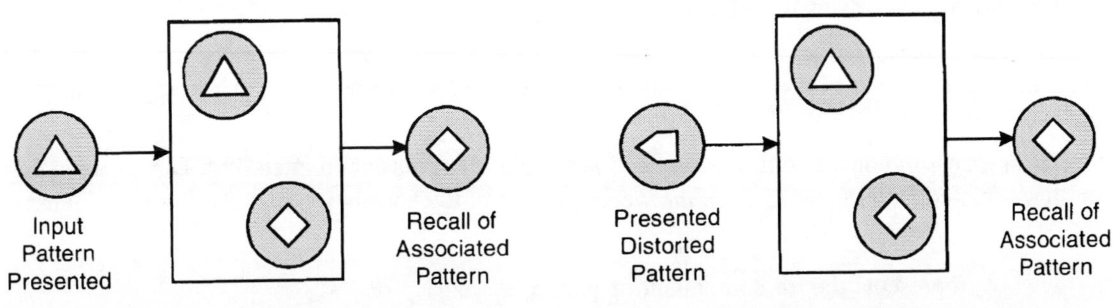

Fig. 6.2. Hetero and Auto Associative Memory Correlators

(2) Hetero Associative Memory : A hetero-associative memory, also known as hetero-associative correlator, is used to retrieve pattern in general, different from the input pattern not only in content but possibly also different in type and format.

Working of Associative Memory

Example : An associative memory is a storehouse of associated patterns which are encoded in some form.

When the storehouse is triggered or excited with a pattern, then the associated pattern pair is recalled or appears at the output. The input could be an exact or distored or partial representation of a stored pattern.

Figure 6.3 shows in illustrates the working of an associated memory.

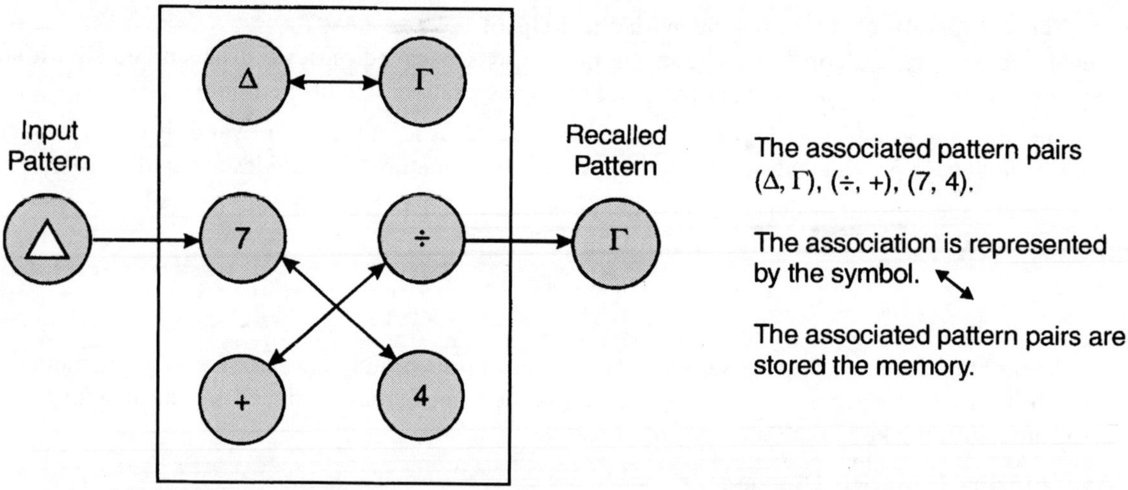

Fig. 6.3. Working of an associated memory.

When the memory is triggered with an input pattern say ú then the associated pattern æ is retrieved automattically.

Related Terms

(1) Encoding or Memorization

Building an associative memory means, constructing a connection weight matrix w such that when an input pattern is presented, and the stored pattern associated with the input pattern is retrieved.

This process of constructing the connection weight matrix is called **encoding**. During encoding, for an associated pattern pair (x_k, y_k), the weight values of the correlation matrix w_k are computed as

$$(w_{ij})_k = (x_i)_k \, (y_j)_k, \text{ where}$$

$(x_i)_k$ represents the ith component of pattern x_k, and

$(y_j)_k$ represents the jth component of pattern y_k

for $i = 1, 2, ..., m$ and $j = 1, 2,..., n$.

Constructing of the connection weight matrix w is accomplished by summing up the individual correlation matrices w_k, *i.e.*,

$$w = \alpha \sum_{k=1}^{p} w_k$$

wherea is the proportionality or normalizing constant.

(2) Retrieval or Recollection

After memorization, the process of retrieving a stored pattern, given an input pattern, is called **decoding.**

Given an input pattern x, the decoding or recollection is accomplished by :

First compute the net **input** to the output units using

$$\text{Input } j = \sum_{i=1}^{m} x_i \, w_{ij}$$

where input j is weighted sum of the input or activation value of node j, for $j = 1, 2, ..., n$.

Then determine the units outputs using a bipolar output function :

$$y_j = \begin{cases} +1 & \text{if input } j \geq \theta \\ -1 & \text{otherwise} \end{cases}$$

where θ_j is the threshold value of output neuron j.

(3) Errors and Noise

The input pattern may contain errors and noise, or may be an **incomplete version** of some previously encoded pattern.

When a corrupted input pattern is presented, the network will retrieve the stored pattern that is closest to actual input pattern.

The presence of noise or errors results only in a more decrease rather than total degradation in the performance of the network.

Thus, associative memories are **robust** and **fault tolerant** because of many processing elements performing highly parallel and distributed computations.

(4) Performance Measures

The **memory capacity** and **content-addressability** are the measures of associative memory performance for correct retrieval. These two performance measures are related to each other.

Memory capacity refers to the maximum number of associated pattern pairs that can be stored and correctly retrieved.

Content addressability is the ability of the network to retrieve the correct stored pattern. If input patterns are mutually orthogonal-perfect retrieval is possible. If the stored input patterns are not mutually orthogonal non-perfect retrieval can happen due to crosstalk among the patterns.

2. Total stored pairs in the memory is P and the associations are defined as

$$X^{(i)} \rightarrow Y^{(i)} \quad \text{for } i = 1, 2, ..., P \qquad ...(6.2)$$

and $$V^{(i)} \neq X^{(i)} \quad \text{for } i = 1, 2, ..., P,$$

The network can be termed as hetero associative memory.

∵ in hetero associative memory, there is association between pairs of 2 ordered sets of vectors $\{X^{(1)}, X^{(2)}, ..., X^{(P)}\}$ and $\{V^{(1)}, V^{(2)}, ..., V^{(P)}\}$ as explained earlier.

If $$X^{(i)} \rightarrow V^{(i)}$$

$$V^i = X^i \quad \text{for } i = 1, 2, ..., P \qquad ...(6.4)$$

Then memory is called auto associative memory.

In auto associative memory, there is association within only one set which is $\{X^{(1)}, X^{(2)}, ..., X^{(P)}\}$.

The realistic application of an auto associative mapping would be recovery of an undistorted prototype vector in response to the distorted prototype key vector.

6.1.1. Different Addressing Modes Which are Used for Memory Data Retrieval

1. Address - addressble memory
2. Content - addressable memory

These 2 addressing modes are commonly used for memory data retrieval. In digital computers, data can be accessed when their correct addresses in the memory are given as shown in Fig. 6.4.

From the Fig. 6.4 it is clear that data have input and output lines, a word line accesses and activates the entire row of binary cells containing word data bits. This activation takes place whenever binary address is decoded by address decoder. The addressed word can be either "read" or replaced during "write" operation.

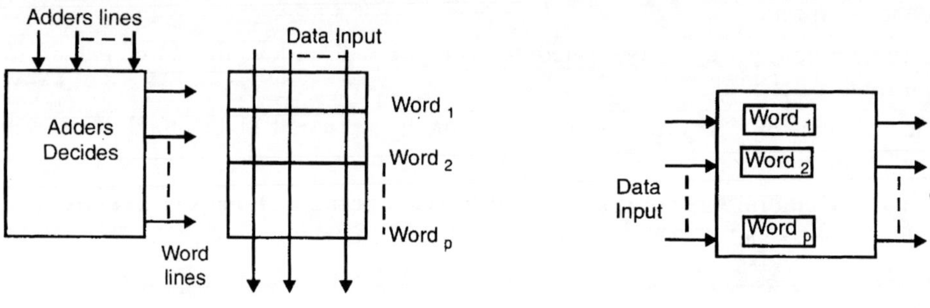

(*a*) Address-addressable memory (*b*) Content addressable memory

Fig. 6.4.

A *content-addressable memory* is a type of memory that allows the recall of data based on the degree of similarity between the input pattern and the patterns stored in memory. It refers to a memory organization in which the memory is accessed by its content as opposed to an explicit address like in the traditional computer memory system. Therefore, this type of memory allows the recall of information based on partial knowledge of its contents.

Suppose we are given a memory of names of several people as shown in the Fig. 6.5. If the given memory is content-addressable, using the erroneous string "Crhistpher Columbos" as key is sufficient to retrieve the correct name "Christopher Colombus." In this sense, this type of memory is robust and fault-tolerant, as this type of memory exhibits some form of error-correction capability.

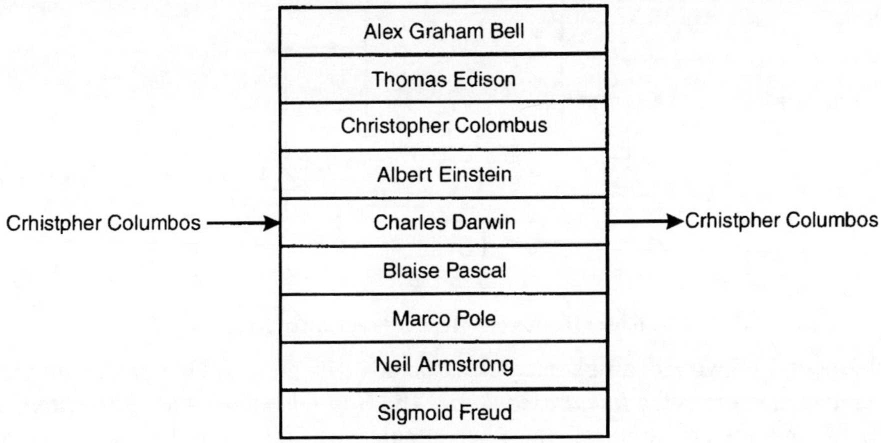

Fig. 6.5.

Another models of associative memory networks are :

1. Static
2. Dynamic memories

As discussed in chapter-2 the feedforward and feedback neutral networks.

Static memory means that there is one feed forward pass only and there is no delay theoretically. Static memories are also termed as non-recurrent memories. Block diagram shown in Fig. 6.6(*a*) and it is clear that

$$V^{(1)} = M_1[X^1] \qquad \text{or} \qquad V^k = M[X^k]$$

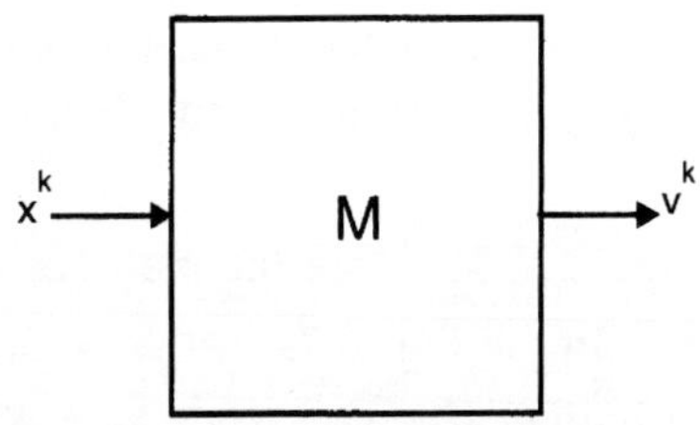

Fig. 6.6(*a*). Static/feedback memory.

Because there is no delay.

Dynamic memory means there exists feedback property and hence time delay. Dynamic memory networks exhibit dynamic evolution means they converge to an equilibrium state according to recursive formula.

$$V^{k+1} = M_2[\underbrace{X^k, V^k}_{\text{pair}}]$$

$k + 1 \Rightarrow$ Time delay of unity.

An the block diagram (Fig. 6.6(*b*)) the delay element in the feedback loop inserts a unity delay Δ.

Auto associative memory based on Hopfield model is on example of recurrent network (organic memory) for which the input X^0 is used to initialize V^0 *i.e.*, $X^0 = V^0$ and input is then removed.

For recurrent hetero-associative memory, total delay is 2Δ because the memory associates pairs of vectors $[X^{(i)}, V^{(i)}]$, $i = 1, 2, ..., P$

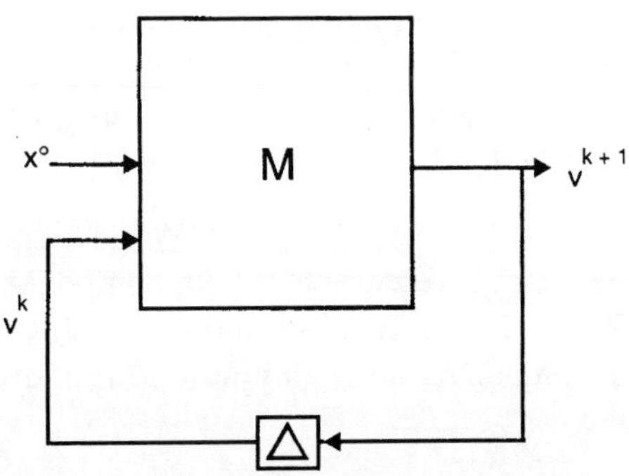

Fig. 6.6(*b*) Auto associative network.

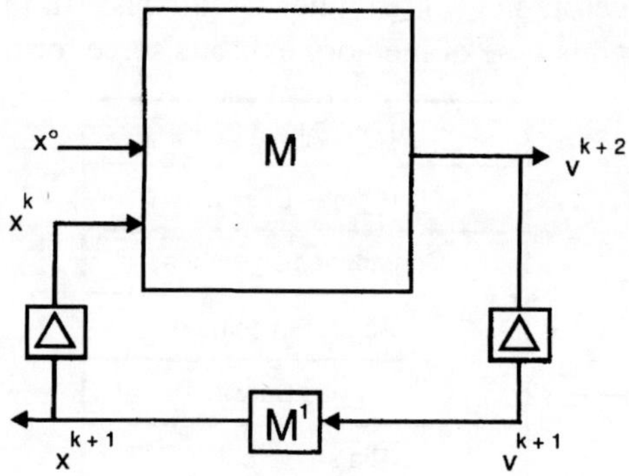

Fig. 6.6(*c*) Recurrent Hetero Associative Network

Artificial neural networks can be used as associative memories. One of the simplest artificial neural associative memory is the *linear associator*. The *Hopfield model* and *bidirectional associative memory* (BAM) models are some of the other popular artificial neural network models used as associative memories.

6.2. LINEAR ASSOCIATOR [MDU, Dec. 2011]

The *linear associator* is one of the simplest and first studied associative memory model. Below is the network architecture of the linear associator. Linear associator is a *feedforward type network* where the output is produced in a single feedforward computation.

For the linear associative memory, an input pattern X is presented and mapped to the output by simply performing the matrix multiplication operation

$$V = WX$$

where X, V, W are the matrices of size $n \times 1$, $m \times 1$ and $m \times n$ respectively.

In the Fig. 6.7, all the m input units are connected to all the n output units via the connection weight matrix $\mathbf{W} = [w_{ij}]_{m \times n}$ where w_{ij} denotes the synaptic strength of the unidirectional connection from the ith input unit to the jth output unit.

It is the connection weight matrix that stores the p different associated pattern pairs $\{(X_k, Y_k) \mid k = 1, 2, ..., p\}$ where $X_k \hat{I} \{-1, +1\}_m$ and $Y_k \hat{I} \{-1, +1\}_n$ in a distributed representation.

Encoding : Building an associative memory is nothing but constructing the connection weight matrix $\boldsymbol{W}$ such that when an input pattern is presented, the stored pattern associated with the input pattern is retrieved. The process of constructing the connection weight matrix is called *encoding*. During encoding the weight values of the correlation matrix $\boldsymbol{W}_k$ for a particular associated pattern pair $(\boldsymbol{X}_k, \boldsymbol{Y}_k)$ are computed as :

$$(w_{ij})_k = (x_i)_k \, (y_j)_k$$

where $(x_i)_k$ represents the ith component of pattern $\boldsymbol{X}_k$, $(y_j)_k$ represents the jth component of pattern $\boldsymbol{Y}_k$ for $i = 1, 2, ..., m$ and $j = 1, 2, ..., n$. Constructing the connection weight matrix $\boldsymbol{W}$ is then accomplished by summing up the individual correlation matrices, *i.e.*,

$$W = \alpha \sum_{k=1}^{p} W_k$$

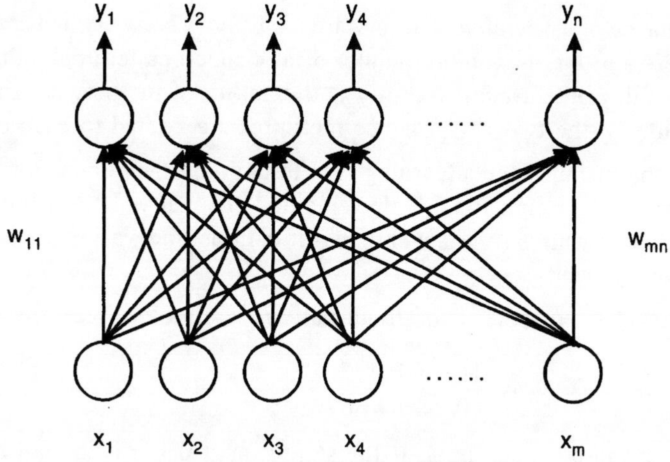

Fig. 6.7. Linear Associator

where α is the proportionality or normalizing constant. α is usually set to $1/p$ to prevent the synaptic values from going too large when there are a number of associated pattern pairs to be memorized. The connection weight matrix construction above simultaneously stores or remembers p different associated pattern pairs in a distributed manner.

Decoding : After encoding or memorization, the network can be used for *retrieval.* The process of retrieving a stored pattern given an input pattern is called *decoding.* Given a stimulus input pattern $\boldsymbol{X}$, decoding or recollection is accomplished by computing the net input to the output units using :

$$\text{input}_j = \sum_{i=1}^{m} x_i w_{ij}$$

where $input_j$ stands for the weighted sum of the input or activation value of node j for $j = 1, 2, ..., n$. Then determine the output of those units using the bipolar output function :

$$y_j = \begin{cases} +1 & \text{if input}_j \geq \theta_j \\ -1 & \text{otherwise} \end{cases}$$

where θ_j is the threshold value of output neuron j. From the foregoing discussions, it can be seen that the output units behave like the *linear threshold units* (McCulloch and Pitts, 1943) and the *perceptrons* (Rosenblatt, 1958) that compute a weighted sum of the input and produces a –1 or +1 depending whether the weighted sum is below or above a certain threshold value.

Important Points:

(1) the input pattern may contain errors and noise, or may be an incomplete version of some previously encoded pattern. Nevertheless, when presented with such a corrupted input pattern, the network will retrieve the stored pattern that is closest to actual input pattern. Therefore, the linear associator (and associative memories in general) is *robust and fault tolerant, i.e.,* the presence of noise or errors results only in a mere decrease rather than total degradation in the performance of the network. Associative memories being robust and fault tolerant are the by products of having a number of processing elements performing highly parallel and distributed computations.

(2) The performance of associative memory are its *memory capacity* and *content-addressability*. Memory capacity refers to the maximum number of associated pattern pairs that can be stored and correctly retrieved while content-addressability is the ability of the network to retrieve the correct stored pattern. Obviously, the two performance measures are related to each other.

(3) It is known that using Hebb's learning rule in building the connection weight matrix of an associative memory yields a significantly low memory capacity. Due to the limitation brought about by using Hebb's learning rule, several modifications and variants have been proposed to maximize the memory capacity.

(4) Perfect retrieval is possible if the input patterns are *mutually orthogonal, i.e.,*

$$X_a X_b = \begin{cases} 1 & \text{if } a = 0 \\ 0 & \text{otherwise} \end{cases}$$

for $a = 1, 2, ..., k$ and $b = 1, 2, ..., k$. If the stored input patterns are not mutually orthogonal, on-perfect retrieval can happen due *crosstalk* among the patterns.

However, even if the input patterns are not mutually orthogonal, accurate retrieval can still be realized if the crosstalk term is small. The degree of correlation among the input patterns gets the limit on the memory capacity and content-addressability of an associative memory.

(5) Another limitation of associative memories is the presence of *spurious memories, i.e.,* meaningless memories or associated pattern pairs other than the original fundamental memories stored. Presence of spurious memories degrades the content-addressability feature of an associative memory. Another characteristic of associative memory is that the storage of the associated pattern pairs $\{(X_k, Y_k) \mid k = 1, 2, ..., p\}$ results in the storage of the associated pattern pairs $\{(X_k^c, Y_k^c) \mid k = 1, 2, ..., p\}$ where c stands for the complement of a vector, *i.e.,* $X_k^c = -X_k$.

(6) Associative memory can be autoassociative or heteroassociative. The difference between autoassociative and heteroassociative memories lies in the retrieved pattern. An autoassociative memory retrieves the same pattern Y given an input pattern X, *i.e.,* $Y = X$. On the other hand, a heteroassociative memory retrieves the stored pattern Y given an input pattern X such that $Y'X$.

(7) An auto associative memory stores the associated pattern pairs $\{(X_k, X_k) \mid k = 1, 2, ..., p\}$. A hetero associative memory is a generalization of an auto associative memory with connection weight matrix equivalent to $\begin{bmatrix} 0 & W \\ W^T & 0 \end{bmatrix}$ and the associated pattern pairs $\{(X_k \mid Y_k, X_k \mid Y_k) \;\; k = 1, 2, \ldots, p\}$

(8) Another implementation of associative memory that differ significantly from the one is the recurrent type networks where the output of the units are fedback to the input units. Recurrent type networks produce their output through recursive computations until they become stable. The popular *Hopfield Model* is of this type of network.

6.3. HOPFIELD NETWORK

The *Hopfield model* was proposed by John Hopfield of the California Institute of Technology during the early 1980s. The dynamics of the Hopfield model is different from that of the *linear associator model* in that it computes its output recursively in time until the system becomes stable. Below is a Hopfield model with six units, where each node is connected to every other node in the network.

Unlike the linear associator model which consists of two layers of processing units, one serving as the input layer while the other as the output layer, the Hopfield model consists of a single layer of processing elements where each unit is connected to every other unit in the network other than itself. The connection weight matrix $\boldsymbol{W}$ of this type of network is square and symmetric, *i.e.*, $w_{ij} = w_{ji}$ for $i, j = 1, 2, ..., m$. Each unit has an extra external input I_i. This extra input leads to a modification in the computation of the net input to the units :

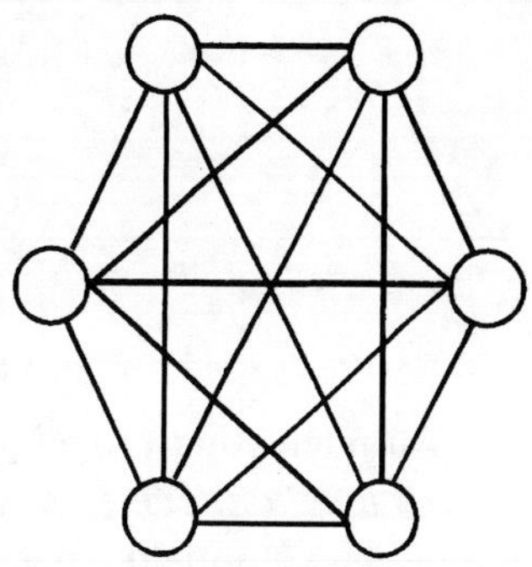

Fig. 6.8. Hopfield Model

$$\text{input}_j = \alpha \sum_{i=1}^{m} x_i w_{ij} + I_j$$

where $j = 1, 2, ..., m$.

Unlike the linear associator, the units in the Hopfield model act as both input and output units. But just like the linear associator, a single associated pattern pair is stored by computing the weight matrix as follows :

$$\boldsymbol{W}_k = \boldsymbol{X}_k^T \boldsymbol{Y}_k$$

where $\boldsymbol{Y}_k = \boldsymbol{X}_k$ to store p different associated pattern pairs. Since the Hopfield model is an autoassociative memory model, patterns, rather than associated pattern pairs, are stored in memory.

After encoding, the network can be used for decoding. Decoding in the Hopfield model is achieved by a collective and recursive relaxation search for a stored pattern given an initial stimulus pattern. Given an input pattern $\boldsymbol{X}$, decoding is accomplished by computing the net input to the units and determining the output of those units using the output function to produce the pattern $\boldsymbol{X}'$. The pattern $\boldsymbol{X}'$ is then fed back to the units as an input pattern to produce the pattern $\boldsymbol{X}''$. The pattern $\boldsymbol{X}''$ is again fedback to the units to produce the pattern $\boldsymbol{X}'''$. The process is repeated until the network stabilizes on a stored pattern where further computations do not change the output of the units.

If the input pattern $\boldsymbol{X}$ is an incomplete pattern or if it contains some distortions, the stored pattern to which the network stabilizes is typically one that is most similar to $\boldsymbol{X}$ without the distortions. This feature is called *pattern completion* and is very useful in many image processing applications.

During decoding, there are several schemes that can be used to update the output of the units. The updating schemes are *synchronous* (or parallel as termed in some literatures), *asynchronous* (or sequential), or a combination of the two (*hybrid*). Using the synchronous updating scheme, the output of the units are updated as a group prior to feeding the output back to the network. On the other hand, using the asynchronous updating scheme, the output of the units are updated in some order (*e.g.*, random or sequential) and the output are then fed back to the network after each unit update. Using the hybrid synchronous-asynchronous updating scheme, subgroups of units are updated synchronously while units in each subgroup updated asynchronously.

Discrete Hopfield Model

In the discrete Hopfield model, the units use a slightly modified bipolar output function where the states of the units, *i.e.*, the output of the units remain the same if the current state is equal to some threshold value :

$$x_1(t+1) = \begin{cases} +1 & \text{if input} > \theta_i \\ x_i(t) & \text{if input} = \theta_i \\ -1 & \text{if input} < \theta_i \end{cases}$$

For $i = 1, 2, \ldots m$ and t denotes the discrete time.

An interesting property of recurrent type networks is that their state can be described by an *energy function*. The energy function is used to prove the stability of recurrent type networks. For the discrete Hopfield model with $w_{ii} = 0$ and $w_{ij} = w_{ji}$ using the asynchronous updating scheme, the energy function E according to Hopfield (1982) is defined as :

$$E = -\frac{1}{2}\sum_{i=1}^{m}\sum_{j=1}^{m} x_i w_{ij} x_j - \sum_{i=1}^{m} x_i I_i + \sum_{i=1}^{m} x_i \theta_i$$

where the local minima of the energy function correspond to the energy of the stored patterns. Hopfield (1982) has shown that the energy of the discrete Hopfield model decreases or remains the same after each unit update. Therefore, the network will eventually converge to a local minimum that corresponds to a stored pattern. The stored pattern to which the network converges depends on the input pattern and the connection weight matrix.

Continuous Hopfield Model

The continuous Hopfield model is just a generalization of the discrete case. Here, the units use a continuous output function such as the sigmoid or hyperbolic tangent function. In the continuous Hopfield model, each unit has an associated capacitor *Ci* and resistance *ri* that model the capacitance and resistance of real neuron's cell membrane, respectively. Thus the state equation of each unit is now

$$C_j \frac{d\, input_j}{dt} = \sum_{i=1}^{m} x_i w_{ij} - \frac{input_j}{R_j} + I_j$$

where

$$\frac{1}{R_j} = \frac{1}{\rho_j} + \sum_{i=1}^{m} w_{ij}$$

Just like in the discrete case, there is an energy function characterizing the continuous Hopfield model. The energy function due to Hopfield (1984) is given by

$$E = -\frac{1}{2}\sum_{i=1}^{m}\sum_{j=1}^{m} x_i w_{ij} x_j - \sum_{i=1}^{m} x_i I_i + \sum_{i=1}^{m}\left(\frac{1}{R_j}\right)\int_0^{x_i} f^{-1}(x)\,dx$$

where f is the output function of the units.

It can be shown that $dE/dt = 0$ when $w_{ij} = w_{ji}$. Therefore, the energy of the continuous Hopfield model decreases or remains the same. The minimum energy of the continuous Hopfield model also exists using analogous computation as that in the discrete Hopfield model.

6.4. BIDIRECTIONAL ASSOCIATIVE MEMORY (BAM) [MDU Dec. 2007, Dec. 2011]

BAM is a hetero associative, content-addressable memory consisting of 2 layers. The network structure of BAM is similar to that of linear associator but connections are bidirectional that is

$$w_{ij} = w_{ji} \qquad \text{for} \begin{cases} i = 1, 2, \ldots., m \\ j = 1, 2, \ldots., n \end{cases}$$

i.e., BAM allows **forward and backward** flow of information between the layers. In BAM the units in both layers serve as both input and output units depending on the direction of propagation.

Propagating the signals from *X* layer to *Y* layer makes the units in *X* layer to act as input units while units in *Y* layer act as output units. Same is true for other direction because memory is bidirectional.

BAM model can perform both auto and hetero associative recall of stored information.

Like is linear associator and hopfield network encoding in BAM can be carried out by using

$$w_k = x_K^T y_K \text{ (to store a single associated pattern pair)}$$

and

$$w = \alpha \sum_{K=1}^{P} w_K \text{ (to store simultaneously several associated pattern pairs)}$$

This process is generally called storage or encoding.

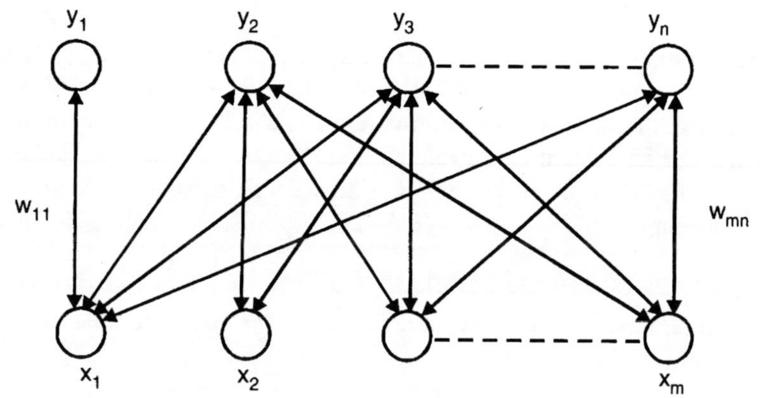

Fig. 6.9. BAM model (arrows are bidirectional).

After encoding, network can be used for decoding. The decoding process is as follows :

1. An input pattern can be applied either on X layer or on the Y layer.
2. When input pattern is applied, network will propagate the input pattern to other layer allowing the units in other layer to compute their output values.
3. Pattern that was produced by other layer is then propagated back to original layer and units in the original layer compute the output values. New pattern that was produced of the original layer is again propagated to other layer.
4. The process is repeated until further propagations and computations do not result in a change in the associates of units in both layers where final pattern pair is one of the stored associated pattern pairs.

5. Final pattern pair that will be produced by the network depends on initial pattern pair and connection weight matrix.

BAM also uses Hebb's learning rule to build the connection weight matrix to store the associated pattern pairs.

Addition and Deletion of Pattern Pairs

Given a set of pattern pairs (X_i, Y_i), for $i = 1, 2, ..., n$ and a sct of correlation matrix M :

- ➡ a new pair (X', Y') can be added or
- ➡ an existing pair (X_j, Y_j) can be deleted from the memory model.

Addition : Add a new pair (X', Y'), to existing correlation matrix M, then the new correlation matrix M_{new} is given by

$$M_{\text{new}} = X_1^T Y_1 + X_1^T Y_1 + ... + X_n^T Y_n + X'^T Y'$$

Deletion : Subtract the matrix corresponding to an existing pair (X_j, Y_j) from the correlation matrix M, them the new correlation matrix M_{new} is given by

$$M_{\text{new}} = M -- \left(X_j^T Y_j\right)$$

Note : The addition and deletion of information is similar to the functioning of the system as a human memory exhibiting learning and forgetfulness.

Energy Function for BAM

A system that changes with tim is a dynamic system. There are two types of dynamics in a neural network. During training phase it iteratively update weights and during production phase it asymptotically converges to the solution patterns. State is a collection of qualitative and qualitative items that characterize the system *e.g.,* weights, data flows. The Energy function (or Lyapunov function) is bounded function of the system state that decreases with time and the system is the minimum energy system.

Let a pair (A, B) defines the state of a BAM

- ➡ To store a pattern, the value of the energy function for that pattern has to occupy a minimum point in the energy landscape.

Let a pair (A, B) defines the state of a BAM

- ➡ to store a pattern, the value of the energy function for that pattern has to occupy a minimum point in the energy landscape.
- ➡ also adding a new pattern must not destroy the previously stored patterns.

The stability of a BAM can be proved by identifying the energy function E with each state (A, B).

For auto-associative memory : the energy function is

$$E(A) = -AM\ AT$$

where M is a matrix corresponds to the local minimum of energy function or $M = \sum_{i=1}^{N} x_i^T y_i$

For auto-associative memory : the energy function is

$$E(A) = -AM\ AT$$

For bidirectional hetero associative memory : the energy function is

$(A, B) = -AM\,BT$; for a particular case $A = B$, it corresponds to Hopfield auto associative function.

We wish to retrieve the nearest of $(A_i\ B_i)$ pair, when any (α, β) pair is presented as initial condition to BAM. The neurons change their states until a bidirectional stable state (A_f, B_f) is reached. Kosko has shown that such stable state is reached for any matrix M when it corresponds to local minimum of the energy function. Each cycle of decoding lowers the energy E if the energy function for any point (α, β) is given by $E = \alpha\, M\, \beta T$.

If the energy $E = A_i\, M\, B_i^T$ evaluated using coordinates of the pair (A_i, B_i) does not constitute a local minimum, then the point cannot be recalled, even though one starts with $\alpha = A_i$. Thus Kosko's encoding method does not ensure that stored pairs are at a local minimum.

Example : Kosko's BAM for Retrieval of Associated Pair

The working of Kosko's BAM for retrieval of associated pair. Start with X3, and hope to retrieve the associated pair Y_3.

Consider $N = 3$ pattern pairs (A_1, B_1), (A_2, B_2), (A_3, B_3) given by

$$A_1 = (1\ 0\ 0\ 0\ 0\ 1) \qquad B_1 = (1\ 1\ 0\ 0\ 0)$$
$$A_2 = (0\ 1\ 1\ 0\ 0\ 0) \qquad B_2 = (1\ 0\ 1\ 0\ 0)$$
$$A_3 = (0\ 0\ 1\ 0\ 1\ 1) \qquad B_3 = (0\ 1\ 1\ 1\ 0)$$

Convert these three binary pattern to bipolar form replacing 0s by –1s.

$$X_1 = (1\ -1\ -1\ -1\ -1\ 1) \qquad Y_1 = (1\ 1\ -1\ -1\ -1)$$
$$X_2 = (-1\ 1\ 1\ -1\ -1\ 1) \qquad Y_2 = (1\ -1\ 1\ -1\ -1)$$
$$X_3 = (-1\ -1\ 1\ -1\ 1\ 1) \qquad Y_3 = (-1\ 1\ 1\ 1\ -1)$$

The correlation matrix M is calculated as 6×5 matrix

$$M = X_1^T Y_1 + X_2^T Y_2 + X_3^T Y_3 = \begin{bmatrix} 1 & 1 & -3 & -1 & 1 \\ 1 & -3 & 1 & -1 & 1 \\ -1 & -1 & 3 & 1 & -1 \\ -1 & -1 & -1 & 1 & 3 \\ -3 & 1 & 1 & 3 & 1 \\ -1 & 3 & -1 & 1 & -1 \end{bmatrix}$$

Suppose we start with $\alpha = X_3$, and we hope to retrieve the associated pair Y_3. The calculations for the retrieval of Y_3 yield :

$$\alpha M = \begin{bmatrix}-1 & -1 & 1 & -1 & 1 & 1\end{bmatrix} (M) = (-6\ \ 6\ \ 6\ \ 6\ \ -6)$$

$$\Phi(\alpha M) = \beta' = (-1\ \ 1\ \ 1\ \ 1\ \ -1)$$

$$\beta' M^T = (-5\ \ -5\ \ 5\ \ -3\ \ 7\ \ 5)$$

$$\Phi(\beta' M^T) = (-1\ \ -1\ \ 1\ \ -1\ \ 1\ \ 1) = \alpha'$$

$$\alpha' M = (-1\ \ -1\ \ 1\ \ -1\ \ 1\ \ 1)\ M = (-6\ \ 6\ \ 6\ \ 6\ \ -6)$$

$$\Phi(\alpha' M) = \beta'' = (-1\ \ 1\ \ 1\ \ 1\ \ 1\ \ -1)$$

This retrieved pattern β' is same as Y_3.

Hence, $(\alpha_f, \beta_f) = (X_3, Y_3)$ is correctly recalled, a desired result.

Example : Incorrect Recall by Kosko's BAM

The working of incorrect recall by Kosko's BAM.

Start with X_2, and hope to retrieve the associated pair Y_2.

Consider $N = 3$ pattern pairs (A_1, B_1), (A_2, B_2), (A_3, B_3) given by

$$A_1 = (1\ 0\ 0\ 1\ 1\ 1\ 0\ 0\ 0) \qquad B_1 = (1\ 1\ 1\ 0\ 0\ 0\ 0\ 1\ 0)$$

$$A_2 = (0\ 1\ 1\ 1\ 0\ 0\ 1\ 1\ 1) \qquad B_2 = (1\ 0\ 0\ 0\ 0\ 0\ 0\ 0\ 1)$$

$$A_3 = (1\ 0\ 1\ 0\ 1\ 1\ 0\ 1\ 1) \qquad B_3 = (0\ 1\ 0\ 1\ 0\ 0\ 1\ 0\ 1)$$

Convert these three binary pattern to bipolar form replacing 0s by –1s.

$$X_1 = (1\ -1\ -1\ 1\ 1\ 1\ -1\ -1\ -1) \qquad Y_1 = (1\ 1\ 1\ -1\ -1\ -1\ -1\ 1\ -1)$$

The correlation matrix M is calculated as 9 × 9 matrix

$$M = X_1^T Y_1 + X_2^T Y_2 + X_3^T Y_3$$

$$= \begin{bmatrix} -1 & 3 & 1 & 1 & -1 & -1 & 1 & 1 & -1 \\ 1 & -3 & -1 & -1 & 1 & 1 & -1 & -1 & 1 \\ -1 & -1 & -3 & 1 & -1 & -1 & 1 & -3 & 3 \\ 3 & -1 & 1 & -3 & -1 & -1 & -3- & 1 & -1 \\ -1 & 3 & 1 & 1 & -1 & -1 & 1 & 1 & -1 \\ -1 & 3 & 1 & 1 & -1 & -1 & 1 & 1 & -1 \\ 1 & -3 & -1 & -1 & 1 & 1 & -1 & -1 & 1 \\ -1 & -1 & -3 & 1 & -1 & -1 & 1 & -3 & 3 \\ -1 & -1 & -3 & 1 & -1 & -1 & 1 & -3 & 3 \end{bmatrix}$$

Start with $\alpha = X_2$, and hope to retrieve the associated pair Y_2. The calculations for the retrieval of Y_2 yield.

$$\alpha M = (5\ -19\ -13\ -5\ 1\ 1\ -5\ -13\ 13)$$

$$F(\alpha M) = (1\ -1\ -1\ -1\ 1\ 1\ -1\ -1\ 1) = \beta'$$

$$\beta' MT = (-11\ 11\ 5\ 5\ -11\ -11\ 11\ 5\ 5)$$

$$F(\beta' MT) = (-1\ 1\ 1\ 1\ -1\ -1\ 1\ 1\ 1) = \alpha'$$

$$\alpha' M = (5\ -19\ -13\ -5\ 1\ 1\ -5\ -13\ 13)$$

$$F(\alpha' M) = (1\ -1\ -1\ -1\ 1\ 1\ -1\ -1\ 1) = \beta''$$

$$= \beta'$$

Here $\beta'' = \beta'$. Hence the cycle terminates with

$$\alpha_F = \alpha' = (-1\ 1\ 1\ 1\ -1\ -1\ 1\ 1\ 1) = X_2$$

$$\beta_F = \beta' = (1\ -1\ -1\ -1\ 1\ 1\ -1\ -1\ 1) \neq X_2$$

But β' is not Y_2. Thus the vector pair (X_2, Y_2) is not recalled correctly by Kosko's decoding process.

Check with Energy Function : Compute the energy functions

for the coordinates of pair (X_2, Y_2), the energy $E_2 = -X_2M\ Y_2T = -71$

for the coordinates of pair (α_F, β_F), the energy $EF = -\alpha FM\ \beta FT = -75$

However, the coordinates of pair (X_2, Y_2) is not at its local minimum can be shown by evaluating the energy E at a point which is "one Hamming distance" way from Y_2. To do this consider a point

$$Y'_z = (1\ \ -1\ \ -1\ \ -1\ \ 1\ \ -1\ \ -1\ \ -1\ \ 1)$$

where the fifth component -1 of Y_2 has been changed to 1. Now

$$E = -X_2 M Y_2'^T = -73$$

which is lower than E_2 confirming the hypothesis that (X_2, Y_2) is not at its local minimum of E.

Note : The correlation matrix M used by Kosko does not guarantee that the energy of a training pair is at its local minimum. Therefore, a pair P_i can be recalled if and only if this pair is at a local minimum of the energy surface.

6.4.1. Discrete Bidirectional Associative Memory

Network propagates an input pattern X to Y layer and Y layer compute the net input using :

Input value is calculated using

$$y_j = \sum_{i=1}^{m} x_i\, w_{ij}$$

Output value is calculated using

$$y_i(t+1) = \begin{cases} +1 & \text{if input } y_i > 0 \\ y_i(t) & \text{if input } y_i = 0 \\ -1 & \text{if input } y_i < 0 \end{cases} \qquad \begin{cases} i = 1, 2, \ldots., m \\ j = 1, 2, \ldots., n \end{cases}$$

pattern Y produced on Y layer is then propagated back to X layer allowing units in X-layer to compute their net input using :

Input value is calculated using

$$x_j = \sum_{i=1}^{n} y_i\, w_{ij}$$

Output value is calculated using

$$x_i(t+1) = \begin{cases} +1 & \text{if input } x_i > 0 \\ x_i(t) & \text{if input } x_i = 0 \\ -1 & \text{if input } x_i < 0 \end{cases}$$

Energy function characterize the state of discrete Bidirectional Associative Memory and given by :

$$E = -\sum_{i=1}^{m} \sum_{j=1}^{n} x_i w_{ij} y_j$$

Energy of discrete BAM is bounded and given below

$$E = -\sum_{i=1}^{m} \sum_{j=1}^{n} |w_{ij}|$$

which guarantees that the network will converge to a local minimum corresponding to a stored associated pattern pair.

Continuous BAM

In this, the units use signoid or hyperbolic tangent output function. The units in X-layer have an extra external input I_i while units in Y-layer have an extra external input J_j for $i = 1, 2, ..., m$ and $j = 1, 2, ..., n$

These extra input lead to a modification in the computation of net input to units :

$$\text{Input } x_i = \sum_{j=1}^{n} y_i w_{ij} + I_i$$

$$\text{Input } y_i = \sum_{j=1}^{m} x_i w_{ij} + j_i$$

Energy function for continuous BAM is given by

$$E = \sum_{i=1}^{m} \sum_{j=1}^{n} f(x_i) f(y_j) w_{ij} - \sum_{i=1}^{m} f(x_i) I_i - \sum_{j=1}^{n} f(y_j) J_j$$

$$+ \sum_{i=1}^{m} \int_0^{x_i} f(x_i) x_i \, dx_i + \sum_{j=1}^{n} \int_0^{y_i} f(y_j) y_j \, dy_j$$

6.5. RECURRENT AUTOASSOCIATIVE MEMORY

In order to produce an improved association there is a need for suppressing the output noise at memory output in linear associative memory. This can be done by thresholding the output and by recycling of the output to input.

The repetitive process of recycling of the output to the input through the network can be performed by a **Recurrent Neural Network (RNN).**

RNN is similar to linear associative memory. But it has feedback is dynamical.

There are two modes of operation of updation.

1. Under the asynchronous update mode, only one neuron is allowed to compute or change state at a time and then all outputs are delayed by a time ú that is produced by unity delay element in the feedback loop.

2. In synchronous mode

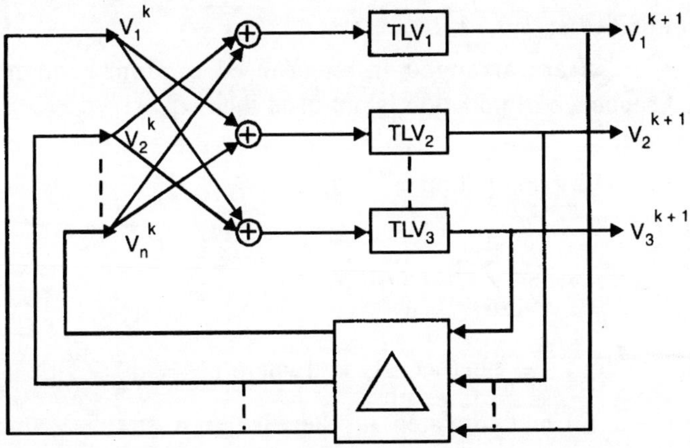

Fig. 6.10. Hopfield Model Autoassociative Memory (Recurrent autoassociative memory)

Explanation of figure dots represents the neurons, TLU threshold logic unit and these are used because associative memories are updated in discrete time.

$\Delta \rightarrow$Unit delay in time.

6.5.1. Storage and Retrieval Algorithm

As explained earlier also storage algorithm is called encoding and the retrieval algorithm is called decoding process.

Given : P bipolar binary vector $\{S^{(1)}, S^{(2)}, ..., S^{(P)}\}$

where S^m is $n \times 1$ for $m = 1, 2, ..., P$

Initializing vector V^0 is $n \times 1$.

Storage Algorithm

Step 1 : Weight matrix W is $(n \times n)$: $W \leftarrow 0$, $m \leftarrow 1$

Step 2 : Vector S^m is stored. Therefore the storage algorithm for calculating the weight matrix is

$$W \leftarrow S^{(m)} S^{(m)t} - I$$

or

$$w_{ij} = (1 - S_{ij}) \sum_{m=1}^{P} s_i^{(m)} s_j^{(m)} \qquad [S_{ij} \rightarrow \text{Kronecker function}]$$

Step 3 : If $m < P$ then $m \leftarrow m + 1$ and go to step 2, otherwise go to step 4.

Step 4 : Storage is complete, output vector W.

Note. If instead of **bipolar** binary vectors, unipolar binary vectors S^m for $m = 1, 2, ..., P$ are used then entries of original unipolar vector S^m are replaced by $2S_i^{(m)} - 1$ where $i = 1, 2, ..., n$

$$\therefore \qquad w_{ij} = (1 - S_{ij}) \sum_{m=1}^{P} (2S_i^{(m)} - 1)(2S_j^{(m)} - 1)$$

Retrieval Algorithm

1. Cycle counter K is initialized, $K \leftarrow 1$. Within the cycle counter i is initialized i $\leftarrow$ 1 and network is initialized as $V \leftarrow V^0$.
2. Integers 1, 2, ..., n are arranged in an ordered random sequences $\alpha_1, \alpha_2, \ldots, \alpha_n$. If randomized sequence of updation is not used then $\alpha_1 = 1, \alpha_2 = 2, \ldots \alpha_m = n$ or $\alpha_i = i$ at each update cycle.
3. Neuron i is updated by computing $V^{new}\ \alpha_i$

$$\text{net}_{\alpha_i} = \sum_{j=1}^{n} W_{\alpha_i j} V_j$$

$$V_{\text{new}\,\alpha_i} = \text{sgn}\,(\text{net}_{\alpha_i}) \qquad [\text{where net} = w_{ij}^{k}\, V_j^{k}\,]$$

4. If $i < n$ then $i \leftarrow i + 1$ and goto step 3, otherwise go to step 5.
5. If $V_{\text{new}\,\alpha_i} = V_{\alpha_i}$ for $i = 1, 2, \ldots, n$

$\Rightarrow$ no updation and retrieval is complete output is K and $V_{\text{new}_1}, V_{\text{new}_2}, V_{\text{new}_n}$ otherwise $K \leftarrow K + 1$, go to step 2.

Working of an auto-correlator

- how to store the patterns
- how to retrieve/recall a pattern from the stored patterns, and
- how to recognize a noisy pattern

(*i*) **How to Store Patterns :** Example

Consider the three bipolar patterns A1, A2, A3 to be stored as an auto-correlator.

$$A_1 = (-1 \quad 1 \quad -1 \quad 1)$$

$$A_2 = (1 \quad 1 \quad 1 \quad -1)$$

$$A_3 = (-1 \quad -1 \quad -1 \quad 1)$$

Note that the outer product of two vectors U and V is

$$U \otimes V = U^T V = \begin{bmatrix} U_1 \\ U_2 \\ U_3 \\ U_4 \end{bmatrix} \begin{bmatrix} V_1 & V_2 & V_3 \end{bmatrix} = \begin{bmatrix} U_1V_1 & U_1V_2 & U_1V_3 \\ U_2V_1 & U_2V_2 & U_2V_3 \\ U_3V_1 & U_3V_2 & U_3V_3 \\ U_4V_1 & U_4V_2 & U_4V_3 \end{bmatrix}$$

Thus, the outer products of each of these three A_1, A_2, A_3 bipolar patterns are

$$\left[A_1^T\right]_{4\times 1} \left[A_1\right]_{1\times 4} = \begin{Bmatrix} 1 & -1 & 1 & -1 \\ -1 & 1 & -1 & 1 \\ 1 & -1 & 1 & -1 \\ -1 & 1 & -1 & 1 \end{Bmatrix}$$

$$[A_2^T]_{4\times 1}[A_2]_{1\times 4} = \begin{Bmatrix} 1 & 1 & 1 & -1 \\ 1 & 1 & 1 & -1 \\ 1 & -1 & 1 & -1 \\ -1 & -1 & -1 & 1 \end{Bmatrix}$$

$$[A_3^T]_{4\times 1}[A_3]_{1\times 4} = \begin{Bmatrix} 1 & 1 & 1 & -1 \\ 1 & 1 & 1 & -1 \\ 1 & 1 & 1 & -1 \\ -1 & -1 & -1 & 1 \end{Bmatrix}$$

Therefore the connection matrix is

$$T = [t_{ij}] = \sum_{i=1}^{3}[A_i^T]_{4\times 1}[A_i]_{1\times 4}\,[t_{ij}] = \sum_{i=1}^{3}[A_i^T]_{4\times 1}[A_i]_{1\times 4} \begin{Bmatrix} 3 & 1 & 3 & -3 \\ 1 & 3 & 1 & -1 \\ 3 & 1 & 3 & -3 \\ -3 & -1 & -3 & 3 \end{Bmatrix}$$

This is how the patterns are stored.

Retrieve the pattern from stored pattern: from previous example.

$$T = [t_{ij}] = \sum_{i=1}^{3}[A_i^T]_{4\times 1}[A_i]_{1\times 4} = \begin{Bmatrix} 3 & 1 & 3 & -3 \\ 1 & 3 & 1 & -1 \\ 3 & 1 & 3 & -3 \\ -3 & -1 & -3 & 3 \end{Bmatrix}$$

and one of the three stored pattern is $A_2 = (1\ \ 1\ \ 1\ \ -1)$

- ➡ Retrieve or recall of this pattern A_2 from the three stored patterns.
- ➡ The recall equation is

$$a_j^{new} = f\left(a_i\, t_{ij}, a_j^{old}\right) \text{ for } \forall_j = 1, 2, \ldots., p$$

Computation for the recall equation A_2 yields $\alpha = \Sigma a_i\, t_{ij}$ and then find b.

$$\alpha = \begin{matrix} \alpha = \Sigma a_i\, t_{i,j=1} \\ \alpha = \Sigma a_i\, t_{i,j=2} \\ \alpha = \Sigma a_i\, t_{i,j=3} \\ \alpha = \Sigma a_i\, t_{i,j=4} \end{matrix} \begin{Bmatrix} 1\times 3 & + & 1\times 1 & + & 1\times 3 & + & -1\times -3 \\ 1\times 1 & + & 1\times 3 & + & 1\times 1 & + & -1\times -1 \\ 1\times 3 & + & 1\times 1 & + & 1\times 3 & + & -1\times -3 \\ 1\times -3 & + & 1\times -1 & + & 1\times -3 & + & -1\times 3 \end{Bmatrix} \begin{matrix} = & 10 & 1 \\ = & 6 & 1 \\ = & 10 & 1 \\ = & -1 & -1 \end{matrix}$$

Therefore $\quad a_j^{new} = f(a_i\, t_{ij}, a_j^{old})$ for $\forall_j = 1, 2, \ldots., p$ is $f(\alpha, \beta)$

$$a_i^{new} = f(10, 10$$

$$a_2^{new} = f(6, 1)$$

$$a_3^{new} = f(10, 1)$$

$$a_4^{new} = f(-1, -1)$$

The values of b is the vector pattern (1, 1, 1, –1) which is A_2.

This is how to retrieve or recall a pattern from the stored patterns.

Similarly, retrieval of vector pattern A_3 as

$$\left(a_1^{new}, a_2^{new}, a_3^{new}, a_4^{new}\right) = (-1, -1, -1, 1) = A_3$$

Recognition of Noisy Patterns

Consider a vector $A' = (1, 1, 1, 1)$ which is a noisy presentation of one among the stored patterns.

- Find the proximity of the noisy vector to the stored patterns using Hamming distance measure.
- Note that the Hamming distance (HD) of a vector X from Y,

  where $X = (x_1, x_2, ..., x_n)$ and $Y = (y_1, y_2, ..., y_n)$ is given by

$$HD(x, y) = \sum_{i=1}^{m} \left| (x_i - y_i) \right|$$

The HDs of A' from each of the stored patterns A_1, A_2, A_3 are

$$HD(A', A_1) = \Sigma\ |(x_1 - y_1)|, |(x_2 - y_2)|, |(x_3 - y_3)|, |(x_4 - y_4)|$$

$$= \Sigma\ |(1 - (-1))|, |(1 - 1)|, |(1 - (-1))|, |(1 - 1)|$$

$$= 4$$

$$HD(A', A_2) = 2$$

$$HD(A', A_3) = 6$$

Therefore, the vector A' is closest to A_2 and so resembles it. In other words the vector A' is a noisy version of vector A_2. Computation of recall equation using vector A' yields

$$\begin{matrix} \alpha = \Sigma a_i\, t_{i,j=1} \\ \alpha = \Sigma a_i\, t_{i,j=2} \\ \alpha = \Sigma a_i\, t_{i,j=3} \\ \alpha = \Sigma a_i\, t_{i,j=4} \end{matrix} \left\{ \begin{matrix} 1\times 3 & + & 1\times 1 & + & 1\times 3 & + & 1\times -3 \\ 1\times 1 & + & 1\times 3 & + & 1\times 1 & + & 1\times -1 \\ 1\times 3 & + & 1\times 1 & + & 1\times 3 & + & 1\times -3 \\ 1\times -3 & + & 1\times -1 & + & 1\times -3 & + & 1\times 3 \end{matrix} \right\} \begin{matrix} = & 4 & 1 \\ = & 4 & 1 \\ = & 4 & 1 \\ = & -4 & -1 \end{matrix}$$

Therefore $$a_j^{new} = f(a_i\, t_{ij} . a_j^{old}) \text{ for } \forall_j = 1, 2, ..., p \text{ is } f(\alpha, \beta)$$

$$a_1^{new} = f(4, 1)$$

$$a_2^{new} = f(4, 1)$$

$$a_3^{new} = f(4, 1)$$

$$a_4^{new} = f(-4, -1)$$

The values of b is the vector pattern $(1,\ 1,\ 1,\ -1)$ which is A_2.

6.6. APPLICATIONS

The associative memories found application in:

1. Medical Diagnosis: The input data is patient's personal information, blood pressure, temperature etc.

- Assisting doctors with their diagnosis by analyzing the reported symptoms.
- Cost estimation etc.

2. Financial:

(*a*) Stock market prediction—predict the future movement of the security using the previous data.

(*b*) Credit rating—assign the credit ratings to companies or individuals based on their financial state.

(*c*) Price forecast—forecast prices of raw materials, products etc.

3. Human Resource Management:

(*a*) Employ selection and hiring—predict on which job an applicant will achieve its best job performance.

(*b*) Staff scheduling—predict staff requirements for restaurants, banks etc.

(*c*) Personal profiling—forecast successful completion of training program, identify employees suitable for certain task.

Summary

- Associative memories are a class of ANN capable of implementing complex associative mapping following are the most important quality criteria for design of neural associative memories.
  1. Very large capacity for storing patterns.
  2. Must be immune to noisy or missing data.
  3. No memory limit cycles which are oscillatory states.
- Due to the parallel processing mode, associative memories can be successfully applied if correctly designed.
- Neural associative memories are useful in a wide range of applications such as
  (1) Content addressable storage
  (2) Search and Retrieve tasks
  (3) Image restoration

(4) Pattern recognition, classification.

(5) Optical signal processing

(*i*) Code faltering, (*ii*) Code mapping, (*iii*) Code shifting, (*iv*) Projecting etc.

➠ The neural associative memory models follow different neural network architectures to memorize information. The network architectures are either single layer or two layers.

(*i*) The **linear associator model**, is a feed forward type network, consists, two layers of processing units, one serving as the input layer while the other as the output layer.

(*ii*) The **Hopfield model**, is a single layer of processing elements where each unit is connected to every other unit in the network other than itself.

(*iii*) The **Bi-directional Associative Memory (BAM) model** is similar to that of linear associator but the connections are bidirectional.

Review Questions

1. State different Addressing Modes Which are Used for Memory Data Retrieval.

2. What is Linear Associator. How it associates various patterns in ANN.

3. Explain the working of BAM.

4. State the difference between continuous and discrete BAM.

5. Write a note on Recurrent associative memory.

❑❑❑

CHAPTER

7

HAMMING NETWORKS AND SELF ORGANIZING NETWORKS

INSIDE THIS CHAPTER

7.1. HAMMING DISTANCE

Hamming Distance (HD) is defined as the distance between two binary codes A and B of the same length as the number of place in which a_i , and b_j differ.

OR

HD can be defined as the lowest number of edges that must be traversed between the two relevant codes

1. Example:

$$A: [1 \quad -1 \quad 1 \quad 1 \quad 1 \quad 1]$$

$$B: [1 \quad 1 \quad -1 \quad -1 \quad -1 \quad 1]$$

=> HD(A, B) = 4

If A and B are bipolar binary components, then the scalar product of A, B :

$$A^tB = \underbrace{[n - HD(A, B)]}_{\text{bits are agreed}} - \underbrace{HD(A, B)}_{\text{bits are different}}$$

$$= n - 2HD(A, B)$$

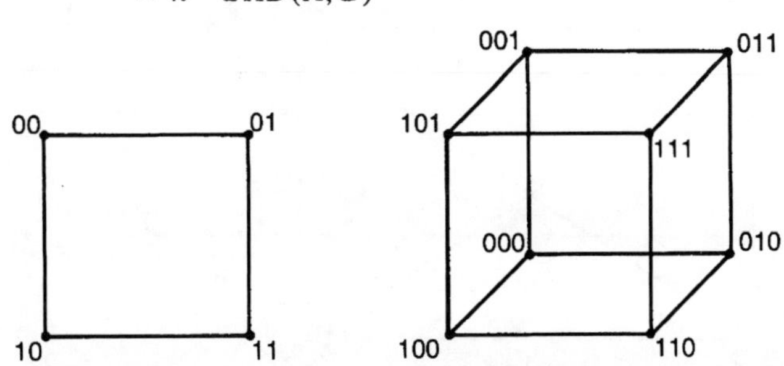

(a) (b)

Fig. 7.1.

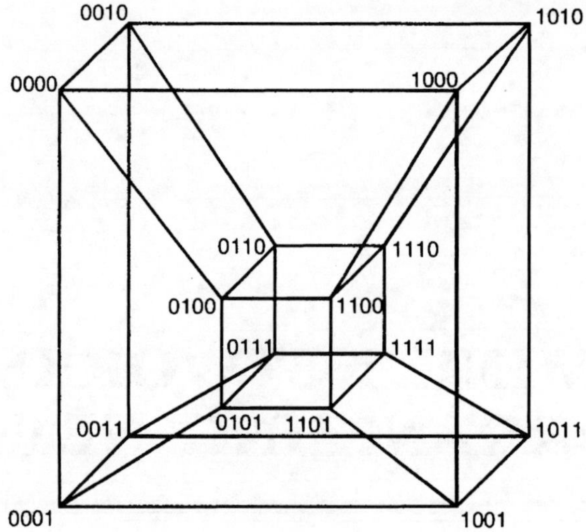

Fig. 7.1(c).

Figure 7.1 (*a*), (*b*), (*c*) shows the interpretation for of hamming distance for codes with 2, 3, 4 bits. As it is clear from the figures's that the connected vertices are differ by only one bit. From Fig. 7.1 (*b*) 000 is connected with 00<u>1</u>(differ in only position 0),100 (differ in only from 000 2nd position), 010 (differ in 1st position).

7.2. HAMMING NET AND MAXNETS

MAXNETS are Developed by Lippmann, 1987. MAXNETs can be used as a subset to pick the node whose input is the largest. MAXNETS are completely interconnected (including self-connection). Symmetric weights are used in this. No training is required in MAXNETS. In MAXNETS Weights are fixed.

For a two layer classifier of binary bipolar vectors, and *p* classes, *p* output neurons, the strongest response of a neuron indicates it has the minimum HD value and the category this neuron represents.

7.2.1. MAXNETs

Architecture of MAXNETs

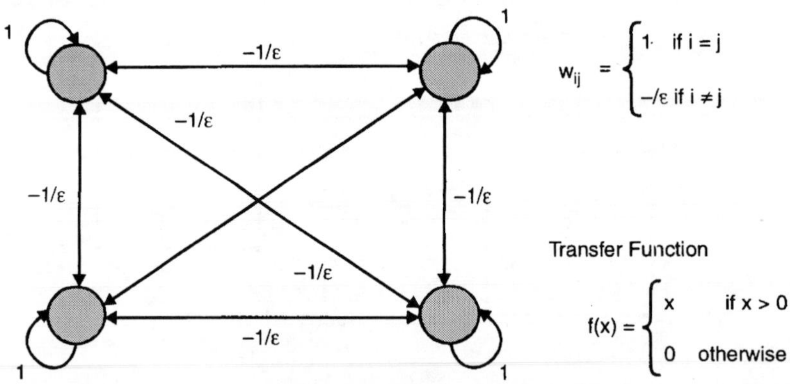

Fig. 7.2. Architecture of MAXNET

Procedure of MAXNET

1. Initialize activations and weights.
2. Update activation of each node.

$$x_j^{new} = f\left(\sum_j w_{ij} x_j^{old}\right)$$

$$x_j^{old} = x_j^{new}$$

3. If more than one node has a nonzero activation, continue; otherwise, stop.

7.2.2. Hamming Net

Hamming Nets are developed by Lippmann, 1987. These are used to determine which of several exemplar vectors is most similar to an input vector. The Exemplar vectors determine the weights of the net and the measure of similarity between the input vector and the stored exemplar vectors is (n – HD between the vectors) where HD is the hamming distance defined earlier.

Weights and Transfer Function of Hamming Net

Bipolar (–1, 1)

$$\begin{cases} w_{ij} = \frac{1}{2} x_i \\ w_b = \frac{n}{2} \end{cases}$$

Binary (0, 1)

$$\begin{cases} w_{ij} = \begin{matrix} 1 \text{ if } x_i = 1 \\ -1 \text{ if } x_i = 0 \end{matrix} \\ w_b = \# \text{ of 0's in vector} \end{cases}$$

Transfer Function (Identity Function)

$$f(x) = \begin{cases} x & \text{if } x > 0 \\ 0 & \text{otherwise} \end{cases}$$

Architecture of Hamming Net

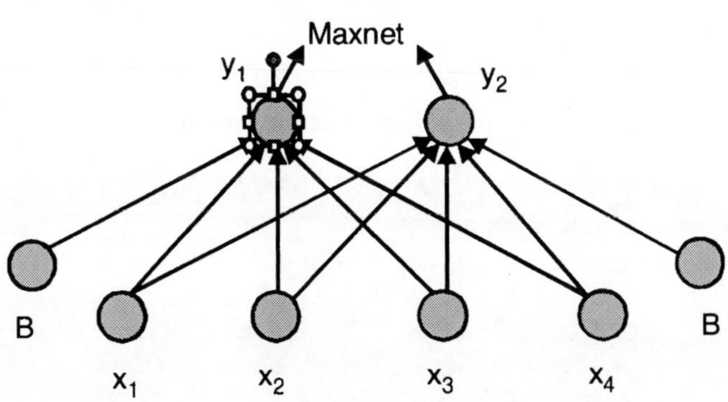

Fig. 7.3. Architecture of Hamming Net

Procedure of the Hamming Net

1. Initialize weights to store the m exemplar vectors
2. For each input vector x
   Compute

$$net_{Y_j} = net_{Y_j} = b_j + \sum_i w_{ij} x_i$$

3. Initialize activation for MAXNET $y_j(0) = net_{Y_j}$
4. MAXNET iterates for find the best match exemplar.

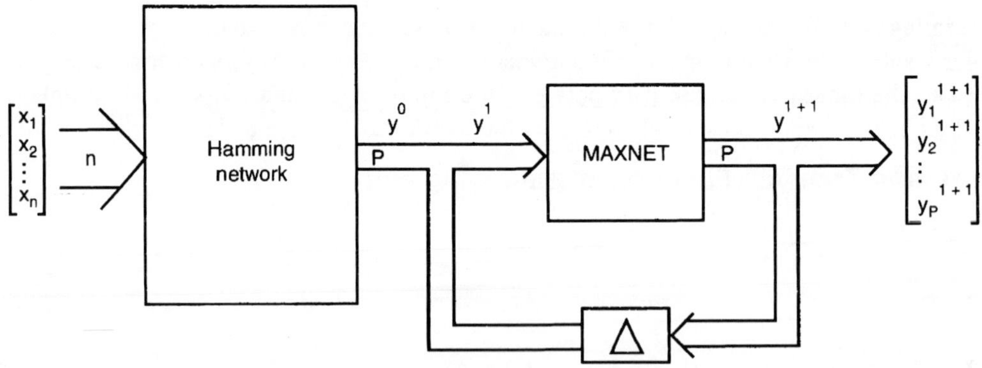

Fig. 7.4. Block Diagram of Minimum HD Classifier

1. From the figure it is clear that, we have input vector $\underline{X}$, p classes => p neurons for output, therefore output vector $\underline{Y} = [y_1, ..., y_p]$.
2. For any output neuron, m, $m = 1, ..., p$, we have
   $\underline{W}_m = [w_{m1}, w_{m2}, ..., w_{mn}]^t$ and $m = 1, 2, ..., p$
   $w_{m1}, w_{m2}, ..., w_{mn}$ are the weights between input $\underline{X}$ and each output neuron.
   Also, assuming that for each class m, one has the prototype vector $\underline{S}^{(m)}$ as the standard to be matched.
3. For classifying p classes, one can say the m'^{th} output is 1 if and only if

$$\underline{X} = \underline{S}^{(m)} \Rightarrow \text{happens only } \underline{W}^{(m)} = \underline{S}^{(m)}$$

4. Output for the classifier are

$$\underline{X}^t\underline{S}^{(1)}, \underline{X}^t\underline{S}^{(2)}, \ldots \underline{X}^t\underline{S}^{(m)}, \ldots \underline{X}^t\underline{S}^{(p)}$$

So when $\underline{X} = \underline{S}^{(m)}$, the m'^{th} output is n and other outputs are smaller than n.

$$\underline{X}^t\,\underline{S}^{(m)} = (n - HD(\underline{X}, \underline{S}^{(m)})) - HD(\underline{X}, \underline{S}^{(m)})$$

$$\therefore \qquad \frac{1}{2}\underline{X}^t\,\underline{S}^{(m)} = \frac{n}{2} - HD(\underline{X}, \underline{S}^{(m)})$$

So the weight matrix:

$$\underline{W_H} = \frac{1}{2}\underline{S}$$

$$\underline{W_H} = \frac{1}{2}\begin{bmatrix} S_1^{(1)} & S_2^{(1)} & \ldots & S_n^{(1)} \\ S_1^{(2)} & S_2^{(2)} & \ldots & S_n^{(2)} \\ \vdots & \vdots & \vdots & \vdots \\ S_1^{(p)} & S_2^{(p)} & \ldots & S_n^{(p)} \end{bmatrix}$$

5. By giving a fixed bias $\frac{n}{2}$ to the input

then $$net_m = \frac{1}{2}\underline{X}^t\,\underline{S}^{(m)} + \frac{n}{2} \quad \text{for } m = 1, 2, \ldots, p$$

or $$net_m = n - HD(\underline{X}, \underline{S}^{(m)})$$

6. To scale the input 0~n to 0~1 down, transfer function that can be applied as

$$f(net_m) = net_m \quad \text{for } m = 1, 2, \ldots, p$$

7. So for the node with the highest output means that the node has smallest HD between input and prototype vectors $\underline{S}^{(1)} \ldots \underline{S}^{(m)}$

i.e., $f(\text{net}_m) = 1$

for other nodes

$$f(\text{net}_m) < 1$$

MAXNETS: The MAXNET operates to suppress values at output instead of output values of Hamming network. MAXNET is employed as a second layer only for the cases where an enhancement of the initial dominant response of m'^{th} node is required.

i.e., the purpose of MAXNET is to let $max\{y_1, \ldots, y_p\}$ equal to 1 and let others equal to 0.

To achieve this, one can let

$$y_i^{k+1} = f\left(y_i^k - \sum_j (\varepsilon \bullet y_j^k)\right)$$

where $i = 1, \ldots, p$; $j = 1, \ldots, p$; $j \neq i$

$\because$ y_i is bounded by $0 \leqq y_i \leqq 1$ for $i = 1, \ldots, p$

↑

Output of Hamming Net

and y_i^{k+1} can only be 1 or 0.

So, ε is bounded by $0 < \varepsilon < \frac{1}{p}$

ε: lateral interaction coefficient and

$$net_k = \underline{W}_M \bullet y^k$$

And the transfer function is

$$f(net) = \begin{cases} 0 & net < 0 \\ net & net \geq 0 \end{cases}$$

Thus, hamming networks only tells which class it will most likely, not to restore distorted pattern.

7.3. UNSUPERVISED LEARNING OF CLUSTERS [MDU Dec. 2011]

As discussed in previous chapters the meaning of learning and different types of learning *i.e.*, supervised and unsupervised. Now we deal with unsupervised learning in more detail. By contrast with SUPERVISED LEARNING or REINFORCEMENT LEARNING, there are no explicit target outputs or environmental evaluations associated with each input; rather the unsupervised learner brings to bear prior biases as to what aspects of the structure of the input should be captured in the output.

Learning is based on clustering of input data. No prior knowledge is assumed to be available regarding an input membership in a particular class. Rather, gradually detected characteristics and a history of training will be used to assist the network in defining classes and possible boundaries between them.

Clustering [MDU Dec. 2009]

Clustering is understood to be the grouping of similar objects and separating of dissimilar ones. The objective of clustering neural networks discussed in this section is to categorize or cluster data. The classes must first be found from the correlations of an input data stream. Since the network actually deals with unlabeled data, the clustering should be followed by labeling clusters with appropriate category names or numbers. This process of providing the category of objects with a label is usually termed as ***calibration.***

Although the algorithm won't have names to assign to these clusters, it can produce them and then use those clusters to assign new examples into one or the other of the clusters. This is a data-driven approach that can work well when there is sufficient data; for instance, social information filtering algorithms, such as those that Amazon.com use to recommend books, are based on the principle of finding similar groups of people and then assigning new users to groups. In some cases, such as with social information filtering, the information about other members of a cluster (such as what books they read) can be sufficient for the algorithm to produce meaningful results. In other cases, it may be the case that the clusters are merely a useful tool for a human analyst. Unfortunately, even unsupervised learning suffers from the problem of overfitting the training data. There's no silver bullet to avoiding the problem because any algorithm that can learn from its inputs needs to be quite powerful.

Suppose we are given a set of patterns without any information as to the number of classes that may be present in the set. The clustering problem in such a case is that of identifying the number of classes according to a certain criterion, and of assigning the membership of the patterns in these classes. The clustering technique presented below :

Assumption : The number of classes is known a priori. The pattern set $\{x_1, x_2, \dots, x_n\}$ is submitted to the input to determine decision functions required to identify possible clusters. Since no information is available from the teacher as far as the desired classifier's responses, we will use the similarity of incoming patterns as the criterion for clustering. To define a cluster, we need to establish a basis for assigning patterns to the domain of particular cluster.

How we will assign a pattern to domain of particular cluster? There are 2 rules for this.

(1) The most common similarity rule is to find the Euclidean distance between 2 patterns x for x_i defined as

$$|X - X_i| = [(X - X_i)^t (X - X_i)]^{1/2} \quad ...(7.1)$$

Smaller the distance, closer the patterns. Using (7.1) the distance between all the pairs are computed. Now how to distinguish the clusters? A distance T can then be chosen to discriminate clusters. The value T is understood as the maximum distance between patterns within a single cluster. Figure 7.5 shows an example of two clusters with a T value chosen to be greater than the typical within-cluster distance but smaller than the between-cluster distance.

Let there are 2 patterns X_1 and X_2 and we have to find which pattern is similar to pattern X. Then using (1) distance is calculated. The value for which the distance is small ,closer the pattern is with x

(2) Another similarity rule is the cosine of the angle between $\boldsymbol{x}$ and $\boldsymbol{x}_i$:

$$\cos \alpha = \frac{(X^t X_i)}{\|X\|.\|X_i\|} \quad ...(7.2)$$

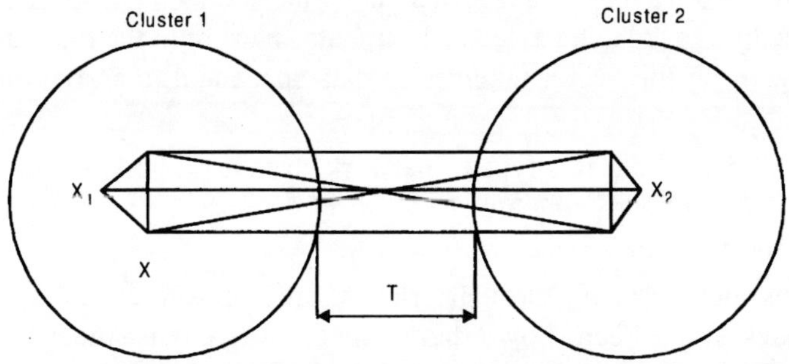

Fig. 7.5. Measures of Similarity between 2 Clusters using Distance

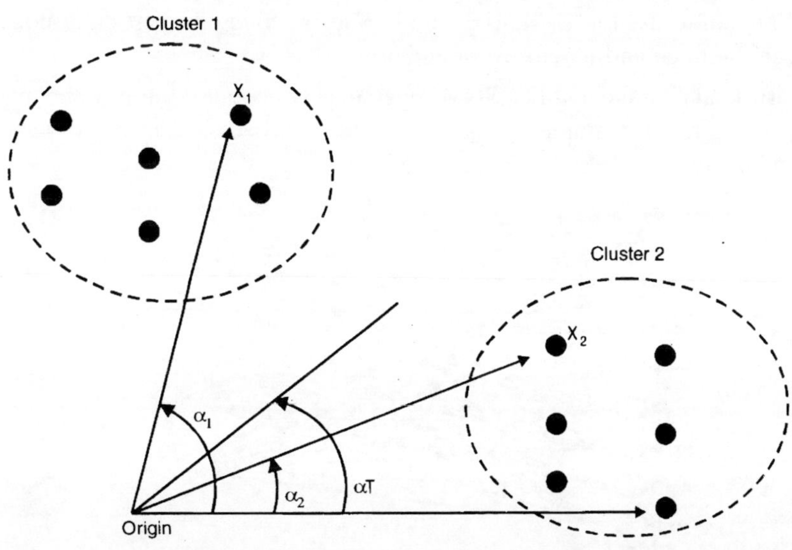

Fig. 7.6. Measures of Similarity Between 2 Clusters using Normalized Scalar Product.

This rule is particularly useful when clusters develop along certain principal and different axes as shown in Figure 7.6. For cos $\alpha_2 <$ cos α_1 pattern x is more similar to x_2 than to x_1. It would thus be natural to group it with the second of the two apparent clusters. To facilitate this decision, the threshold angle α_t can be chosen to define the minimum angular cluster distance. It should be noted, however, that the measure defined in (7.2) should be used according to certain additional qualifications. If the angular similarity criterion (7.2) is to be efficient, vectors x_1, x_2, and x should be of comparable, or better, identical lengths.

7.4. SELF ORGANIZATION MODELS

The term self organization refers to the ability to learn and organize the information without being given the correct answers for input patterns. The self organized networks perform **unsupervised** learning. This process is same as some neurobiological phenomena such **as how a baby learns since it does not know what is correct.**

A self-organizing map consists of components called nodes or neurons. Associated with each node is a weight vector of the same dimension as the input data vectors and a position in the map space. The usual arrangement of nodes is a regular spacing in a hexagonal or rectangular grid. The self-organizing map describes a mapping from a higher dimensional input space to a lower dimensional map space. The procedure for placing a vector from data space onto the map is to find the node with the closest weight vector to the vector taken from data space and to assign the map coordinates of this node to our vector.

While it is typical to consider this type of network structure as related to feedforward networks where the nodes are visualized as being attached, this type of architecture is fundamentally different in arrangement and motivation.

Useful extensions include using toroidal grids where opposite edges are connected and using large numbers of nodes. It has been shown that while self-organizing maps with a small number of nodes behave in a way that is similar to K-means, larger self-organizing maps rearrange data in a way that is fundamentally topological in character.

It is also common to use the U-Matrix. The U-Matrix value of a particular node is the average distance between the node and its closest neighbors.

Like most artificial neural networks, SOMs operate in two modes: training and mapping. Training builds the map using input examples. It is a competitive process, also called vector quantization. Mapping automatically classifies a new input vector.

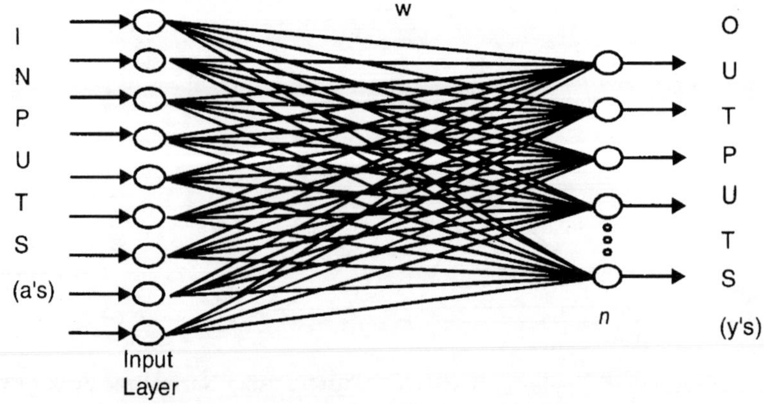

Fig. 7.7. Architecture of Self Organising Model

Procedure of Self Organizing Models

Initialize weights uniformly and normalize to unit length

Normalize inputs to unit length

Present an input vector x

calculate Euclidean distance between x and all Kohonen neurons

select winning output neuron j (with the smallest distance)

update the winning neuron $w^{new} = w^{old} + \eta[a - w^{old}]$

re-normalize weights to j (sometimes skipped)

present next training vector

Method

1. Normalize input vectors, **a**, by: $a'_{ki} = \dfrac{a_{ki}}{\sqrt{\sum_i a_{ki}^2}}$
2. Normalize weight vectors, **w**, by: $w'_{ji} = \dfrac{w_{ji}}{\sqrt{\sum_i w_{ji}^2}}$
3. Calculate distance from **a** to each **w** by :
$$d_j = \sqrt{\sum_i (a'_{ki} - w'_{ji})^2}$$
4. Min d wins (this is the winning neuron)
5. Update **w** of the min d neuron by :
$$w_{new\,ji} = w'_{old\,ji} + \eta(a'_{ki} - w'_{old\,ji})$$
6. Return to 2 and repeat for all input vectors **a**
7. Reduce η if applicable.
8. Repeat until weights converge (stop changing).

Learning Algorithm Overview

A SOM does not need a target output to be specified unlike many other types of network. Instead, where the node weights match the input vector, that area of the lattice is selectively optimized to more closely resemble the data for the class the input vector is a member of. From an initial distribution of random weights, and over many iterations, the SOM eventually settles into a map of stable zones. Each zone is effectively a feature classifier.

Training occurs in several steps and over many iterations:

1. Each node's weights are initialized.
2. A vector is chosen at random from the set of training data and presented to the lattice.
3. Every node is examined to calculate which one's weights are most like the input vector. The winning node is commonly known as the Best Matching Unit (BMU).

4. The radius of the neighbourhood of the BMU is now calculated. This is a value that starts large, typically set to the 'radius' of the lattice, but diminishes each time-step. Any nodes found within this radius are deemed to be inside the BMU's neighbourhood.
5. Each neighbouring node's (the nodes found in step 4) weights are adjusted to make them more like the input vector. The closer a node is to the BMU, the more its weights get altered.
6. Repeat step 2 for N iterations.

7.4.1. Kohonen Self Organizing Maps [MDU May 2008]

Kohonen SOMs (Self-Organizing Maps) are unsupervised learning neural networks widely used to reduce dimensionality of input space preserving its topological structure. A typical Kohonen SOM architecture is shown below. It consists of an input layer connected to a output layer (**Two-dimensional Kohonen layer**) via a **Kohonen Connector** consisting of **Kohonen Synapses**. Each neuron in a Kohonen Layer is associated with a unique set of co-ordinates in two-dimensional space, and hence is referred to as a **Position Neuron**. The input layer with 'n' input neurons is fed with n-dimensional input data one by one. The output layer organizes itself to represent the inputs. Hence the name 'self-organizing map'.

The objective of SOM training is to ensure that different parts of the network respond similarly to similar input vectors. So, the training mainly involves analysing the behaviour of the network for a training sample and adjusting the weights of synapses to ensure that the network exhibits a similar behaviour for a similar input. The training procedure involves following steps.

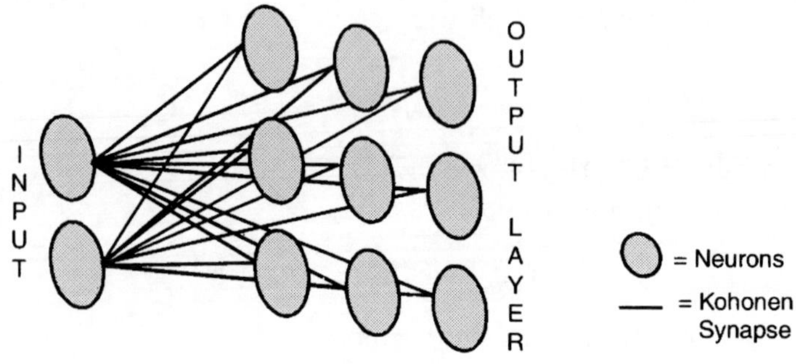

Fig. 7.8.

- Initialize the weights to small random values.
- Choose a random training sample, assign the input vector to the input neurons and run the network.
- The output of a neuron in output layer represents the similarity between its weight vector (source synapse weights) and the input vector. The output neuron which has the highest output value is declared as the **winner neuron** for current input.
- Calculate the distance of each output neuron from the winner using a Neighborhood function
- Update the weights of synapses using following formulae. (This adjusts the weights corresponding to winner and its neighbors such that, a neuron close to the current winner wins for a similar input)

Similarity = Difference between source neurons output and the synapse weight
Weight Change = (Learning Rate) * (Neighborhood value of target neuron) * Similarity

➡ Similarly, train all samples in a random order. This completes one **training cycle** (or Training Epoch)

➡ Repeat the steps to complete the specified number of training epochs.

The trained SOM maps any input vector to a winner neuron, which can be interpreted as the position of vector in two-dimensional space.

Various Self-Organizing Models are :

Kohonen Network, Hebbian Learning, Competitive learning, Adaptive Resonance Theory.

7.5. WINNER-TAKE-ALL-LEARNING (KOHONEN NETWORK)

[MDU Dec. 2007, May 2009]

(1) It consists of single layer of nodes plus an input layer.

(2) Each layer receives inputs from environment and from other nodes within the layer.

(3) Important point about this network is that both weight vector and input vectors are normalized to a constant value before learning

$$\hat{W}_i = \frac{(W_i)}{|W_i|} \qquad \text{for } i = 1, 2, \ldots P$$

(4) Each node computes by taking the dot product of its weight vector and input vector. $O_j = X.W_j$ where O_j is output or activation level of unit j.

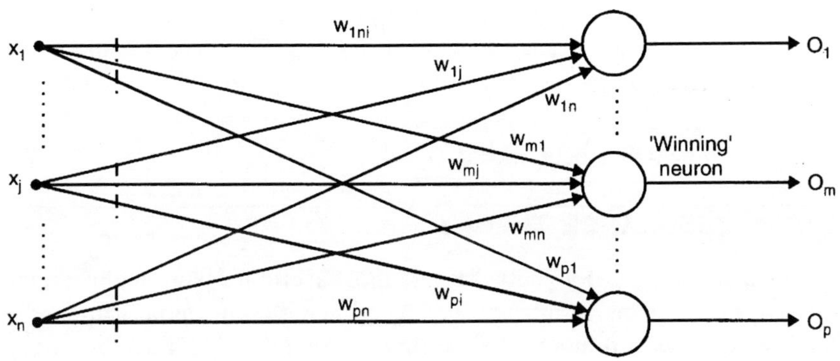

Fig. 7.9.

(5) Node with the largest activation level is declared the winner in the competition. This **winning** node is the only node that will generate an output signal, the activation level of other neurons is suppressed to zero. The Winning node with largest net_i is allowed to change the weight that is weight adjustment is for winning neuron only while weights of other neurons remain unaffected. The weight adjustment criteria for this mode of training is such that

$$|X - W_m| = \min\{|X - W_i|\} \qquad \ldots(7.3)$$

where $i = 1, 2, \ldots P$ and m is winning neuron no.

min represents the minimum. As explained earlier also in clustering that smaller the distance more close the patterns are and finding the minimum of P distances corresponds to finding maximum among the p scalar products.

$$W_m^t X = \max (W_i^t)X \quad \text{where } i = 1, 2, ..., P$$

The left hand side of above equation can be written as

$$|X - W_m| = (X^t X - 2W_m^t X - 1)^{1/2} \qquad ...(7.4)$$

It is clear that searching for the minimum of p distances as on right hand side of (7.3) Corresponds to finding maximum among p scalar products

$$W_m^{\,t} X = \max(W_i^{\,t} X) \qquad ...(7.5)$$

For $i = 1, 2, ..., p$

After the winning neuron has been identified and declared a winner, its weights must be adjusted so that the distance is reduced in the current training step. Thus $|X - W_m|$ must be reduced, preferably along the gradient direction in the weight space $w_{ml}, w_{m2}, ..., w_{mn}$

$$\nabla w_m \| x - w_m \|^2 = -2(x - w_m) \qquad ...(7.6)$$

It seems reasonable to reward the weights of the $X - W_m$ winning neuron with an increment of weight in the negative gradient direction, thus in the direction x - w_m. We thus have

$$\Delta \hat{w}_m' = \alpha(x - \Delta \hat{w}_m) \qquad ...(7.7)$$

where α is a small learning constant selected heuristically, usually between 0.1 and 0.7.

Learning according to Eqs. (7.6) and (7.7) is called Winner –take –All learning and it is a common competitive and unsupervised learning technique. The winning neuron with the largest net is rewarded with a weight adjustment, while the weights of others remain unaffected.

Another modification of the winner-take-all learning rule is that both the winners' and losers' weights are adjusted in proportion to their level of responses. This may be called leaky competitive learning and should provide more subtle learning in the cases for which clusters may be hard to distinguish.

Other type of self organizing network is

7.6. ADAPTIVE RESONANCE THEORY

Adaptive resonance theory (ART) networks are most useful for pattern clustering, classification and recognition. They can also perform pattern association with some modifications. These networks work on binary or analog- valued input. Their ability to generalize is limited because ART networks can lack the hidden layer of neurons which perform feature recognition or pattern recognition in the backpropagation network.

The basic ART system is an unsupervised learning model. It typically consists of a comparison field and a recognition field composed of neurons, a vigilance parameter, and a reset module. The vigilance parameter has considerable influence on the system: higher vigilance produces highly detailed memories (many, fine-grained categories), while lower vigilance results in more general memories (fewer, more-general categories). The comparison field takes an input vector (a one-dimensional array of values) and transfers it to its best match in the recognition field. Its best match is the single neuron whose set of weights (weight vector) most closely matches the input vector. Each recognition field neuron outputs a negative signal (proportional to that neuron's quality of match to the input vector) to each of the other recognition field neurons and inhibits their output accordingly. In this way the recognition field exhibits lateral inhibition, allowing each neuron in it to represent a category to which input vectors are classified. After the input vector is classified, the reset module compares the strength of the recognition match to the vigilance parameter. If the vigilance threshold is met, training commences. Otherwise, if the match level does not meet the vigilance parameter, the firing recognition neuron is inhibited until a new input vector is applied; training commences only upon completion of a search procedure. In the search procedure, recognition neurons are disabled one by one by the reset function until the vigilance parameter is satisfied by a recognition match. If no committed recognition neuron's match meets the vigilance threshold, then an uncommitted neuron is committed and adjusted towards matching the input vector.

Training

There are two basic methods of training ART-based neural networks: slow and fast. In the slow learning method, the degree of training of the recognition neuron's weights towards the input vector is calculated to continuous values with differential equations and is thus dependent on the length of time the input vector is presented. With fast learning, algebraic equations are used to calculate degree of weight adjustments to be made, and binary values are used. While fast learning is effective and efficient for a variety of tasks, the slow learning method is more biologically plausible and can be used with continuous-time networks (*i.e.,* when the input vector can vary continuously).

ART 1

ART 1is the simplest variety of ART networks, accepting only binary inputs.

ART 2

ART 2 extends network capabilities to support continuous inputs.

ART 2-A

ART 2-A is a streamlined form of ART-2 with a drastically accelerated runtime, and with qualitative results being only rarely inferior to the full ART-2 implementation.

ART 3

ART 3 builds on ART-2 by simulating rudimentary neurotransmitter regulation of synaptic activity by incorporating simulated sodium (Na+) and calcium (Ca2+) ion concentrations into the system's equations, which results in a more physiologically realistic means of partially inhibiting categories that trigger mismatch resets.

Fuzzy ART

Fuzzy ART implements fuzzy logic into ART's pattern recognition, thus enhancing generalizability. An optional (and very useful) feature of fuzzy ART is complement coding, a means of incorporating the absence of features into pattern classifications, which goes a long way towards preventing inefficient and unnecessary category proliferation.

ARTMAP

ARTMAP, also known as **Predictive ART**, combines two slightly modified ART-1 or ART-2 units into a supervised learning structure where the first unit takes the input data and the second unit takes the correct output data, then used to make the minimum possible adjustment of the vigilance parameter in the first unit in order to make the correct classification.

Fuzzy ARTMAP

Fuzzy ARTMAP is merely ARTMAP using fuzzy ART units, resulting in a corresponding increase in efficacy.

Remaining types are already explained in chapter 2.

7.7. RECALL MODE

The network trained in the winner-take-all mode responds instantaneously during feedforward recall at all p neuron outputs. The response is computed according to $y = M\,[WX]$, where M is non linear diagonal operator. Note that the layer now performs as a filter of the input vectors such that the largest output neuron is found as follows :

$$y_m = \max(y_1, y_2, ..., y_p)$$

and the input is identified as belonging to cluster m. In general, the neurons' activation functions should be continuous in this network. For some applications, however, $y_m = 1$ and $y_i = 0$, $i \neq m$ must be set in the recall mode of the clustering layer. In this way, for example, the weights of the following layer can be fanned out from the activated node of this previous layer while other nodes remain suppressed. Before a one-to-one vector-to-cluster mapping can be made after the network is trained in the unsupervised mode, it needs to be calibrated in a supervised environment. The calibration involves the teacher applying a sequence of p best matching class /cluster inputs and labeling the output nodes 1, 2, ..., p, respectively, according to their observed responses. Obviously, the calibrating labels assigned to the physical neurons of the layer would vary from training to training depending on the sequence of data within the training set, the training parameters, and the initial weights. Once the clustering network is labeled, it can perform as a cluster classifier in a local representation.

7.8. INITIALIZATION OF WEIGHTS

The weights of the network to be trained are initialized at small random values. The initialization affects the complete solution. If all the weights start out with equal weight values and if the solution requires that unequal weights be developed, the network may not be trained properly. Artificial neural networks typically start out with randomized weights for all their neurons. This means that they don't "know" anything and must be trained to solve the particular problem for which they are intended. Broadly speaking, there are two methods for training an ANN, depending on the problem it must solve.

A self-organizing ANN (often called a Kohonen after its inventor) is exposed to large amounts of data and tends to discover patterns and relationships in that data. Researchers often use this type to analyze experimental data.

A back-propagation ANN, conversely, is trained by humans to perform specific tasks. During the training period, the teacher evaluates whether the ANN's output is correct. If it's correct, the neural weightings that produced that output are reinforced; if the output is incorrect, those weightings responsible are diminished. This type is most often used for cognitive research and for problem-solving applications.

Implemented on a single computer, an artificial neural network is typically slower than a more traditional algorithmic solution. The ANN's parallel nature, however, allows it to be built using multiple processors, giving it a great speed advantage at very little development cost. The parallel architecture also allows ANNs to process very large amounts of data very efficiently. When dealing with large, continuous streams of information, such as speech recognition or machine sensor data, ANNs can operate considerably faster than their linear counterparts.

One of the weight selection methods developed for training the network is called convex combination. In this method all weight vectors are initialized at the value

$$w_i^0 = \left[\frac{1}{(n)^{1/2}}\right][1\ 1\ 1 \ldots p]' \qquad \text{for } i = 1, 2, \ldots, p$$

The learning starts with a very low α value. This forces the weight vectors at the beginning of learning to be close to the input vectors and to have equal lengths. As learning progresses, α is slowly increased. This allows for the gradual separation of weights according to the input clusters used for training. This procedure improves the chances for successful training, but does slow down the process. The simple winner-take-all training algorithm activates only one output neuron for each input vector; it thus provides local representation. The training approach can be modified, however, to a form in which ***K*** winners in the competition are considered and awarded weight increases accordingly. Such learning is called ***multiple-winner unsupervised learning.*** The winning ***K*** neurons would be the ones best matching the input vector, instead of a single one as in the winnertake-all mode. The outputs of winning neurons can be set so that they sum to unity. This interpolation process usually leads to an increased mapping accuracy in applications using the winner-take-all layer as the first layer. A network using this approach is said to be operating in an interpolative mode.

If training continues beyond a certain low error level, it will result in undesirable drift of weights. This causes error to increase and the quality of mapping by the network decreases. Therefore the choice of initial weights is however only one of several factors affecting training of error towards an acceptable error minimization.

7.9. SEPARABILITY LIMITATIONS [MDU, Dec. 2007, Dec. 2011]

A Single Layer Perceptron consists of an input layer and an output layer. The activation function applied is hard-limiting function as discussed in chapter 2. An output unit will assume the value 1 if the sum of weighted inputs is greater than its threshold. *i.e.*,

$$W_{ji}X_i > W_j \qquad \text{where } W_{ji} \text{ is weight from unit } i \text{ to unit } j.$$

X_i is the input from unit i and Δ_j is the threshold on unit j

Let there are 2 classes A and B. If $W_{ji}X_i > \Delta_j$ then object will be classified as Class A otherwise Class B

Suppose there are n inputs then the equation $W_{ji}X_i = \Delta_j$ where $i = 1, 2 \ldots n$ forms a hyperplane, dividing the spacc in 2 halves.

Linear Separability : When a linear hyperplane exists to place the instances of one class on one side and those of other class on the other side of plane. Then this is called Linear Separability. For example in case of OR gate we can draw a linear hyperplane to separate the inputs whose corresponding outputs are 1 on one side and the inputs whose corresponding outputs are 0 on other side of the plane.

TABLE 7.1. OR Gate.

Input *A*	Input *B*	Output
0	0	0
0	1	1
1	0	1
1	1	1

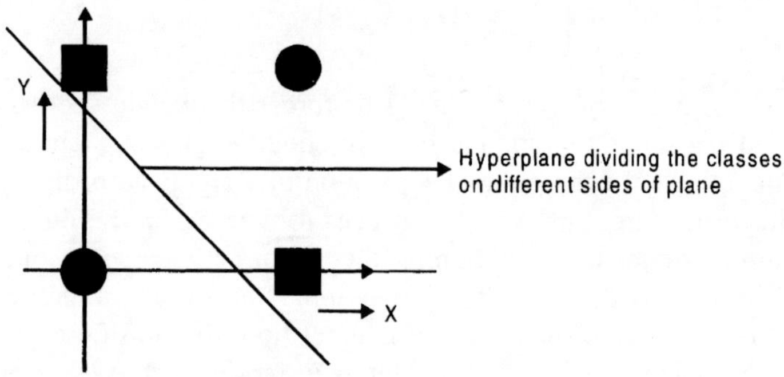

Fig. 7.10. The OR Gate.

But all the classification problems are not linear separable. For example the EX-OR problem is not linear separable because EX-OR is non equality gate *i.e.*,

TABLE 7.2. The EX-OR Gate.

Input *A*	Input *B*	Output
0	0	0
0	1	1
1	0	1
1	1	0

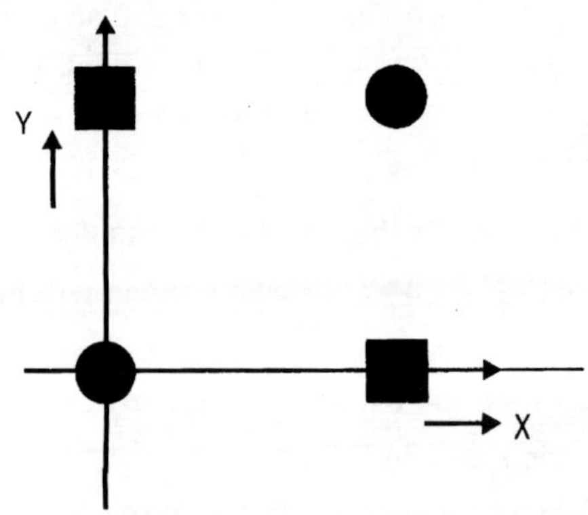

Fig. 7.11. The EX-OR Function.

In this case we can't draw any hyperplane that divides the space so that different classes are on different sides. To cope up with a problem which is not linear separable, a Multilayer Perceptron is required.

Multilayer Perceptron is a feedforward neural network with at least one hidden layer. It can deal with nonlinear classification problems.

Uses of SOM [MDU, Dec. 2008]

SOMs have been applied in many areas. Here are just some of them.
Bibliographic classification.
Image browsing systems.
Medical Diagnosis.
Interpreting seismic activity.
Speech recognition (this is what Kohonen used them for initially).
Data compression.
Separating sound sources.
Environmental modelling.
Vampire classification!

7.10. IMPORTANT CONCEPTS

Differentiate Between Competitive Learning and Connectionist Learning Algorithms [MDU, May 2010]

Competitive Learning: Competitive Learning Networks

A basic **competitive learning** network has one layer of input neurons and one layer of output neurons. An input pattern **x** is a sample point in the *n*-dimensional real or binary vector space. Binary-valued (1 or 0) *local representations* are more often used for the output nodes. That is, there are as many output neurons as the number of classes and each output node represents a pattern category.

A competitive learning network comprises the feedforward excitatory network(s) and the lateral inhibitory network(s). The feedforward network usually implements an excitatory **Hebbian learning rule**. It consist of when an input cell persistently participates in firing an output cell , the input cell's influence firing that output cell is increased. The lateral competitive network is inhibitory in nature. The network serves the important role of selecting the winner, often via a competitive learning process, highlighting the **"winner-take-all"** schema. In a winner-take-all circuit, the output unit receiving the largest input is assigned a full value (*e.g.,* 1), whereas all other units are suppressed to a 0 value. The winner-take-all circuit is usually implemented by a (digital or analog) MAXNET network. Another example of a lateral network is Kohonen's self-organising feature map. By allowing the output nodes to interact via the lateral network, the neural model can be trained to preserve certain topological ordering.

Unsupervised classification procedures are often based on some kind of clustering strategy, which forms groups of similar patterns. The clustering technique is very useful for pattern classification problems. Furthermore, it plays a pivotal role in many competitive learning networks. For a clustering procedure, it is necessary to define a similarity measure to be used for evaluating how close the patterns are.

Basic Competitive Learning Networks

Using no supervision from any teacher, unsupervised networks adapt the weights and verify the results only on the input patterns. One popular scheme for such adaptation is the competitive learning rule, which allows the units to compete for the exclusive right to respond to a particular input pattern. It can be view as a sophisticated clustering technique, whose objective is to divide a set of input patterns into a number of clusters such that the patterns of the same cluster exhibit a certain degree of similarity. The training rules are often the Hebbian rule for the feedforward network and the winner-take-all (WTA) rule for the lateral network.

Connectionist Learning Algorithms

Learning involves improvement in performance. ANNs ability to learn from examples makes them attractive and exciting.

In order to understand the learning process, we need a model of the computation according to which the network operates and we must know what information is available to the network. *The model of the computation is inductive learning from available data examples.*

Connectionist learning typically involves the manipulation of connection weights in a single network of units.

The aim of the learning is to reach a point where the network produces certain types of input/output behaviour. This normally involves *systematically updating the weights on possible connections between units.*

The network weights are updated by *learning rules*, which learning rules govern the learning process. *A learning algorithm refers to the procedure in which learning rules are used for adjusting the network weights.*

In the connectionist learning scenario, the learner is the weight updating procedure and the target representation is the network with a certain configuration of weights.

If the architecture of the network is fixed the hypothesis space is the space of possible weight configurations and a single hypothesis is a particular configuration of weights. If the architecture is not fixed, the hypothesis space is made up of all possible architecture/weight configuration combinations.

There are two learning paradigms:

Supervised learning : When there is a tutor, or a domain expert which gives the learner immediate feedback about the appropriateness of its behaviour.

The supervised learning task is:

Given : *A set of examples and their associated outcomes*;

Determine : *A general description for each outcome that matches only its examples.*

Unsupervised learning :

There are two learning approaches:

- *Incremental learning* : When the examples are processed one at a time;
- *Batch learning* : When a large set of examples are processed at once.

Different network architectures require appropriate learning algorithms.

There are four basic types of *learning algorithms depending on the rules*:

- Error-correction rules
- Boltzmann rule
- Hebbian rule
- Competitive rule

Neural Networks and their Applications

Paradigm	Learning Rule	Network Architecture	Learning Algorithm	Task
Supervised	Error-correction	Single-layer perceptron Multilayer perceptron	Perceptron learning Backpropagation	Pattern classification Regression Time-series Modeling
	Boltzmann	Recurrent	Boltzmann learning	Pattern classification
	Hebbian	Multilayer feed-forward	Linear discriminant analysis	Pattern classification Data Mining
	Competitive	Competitive	Learning vector quantization	Data compression
Unsupervised	Error-correction	Multilayer feed-forward	Simmon's projection	Data analysis
	Hebbian	Feed-forward competitive	Principal component analysis	Data compression Data analysis
	Competitive	Self-organizing maps	Kohonen SOM	Categorization Data analysis

Summary

- Self-organization, which uses the scalar product of similarity measure for developing clusters has been discussed in this chapter. Self-organizing arrays have been able to translate clusters in n-dimensional input pattern space into clustering in low dimensional feature, or image, space.
- The response of a self-organizing neural network can be used to encode input in fewer dimensions while keeping as much of the vital input information as possible. A space with a high dimensionality of input data is generally expected to be mostly empty, and the nonempty parts of the space can be identified and projected into regular array of neurons responding with images that are easy to perceive.
- The competitive learning rule enables the unsupervised training process to work. Winner-take-all learning and an adjustable neighborhood definition have been used for weight formation.
- The networks covered in this chapter have found many applications. They include minimum-distance vector classification, vector quantization, auto- and heteroassociative memory, classification of noisy vectors into clusters, and cluster synthesis. The class of unsupervised learning neural networks is particularly well suited to applications that require data quantization. Examples where the capability of data quantization is useful include statistical analysis, codebook communication, and transmission and data compression. The idea of data quantization is to categorize a given set of random input vectors through a finite

number of classes. The index of the class is found as a number of the output neuron that fires.

➡ Finally, self-organizing feature maps can be applied for extracting the hidden features of multi-dimensional input patterns and their probability of occurrence. The inputs are mapped to the line or plane of output units so that neighborhood relations are transferred from the input space to the neuron output space. This is especially useful for detection of hidden relationships that are difficult to group, organize or categorize in a sensible manner.

Review Questions

1. State and explain the meaning of self-organizing networks.
2. Explain the ways to measure the similarity between the patterns.
3. Is classification and recognition same ? Explain.
4. Explain Supervised and Unsupervised learning with its advantages and disadvantages.
5. How unsupervised learning is useful in clustering the patterns ?
6. Discuss 2 rules for assigning a pattern to domain of particular cluster ?
7. Explain Winner-Take-All-Learning method.
8. Define lateral inhibition.
9. Define leaky competitive learning.
10. Write a note on ART network.

❑❑❑

CHAPTER

8

HYBRID EXPERT SYSTEMS

INSIDE THIS CHAPTER

An expert system refers to a computer system which exhibits the human expert's intelligence. An expert system handles real-world problems requiring the expert's involvement, uses a computer model of expert knowledge and expert reasoning and is comparable with or even superior to a human expert in performance (accuracy and efficiency). MYCIN and DENDRAL are two expert systems in history.

Research on the expert systems has slowed down relative to their development since mid eighties. This is in part due to fact that powerful knowledge-based techniques like the rule-based technique have been mature and the concept of expert systems has merged into many disciplines. On the other hand the neural network approach, which resurged in the last decade, seems to have opened a new direction for expert system development, called the integration of neural networks and conventional expert systems. Therefore the main focus of this chapter to explore integration techniques.

8.1. INTRODUCTION

The basic building blocks of the expert system are shown in the Fig. 8.1. An expert system typically consists of a knowledge base, an inference engine, a user interface and an explanation facility. The knowledge base stores the domain knowledge, and the inference engine reasons with this knowledge for solving the problems. Expert system is connected with conventional software such as database management system. Power of the expert system derives from the knowledge It possesses rather than from the inference methods it employ. The terms "expert systems" and "knowledge-based systems "are sometimes used interchangeably.

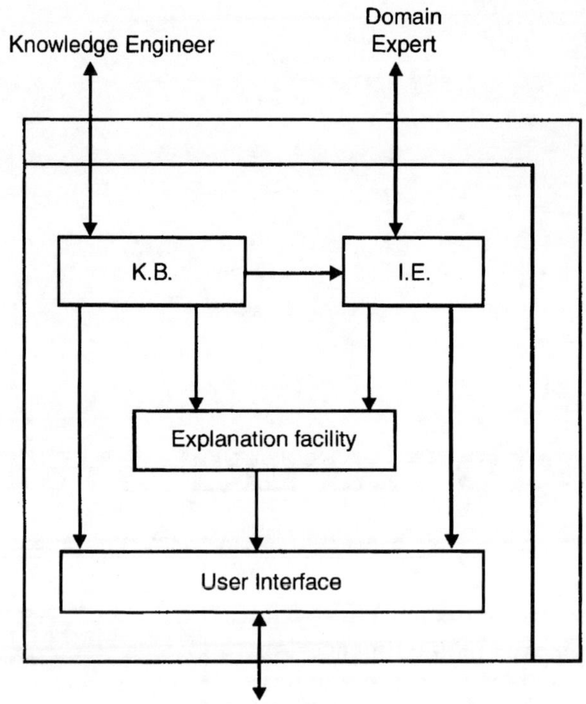

Fig. 8.1. Basic Elements of an Expert System

8.2. KNOWLEDGE ACQUISITION

The domain expertise that needs to be transferred to an expert system is a collection of definitions, relations, specialized facts, procedures and assumptions. The transfer of the knowledge from some knowledge source to a computer system is called **knowledge acquisition**. To acquire knowledge from human experts is known as knowledge engineering. And to extract the human expert's knowledge via interviews or tools is called **knowledge elicitation.** The three models of knowledge acquisition defined by Buchanan and Shortliffe are

(*a*) **Handcrafting:** Code knowledge into program directly.

(*b*) **Knowledge Engineering:** Work with an expert system to organize his/her knowledge in a suitable form for an expert system to use.

(*c*) **Machine Learning:** Extract the knowledge from training examples.

Knowledge acquisition can be divided into following five stages:

(1) **Identification :** Define an appropriate problem and determine the characteristics.

(2) **Conceptualization :** Find the concepts (objects, relations, information etc.) to represent the knowledge.

(3) **Formalization :** Choose a knowledge representation method and an inference mechanism.

(4) **Implementation :** Formulate knowledge in the chosen formalism (rules, frames etc.)

(5) **Testing :** Verify the knowledge and validate the system.

The whole process is iterative. Throughout the process, the knowledge engineer works with the domain expert closely. Knowledge engineering tools automate the knowledge acquisition. Another advantage of using tools is rapid prototyping.

Various expert system tools are available now days are ART, KEE, LOOPS, OPS5 and so on. Tools at the high end provide a mixed environment (combination of different knowledge representation languages) and graphical interfaces and are suitable for large-scale applications. Tools at the lower end are much less versatile and for small- scale applications.

8.3. KNOWLEDGE REPRESENTATION

Knowledge representation is a major issue in building expert system. It is concerned with both the storage of knowledge in proper data structure and the use of knowledge in intelligent processes.

There are 4 levels of knowledge representation:

(*a*) The first level is the **implementation** level, which concerns the possibility of building a computer program for the knowledge representation language.

(*b*) The 2nd level is the **logic** level, which concerns the logic properties of the knowledge representation language, such as meaning of the expressions and the soundness of the associated inference procedures.

(*c*) The 3rd level is the **epistemological** level, which concerns the knowledge structure (*e.g.*, semantics) and the inference strategy of the knowledge representation language.

(*d*) The 4th level is the **conceptual** language, which concerns the actual primitives of the knowledge representation language.

This knowledge representation hierarchy roughly corresponds to the process of the knowledge acquisition described in the last section. Knowledge analysis described in knowledge acquisition corresponds to the logic level of knowledge representation.

Various knowledge representation schemes are:

8.3.1. Logic

In logic, 2 commonly applied rules of inference are:

(*a*) MODUS PONENS : If "A implies B" and "A is true," then "B is true".

(*b*) RESOLUTION : If "A is false or B is true" and "A is true," then "B is true".

The advantage of logic representation includes generality, naturalness, preciseness, flexibility and modularity.

The major disadvantage lies in the separation of representation and utilization and the inefficiency for the inference.

8.3.2. Production Rules

The knowledge base of such systems consists of the rules called productions. Each production rule is put in the form of condition –action pair, *e.g.*," If A and B, then C." When the condition is satisfied, the action is executed.

The advantage of production systems includes modularity, uniformity and natural-ness. Rules have been found quite useful for providing the explanations as to why a question is asked, how a conclusion is reached, and what strategy is used by an expert system.

The disadvantages are the inefficiency of the execution if good control knowledge is unavailable and the difficulty of representing the algorithmic knowledge.

8.3.3. Semantic Nets

A net consists of nodes and link between the nodes. Nodes may be the objects, events, concepts. Links specify the relationship between the nodes.

Inference can be made in semantic nets by

(*a*) **Intersection search :** This is an inference procedure which propagates information from nodes under query through the network and sees where the propagation intersects. The relationship among the objects can be found in this way.

(*b*) **Properly inheritance :** The semantic net allows property inheritance, in which specific object inherits the property from more general objects or classes.

(*c*) **Graph-based matching :** Semantic nets have been used as directed graphs to represent the structural objects in machine learning. In this representation, 2 objects can be compared for similarity or non similarity.

The advantage of the semantic net representation lies in the explicit and succinct association between objects and concepts. However, without extending its formalism, it is difficult to represent quantification, disjunction, and implication.

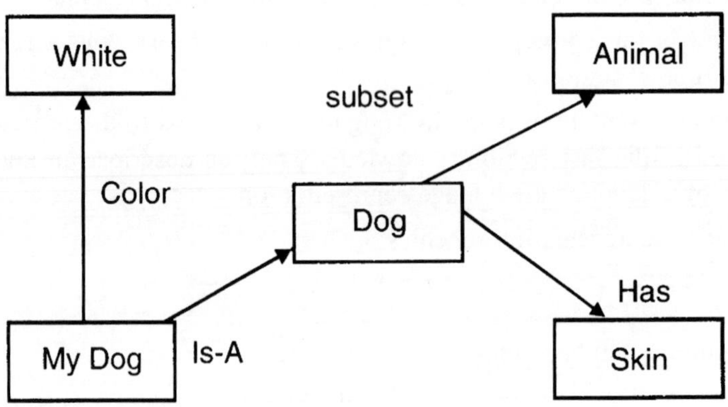

Fig. 8.2. Semantic Nets

8.3.4. Frames

A frame is a collection of slots that characterize an object. Each slot may be filled with a value, a default, another frame, or procedures. Embedding the procedures within a frame is called procedural attachment. Since both procedural and declarative representations have pros and cons, frames are intended to combine their advantages.

The frames representation has several advantages. It is natural for representing the structural objects. It is more efficient than logic. The default reasoning in frames is decidable .in logic the default reasoning is undecidable.

The disadvantage of frames representation is that it lacks a clear semantics. It has the same expressive limitations as semantic net representation.

8.3.5. Connectionists

The long-term knowledge of a connectionist network is encoded as a set of weights on the connections between the units. This approach is used in Neural Networks.

8.3.6. Analogical Representation

In this representation, the structure of representation resembles and corresponds to the structure of the object being represented. Examples include maps, electrical circuits and photographs.

The advantages include naturalness, ease of interpretation and observability of complex relationships.

The disadvantage lies in the limitation to express the domain knowledge.

8.4. WHY REASONING UNDER UNCERTAINTY

(1) In many problem domains it isn't possible to create complete, consistent models of the world. Therefore agents (and people) must act in uncertain worlds (which the real world is).

(2) We want an agent to make rational decisions even when there is not enough information to prove that an action will work.

(3) Some of the reasons for reasoning under uncertainty:

(*a*) **True uncertainty**: For example, flipping a coin.

(*b*) **Theoretical ignorance**: There is no complete theory which is known about the problem domain. For example, medical diagnosis.

(*c*) **Laziness:** The space of relevant factors is very large, and would require too much work to list the complete set of antecedents and consequents. Furthermore, it would be too hard to use the enormous rules that resulted.

8.4.1. Bayes's Rule

Bayes's Rule is the basis for probabilistic reasoning because given a prior model of the world in the form of $P(A)$ and a new piece of evidence B, Bayes's Rule says how the new piece of evidence decreases ignorance about the world by defining $P(A|B)$.

8.4.2. Dempster-Shafer's Theory

The Dempster-Shafer theory is a mathematical theory of evidence based on belief functions and plausible reasoning, which is used to combine separate pieces of information (evidence) to calculate the probability of an event. This theory is a generalization of Bayesian theory of subjective probability. The D-S theory handles the disjunction of hypotheses and can provide hierarchical diagnosis.

8.4.3. Neural Networks

Neural networks not only capable of deducing the useful information but also capable of inducing the knowledge from the data. This capability is due to storage of information in a large number of connection weights and use of heuristic knowledge to adjust the weights properly. Therefore NN can take batter decisions than any other probability density functions.

8.5. HYBRID EXPERT SYSTEMS

Expert networks refer to neural networks used as experts in a particular domain. A major weakness of these systems is that they can't justify their responses as the traditional expert systems do. Some solutions have been proposed. One of the solutions is to build a hybrid system combining neural networks and rule-based techniques.

Five integration techniques have been identified as follows:

8.5.1. Completely Overlapped

In this architecture, the system is both knowledge-based system and a neural network. It has a dual nature. The system optimizes its performance by combining the strengths of the two forms. Depending on the need, it can be presented to the user as a traditional expert system or as a neural network. One form can be converted to other through inherent transition mechanisms. Therefore only one form has to be stored.

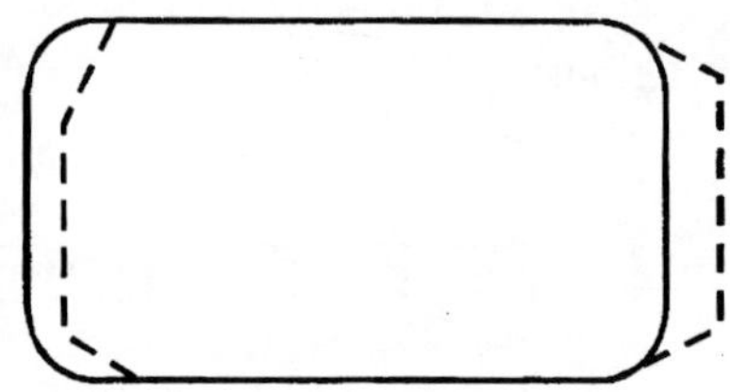

Fig. 8.3. Completely Overlapped Systems

8.5.2. Partially Overlapped

The system is a hybrid of K.B. system and a NN, exhibiting the features of both. The components share some but not all of their own internal variables or data structures. They often communicate through computer internal memory rather than external data files. An expert network augmented with explanation capability is a partially overlapped system. Gallant's connectionist expert system is an example.

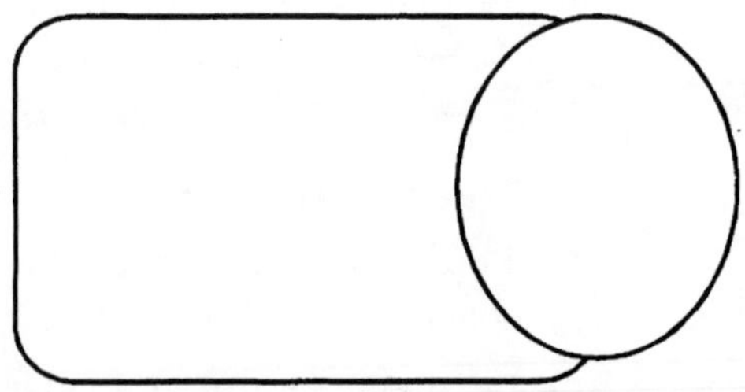

Fig. 8.4. Partially Overlapped Systems

8.5.3. Parallel

A K.B. and NN work in parallel to solve a common problem. Both can be stand-alone systems. The 2 components do not share their own internal variables. They communicate through input, output devices such as data files. For example, in a medical diagnostic system, NN analyzes the signal and images, and a K.B. system interprets the clinical symptoms. And the results are combined.

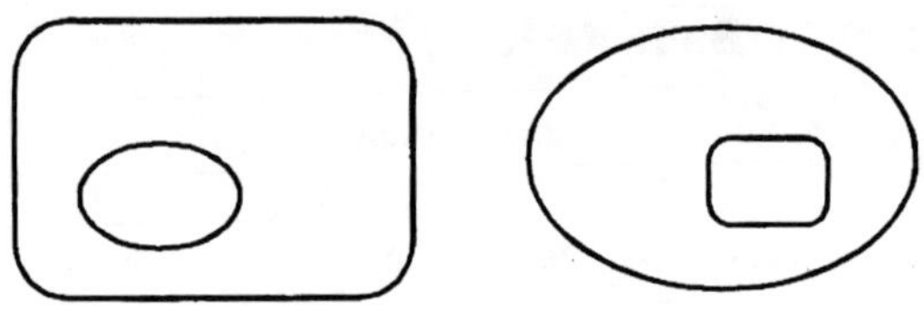

Fig. 8.5. Parallel Systems

8.5.4. Sequential

A K.B. system and NN operate in sequence to solve a particular problem. Again both are stand-alone systems and do not share the internal variables. The output of one component is passed on to other for further processing. *e.g.*, a NN is used as a preprocessor for filtering the noise and transforming the signals to symbols, which are subsequently processed by an expert system.

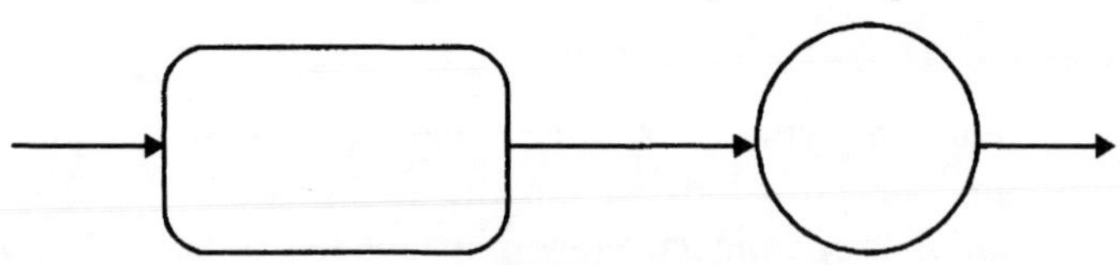

Fig. 8.6. Sequential Systems

8.5.5. Embedded

In this integration, either a K.B. component is embedded within a NN or vice-versa. Inherent information exchange is expected. However, this architecture differs from the partially overlapped architectures in that the system's external feature is determined by the host component only.

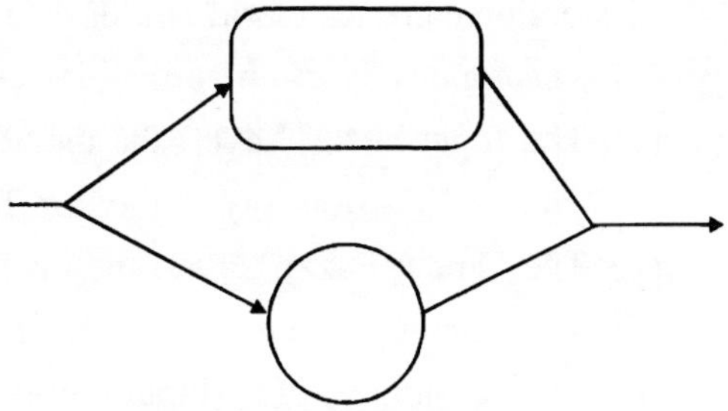

Fig. 8.7. Embedded Systems

The Integration techniques can be categorized according to the nature of coupling:

(1) **Fully Coupled:** Corresponding to the completely overlapped architecture.

(2) **Tightly Coupled:** Including the partially overlapped and embedded architectures.

(3) **Loosely Coupled:** The parallel and sequential architecture.

8.6. FUZZY LOGIC AND NEURAL NETWORKS

Based on Zadeh's fuzzy set theory, fuzzy logic views each predicate as a fuzzy set. Fuzzy neural networks are well known for their ability to handle the fuzzy nature of inference involving symbols. In fuzzy logic, a linguistic variable like "size" can have several linguistic values like "small", "medium", or "large". Each linguistic value is viewed as a fuzzy set associated with a membership function. The degree of membership can be interpreted as the degree of possibility, which evades the requirement of satisfying the probability axioms.

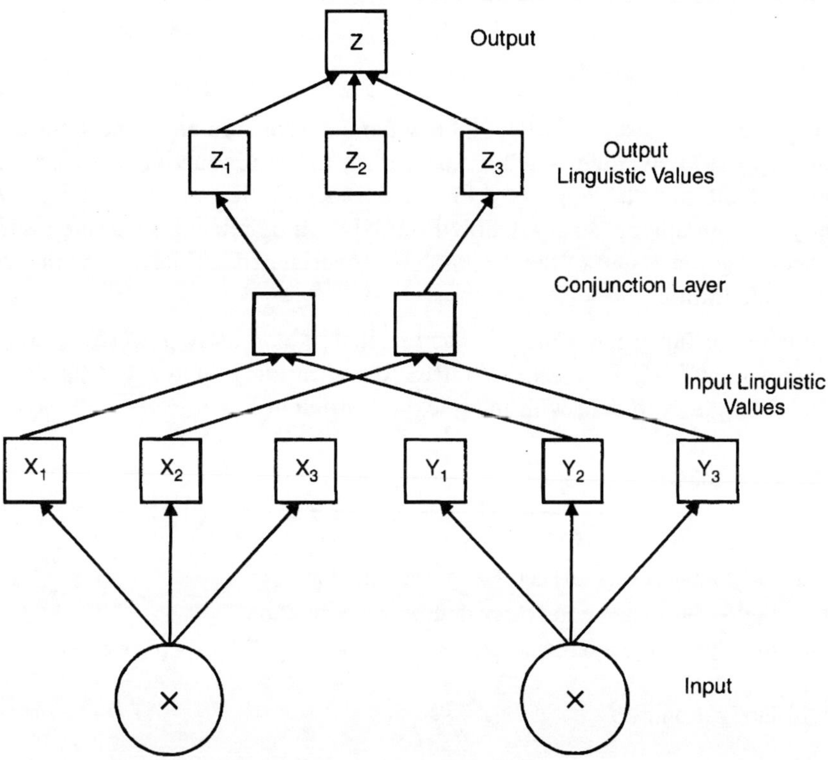

Fig. 8.8. A Fuzzy Neural Network Based on Fuzzy Rules

The architecture of hybrid approach of neural network and fuzzy logic consists of 5 layers:

(1) The Input Layer: no activation function

(2) The Input Fuzzy Layer: the membership function

(3) The Conjunction Layer: the min function

(4) The Output Fuzzy Layer: the max function

(5) THE Output Layer: $F(a) = a/\Sigma X_i \sigma_{ji}$

The activation level of an input unit is the value of certain input variable in the given instance. The input value is passed onto the fuzzy set units, which then translate the value into a degree of membership as the activation level of a fuzzy set unit.

The Conjunction Layer will take the minimum of the inputs it receives from the input fuzzy units shown in Fig. 8.8. The Output Fuzzy Layer will collect the information from 1 or more conjunction units. There exists variation at this point. An Output Fuzzy Layer may take the maximum or sum of the inputs.

The Output Layer generates the final result by integrating the information from output fuzzy set units. How the output unit calculates its activation level depends on the defuzzification scheme.

The fuzzy neural network can learn by adjusting the weights. Different training strategies include:

- Backpropagation
- Reinforcement
- Statistical methods such as random weight change combined with annealing.

Summary

- Expert systems consists of K.B. which separate from inference and control components, contains expert knowledge coded in some form as the production rules, frames etc. In this chapter, we studied the important knowledge acquisition and knowledge representation techniques in an attempt to study how ES& NN can be related. Five integration techniques have been discussed. Out of these completely overlapped architecture is the most advanced form of integration.
- Combination of fuzzy logic and NN has resulted in extremely powerful computation model known as FUZZY NEURAL NETWORK. Representing a linguistic value as a fuzzy set has enabled the system to deal with many expert problems.

Review Questions

1. Explain the architecture of Hybrid Expert Systems.
2. What are different levels of knowledge acquisition ?
3. What are different schemes of knowledge representation ?
4. Explain different integration schemes of building a hybrid expert systems and which one is better.
5. What is Fuzzy Logic ?

❑❑❑

CHAPTER

9

BASICS OF FUZZY LOGIC

INSIDE THIS CHAPTER

9.1. INTRODUCTION

Fuzzy logic was developed by Lotfi A. Zadeh in the 1960s in order to provide mathematical rules and functions which permitted natural language queries. Fuzzy logic provides a means of calculating intermediate values between absolute true and absolute false with resulting values ranging between 0.0 and 1.0. With fuzzy logic, it is possible to calculate the degree to which an item is a member. For example, if a person is 0.53 of tallness, they are "rather tall."

FUZZY set theory exhibits its potential for effective solving the uncertainty in the problem. Fuzziness means 'vagueness'.

9.2. FUZZY VS. CRISP

Fuzzy logic is a super set of conventional (or Boolean) logic and contains similarities and differences with Boolean logic. Fuzzy logic is similar to Boolean logic, in that Boolean logic results are returned by fuzzy logic operations when all fuzzy memberships are restricted to 0 and 1. Fuzzy logic differs from Boolean logic in that it is permissive of natural language queries and is more like human thinking; it is based on degrees of truth.

Fuzzy Logic deals with those imprecise conditions about which a true/false value cannot be determined. Much of this has to do with the vagueness and ambiguity that can be found in everyday life. For example, Consider two situations:

1. Is sun rises in the east.
2. Is ram honest?

The answer to the first condition would be either YES *i.e.*, 1 or NO *i.e.*, 0. Such a logic which demands a binary (0/1) type of handling is termed as crisp. "Temperature is 32°C", "The sun sets in the west" are the examples of crisp sets.

On the other hand the answer to the 2nd query needs not to be a definite YES or NO. There are variety of answers to this query depending on person to person like "extremely honest", "very honest", dishonest, extremely dishonest etc. can be generated. Suppose the value of extremely honest is 1 and value of extremely dishonest is 0, therefore other answers can take values between 1 and 0. Such a situation is termed as fuzzy.

Other example is : "Jane is old". The notion central to fuzzy systems is that truth values (in fuzzy logic) or **membership values** (in fuzzy sets) are indicated by a value on the range [0.0, 1.0], with 0.0 representing absolute Falseness and 1.0 representing absolute Truth. For example, let us take the statement:

"Jane is old."

If Jane's age is 75, we might assign the statement the truth value of 0.80. The statement could be translated into set terminology as follows:

"Jane is a member of the set of old people."

This statement would be rendered symbolically with fuzzy sets as:

mOLD(Jane) = 0.80

Where *m* is the membership function, operating in this case on the fuzzy set of old people, which returns a value between 0.0 and 1.0.

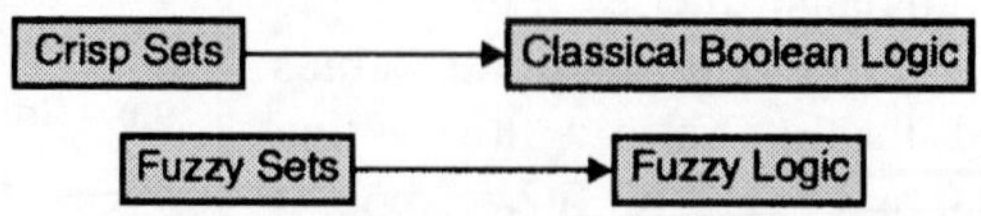

Fig. 9.1. Fuzzy Sets and Crisp Sets

Definition

Fuzzy logic is a branch of fuzzy set theory, which deals with the representation and inference from knowledge. Fuzzy logic, unlike other logical systems, deals with imprecise or uncertain knowledge. In this narrow and perhaps correct sense, fuzzy logic is just one of the branches of fuzzy set theory.

More Definitions

- Fuzzy logic is a set of mathematical principles for knowledge representation based on degrees of membership.
- Unlike two-valued Boolean logic, fuzzy logic is multi-valued. It deals with degrees of membership and degrees of truth.
- Fuzzy logic uses the continuum of logical values between 0 (completely false) and 1 (completely true). Instead of just black and white, it employs the spectrum of colours, accepting that things can be partly true and partly false at the same time.

Now the question arises Why do we need Fuzzy logic?

Natural language abounds with vague and imprecise concepts, such as "Ram is tall," or "It is very hot today." Such statements are difficult to translate into more precise language without losing some of their semantic value: for example, the statement "Ram's height is 152 cm." does not explicitly state that he is tall, and the statement "Ram's height is 1.2 standard deviations above the mean height for women of her age in her culture" is full with difficulties: would a man 1.1999999 standard deviations above the mean be tall? Which culture does Ram belong to, and how is membership in it defined?

The excess of precision and certainty in engineering and scientific research and development is often providing unrealizable solutions. Fuzzy logic, based on the notion of relative graded membership, can deal with information arising from computational perception and cognition that is uncertain, imprecise, vague, partially true, or without sharp boundaries. Thus this is the need of FUZZY LOGIC. Fuzzy logic allows for the inclusion of vague human assessments in computing problems. Also, it provides an effective means for conflict resolution of multiple criteria and better assessment of options. New computing methods based on fuzzy logic can lead to greater adaptability, tractability, robustness, and a lower cost solution in the development of intelligent systems for decision making, identification, recognition, optimization, and control.

9.3. WHY USE FUZZY LOGIC?

FL offers several unique features that make it a particularly good choice for many control problems. Here is a list of general observations about fuzzy logic:

1. Fuzzy logic is conceptually easy to understand.

   The mathematical concepts behind fuzzy reasoning are very simple. Fuzzy logic is a more intuitive approach without the far-reaching complexity.

2. Fuzzy logic is flexible.

   With any given system, it is easy to layer on more functionality without starting again from scratch.

3. Fuzzy logic is tolerant of imprecise data.

   Everything is imprecise if you look closely enough, but more than that, most things are imprecise even on careful inspection. Fuzzy reasoning builds this understanding into the process rather than tacking it onto the end.

4. Fuzzy logic can model nonlinear functions of arbitrary complexity.

   You can create a fuzzy system to match any set of input-output data. This process is made particularly easy by adaptive techniques like Adaptive Neuro-Fuzzy Inference Systems (ANFIS), which are available in Fuzzy Logic Toolbox software.

5. Fuzzy logic can be built on top of the experience of experts.

   In direct contrast to neural networks, which take training data and generate opaque, impenetrable models, fuzzy logic lets you rely on the experience of people who already understand your system.

6. Fuzzy logic can be blended with conventional control techniques.

   Fuzzy systems don't necessarily replace conventional control methods. In many cases fuzzy systems augment them and simplify their implementation.

7. Fuzzy logic is based on natural language.

   The basis for fuzzy logic is the basis for human communication. This observation underpins many of the other statements about fuzzy logic. Because fuzzy logic is built on the structures of qualitative description used in everyday language, fuzzy logic is easy to use.

8. It is inherently robust since it does not require precise, noise-free inputs and can be programmed to fail safely if a feedback sensor quits or is destroyed. The output control is a smooth control function despite a wide range of input variations.
9. Since the FL controller processes user-defined rules governing the target control system, it can be modified and tweaked easily to improve or drastically alter system performance. New sensors can easily be incorporated into the system simply by generating appropriate governing rules.
10. FL is not limited to a few feedback inputs and one or two control outputs, nor is it necessary to measure or compute rate-of-change parameters in order for it to be implemented. Any sensor data that provides some indication of a system's actions and reactions is sufficient. This allows the sensors to be inexpensive and imprecise thus keeping the overall system cost and complexity low.
11. Because of the rule-based operation, any reasonable number of inputs can be processed (1-8 or more) and numerous outputs (1-4 or more) generated, although defining the rule base quickly becomes complex if too many inputs and outputs are chosen for a single implementation since rules defining their interrelations must also be defined. It would be better to break the control system into smaller chunks and use several smaller FL controllers distributed on the system, cach with more limited responsibilities.
12. FL can control nonlinear systems that would be difficult or impossible to model mathematically. This opens doors for control systems that would normally be deemed unfeasible for automation.

When Not to Use Fuzzy Logic

Fuzzy logic is not a cure-all. When should you not use fuzzy logic? The safest statement is the first one made in this introduction: fuzzy logic is a convenient way to map an input space to an output space. If you find it's not convenient, try something else. If a simpler solution already exists, use it. Fuzzy logic is the codification of common sense — use common sense when you implement it and you will probably make the right decision. Many controllers, for example, do a fine job without using fuzzy logic. However, if you take the time to become familiar with fuzzy logic, you'll see it can be a very powerful tool for dealing quickly and efficiently with imprecision and nonlinearity.

9.4. FUZZY LOGIC VS. PROBABILITY

Fuzzy logic may appear similar to probability and statistics as well. However, fuzzy logic is different than probability even though the results appear similar. The probability statement, "There is a 70% chance that Bill is tall" supposes that Bill is either tall or he is not. There is a 70% chance that we know which set Bill belongs. The fuzzy logic statement, "Bill's degree of membership in the set of tall people is .80 " supposes that Bill is rather tall. The fuzzy logic answer determines not only the set which Bill belongs, but also to what degree he is a member. There are no probability statements that pertain to fuzzy logic. Fuzzy logic deals with the degree of membership.

9.5. CRISP SETS

Classical sets are also called 'crisp' sets so as to distinguish them from fuzzy sets. In fact, the crisp sets can be taken as special cases of fuzzy sets. Let A be a crisp set defined over the universe X.

Then for any element x in X, either x is a member of A or *not*. In fuzzy set theory, this property is generalized. Therefore, in a fuzzy set, it is not necessary that x is *a full member* of the set or *not a member*. It can be a *partial member* of the sets.

The generalization is performed as follows: For any crisp set A, it is possible to define a characteristic function $\mu_X = \{0,1\}$, *i.e.*, the characteristic function takes either of the values 0 or 1 in the classical set. For a fuzzy set, the characteristic function can take any value between zero and one.

9.5.1. Set

A set is a well defined collection of objects. Well defined means the object either belongs to or does not belong to the set. The elements or members of a set can be anything: numbers, people, letters of the alphabet, other sets, and so on. Sets are conventionally denoted with capital letters. Sets A and B are equal if and only if they have precisely the same elements.

As set in certain contexts may be associated with its universal set from which it is derived. Given a set A whose objects are $a_1, a_2, ..., a_n$, then $A = \{a_1, a_2, ..., a_n\}$. Here $a_1, a_2, ..., a_n$ are called members of the set. This form of representing is a set is known as list form.

There are two ways of describing, or specifying the members of a set. One way is by intensional definition, using a rule or semantic description:

A is the set whose members are the first four positive integers.

B is the set of colors of the French flag.

The second way is by extension—that is, listing each member of the set. An extensional definition is denoted by enclosing the list of members in brackets:

$$C = \{4, 2, 1, 3\}$$

$$D = \{\text{blue, white, red}\}$$

All set operations preserve the property that each element of the set is unique. The order in which the elements of a set are listed is irrelevant, unlike a sequence or tuple. For example,

$$\{6, 11\} = \{11, 6\} = \{11, 11, 6, 11\},$$

A set may also be defined based on the properties the members have to satisfy. In such a case, a set is defined as.

$$A = \{x \mid P(x)\}$$

where $P(x)$ stands for the property P to be satisfied by the member x. This is read as '*A* is the set of all X such that $P(x)$ is satisfied.

For example: 1. The set F of the twenty smallest integers that are four less than *perfect squares* can be denoted:

$$F = \{n^2 - 4 : n \text{ is an integer; and } 0 = n = 19\}$$

In this notation, the *colon* (":") means "such that", and the description can be interpreted as "*F* is the set of all numbers of the form $n^2 - 4$, such that n is a whole number in the range from 0 to 19 inclusive." Sometimes the *vertical bar* ("|") is used instead of the colon.

2. $A = \{x \mid x \text{ is an odd number}\}$

9.5.2. Membership

The key relation between sets is *membership* – when one set is an element of another. If a is a member of B, this is denoted $a \in B$, while if c is not a member of B then $c \notin B$. For example, with respect to the sets $A = \{1, 2, 3, 4\}$, $B = \{\text{blue, white, red}\}$, and $F = \{n^2 - 4 : n \text{ is an integer; and } 0 = n = 19\}$ defined above,

$4 \in A$ and $285 \in F$; but

$9 \notin F$ and green $\notin B$.

9.5.3. Subset

If every member of set A is also a member of set B, then A is said to be a *subset* of B, written $A \subseteq B$ (also pronounced *A is contained in B*). Equivalently, we can write $B \supseteq A$, read as *B is a superset of A*, *B includes A*, or *B contains A*. The relationship between sets established by $\subseteq$ is called *inclusion* or *containment*.

If A is a subset of, but not equal to B, then A is called a *proper subset* of B, written $A \subsetneq B$ (*A is a proper subset of B*) or $B \supsetneq A$ (*B is a proper superset of A*).

Note that the expressions $A \subset B$ and $B \supset A$ are used differently by different authors; some authors use them to mean the same as $A \subseteq B$ (respectively $B \supseteq A$), whereas other use them to mean the same as $A \subsetneq B$ (respectively $B \supsetneq A$).

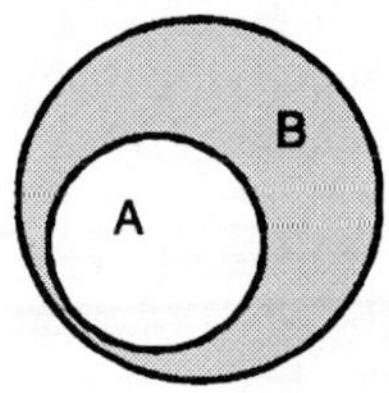

Fig. 9.2. *A* is a subset of *B*

Example:

➡ The set of all men is a proper subset of the set of all people.

➡ $\{1, 3\} \subsetneq \{1, 2, 3, 4\}$.

➡ $\{1, 2, 3, 4\} \subseteq \{1, 2, 3, 4\}$.

The empty set is a subset of every set and every set is a subset of itself:

➡ $\emptyset \subseteq A$.

➡ $A \subseteq A$.

An obvious but useful identity, which can often be used to show that two seemingly different sets are equal:

➡ $A = B$ if and only if $A \subseteq B$ and $B \subseteq A$.

9.5.4. Power Sets

The power set of a set S is the set of all subsets of S. This includes the subsets formed from all the members of S and the empty set. If a finite set S has cardinality n then the power set of S has cardinality 2^n. The power set can be written as $P(S)$.

If S is an infinite (either countable or uncountable) set then the power set of S is always uncountable. Moreover, if S is a set, then there is never a bijection from S onto $P(S)$. In other words, the power set of S is always strictly "bigger" than S.

As an example, the power set of $\{1, 2, 3\}$ is $\{\{1, 2, 3\}, \{1, 2\}, \{1, 3\}, \{2, 3\}, \{1\}, \{2\}, \{3\}, \emptyset\}$. The cardinality of the original set is 3, and the cardinality of the power set is $2^3 = 8$. This relationship is one of the reasons for the terminology *power set*.

9.5.5. Cardinality

There is a unique set with no members and zero cardinality, which is called the *empty set* (or the *null set*) and is denoted by the symbol ? (other notations are used; see empty set). For example, the set of all three-sided squares has zero members and thus is the empty set.

Some sets have infinite cardinality. The set *N* of natural numbers, for instance, is infinite. Some infinite cardinalities are greater than others. For instance, the set of real numbers has greater cardinality than the set of natural numbers. However, it can be shown that the cardinality of (which is to say, the number of points on) a straight line is the same as the cardinality of any segment of that line, of the entire plane, and indeed of any finite-dimensional Euclidean space.

Special sets of numbers include:

- **P**, denoting the set of all primes: $\mathbf{P} = \{2, 3, 5, 7, 11, 13, 17, ...\}$.
- **N**, denoting the set of all natural numbers: $\mathbf{N} = \{1, 2, 3, ...\}$.
- **Z**, denoting the set of all integers (whether positive, negative or zero): $\mathbf{Z} = \{..., -2, -1, 0, 1, 2, ...\}$.
- **Q**, denoting the set of all rational numbers (that is, the set of all proper and improper fractions): $\mathbf{Q} = \{a/b : a, b \in \mathbf{Z}, b \neq 0\}$. For example, $1/4 \in \mathbf{Q}$ and $11/6 \in \mathbf{Q}$. All integers are in this set since every integer a can be expressed as the fraction $a/1$ ($\mathbf{Z} \subsetneq \mathbf{Q}$).
- **R**, denoting the set of all real numbers. This set includes all rational numbers, together with all irrational numbers (that is, numbers which cannot be rewritten as fractions, such as π, e, and $\sqrt{2}$, as well as numbers that cannot be defined).
- **C**, denoting the set of all complex numbers: $\mathbf{C} = \{a + bi : a, b \in \mathbf{R}\}$. For example, $1 + 2i \in \mathbf{C}$.
- **H**, denoting the set of all quaternions: $\mathbf{H} = \{a + bi + cj + dk : a, b, c, d \in \mathbf{R}\}$. For example, $1 + i + 2j - k \in \mathbf{H}$.

Each of the above sets of numbers has an infinite number of elements, and each can be considered to be a proper subset of the sets listed below it. The primes are used less frequently than the others outside of number theory and related fields.

9.5.6. Basic Operations

There are several fundamental operations for constructing new sets from given sets.

Union

Two sets can be "added" together. The *union* of *A* and *B*, denoted by $A \cup B$, is the set of all things which are members of either *A* or *B*.

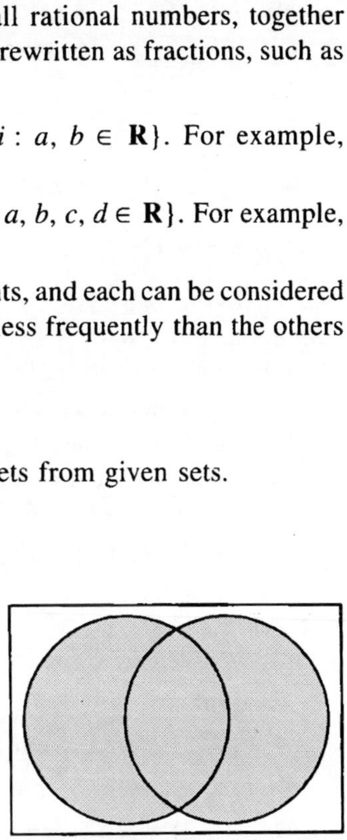

Fig. 9.3.

Examples:

- $\{1, 2\} \cup \{\text{red, white}\} = \{1, 2, \text{red, white}\}$.
- $\{1, 2, \text{green}\} \cup \{\text{red, white, green}\} = \{1, 2, \text{red, white, green}\}$.
- $\{1, 2\} \cup \{1, 2\} = \{1, 2\}$.

Some basic properties of unions:

- $A \cup B = B \cup A$.
- $A \cup (B \cup C) = (A \cup B) \cup C$.

- $A \cup (A \cup B)$.
- $A \cup A = A$.
- $A \cup \emptyset = A$.
- $A \subseteq B$ if and only if $A \cup B = B$.

Intersection

A new set can also be constructed by determining which members two sets have "in common". The *intersection* of A and B, denoted by $A \cap B$, is the set of all things which are members of both A and B. If $A \cap B = \emptyset$, then A and B are said to be *disjoint*.

The **intersection** of A and B, denoted $A \cap B$.

Examples:

- {1, 2} ∩ {red, white} = ∅.
- {1, 2, green} ∩ {red, white, green} = {green}.
- {1, 2} ∩ {1, 2} = {1, 2}.

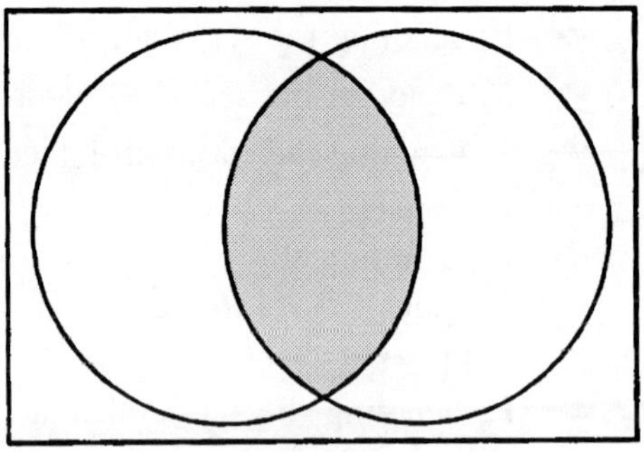

Fig. 9.4.

Some basic properties of intersections:

- $A \cap B = B \cap A$.
- $A \cap (B \cap C) = (A \cap B) \cap C$.
- $A \cap B \subseteq A$.
- $A \cap A = A$.
- $A \cap \emptyset = \emptyset$.
- $A \subseteq B$ if and only if $A \cap B = A$.

Complement

Two sets can also be "subtracted". The *relative complement* of B in A (also called the *set-theoretic difference* of A and B), denoted by $A \setminus B$ (or $A - B$), is the set of all elements which are members of A but not members of B. Note that it is valid to "subtract" members of a set that are not in the set, such as removing the element *green* from the set {1, 2, 3}; doing so has no effect.

In certain settings all sets under discussion are considered to be subsets of a given universal set U. In such cases, $U \setminus A$ is called the *absolute complement* or simply *complement* of A, and is denoted by A'.

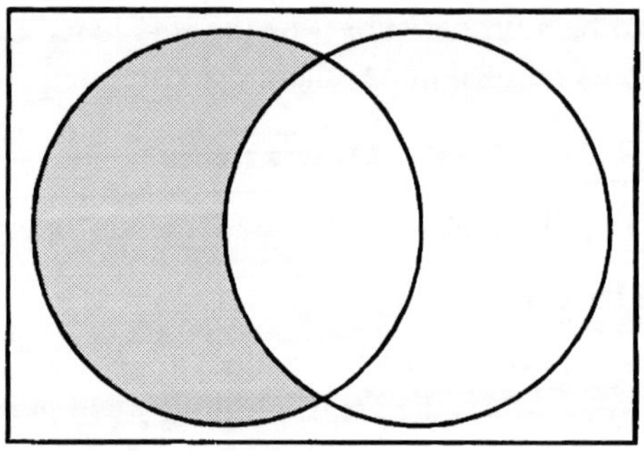

Fig. 9.5. The Relative Component of B in A

Examples:

- {1, 2} \ {red, white} = {1, 2}.
- {1, 2, green} \ {red, white, green} = {1, 2}.
- {1, 2} \ {1, 2} = ∅.
- {1, 2, 3, 4} \ {1, 3} = {2, 4}.
- If U is the set of integers, E is the set of even integers, and O is the set of odd integers, then $E' = O$.

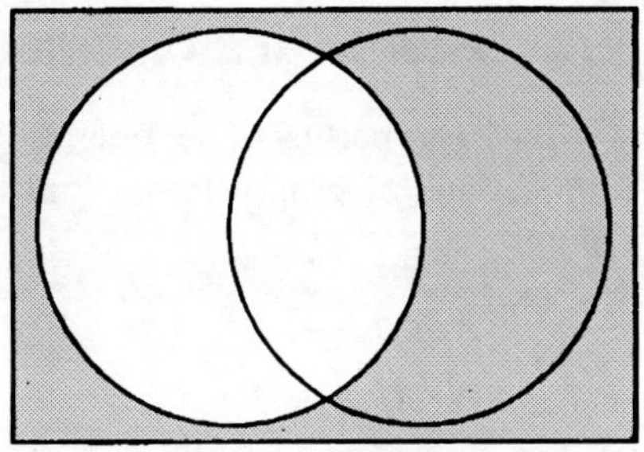

Fig. 9.6. The complement of A in U

Some basic properties of complements:

- $A \setminus B \neq B \setminus A$.
- $A \cup A' = U$.
- $A \cap A' = \emptyset$.
- $(A')' = A$.
- $A \setminus A = \emptyset$.
- $U' = \emptyset$ and $\emptyset' = U$.
- $A \setminus B = A \cap B'$.

An extension of the complement is the symmetric difference, defined for sets A, B as

$$A \Delta B = (A \setminus B) \cup (B \setminus A).$$

For example, the symmetric difference of {7,8,9,10} and {9,10,11,12} is the set {7,8,11,12}.

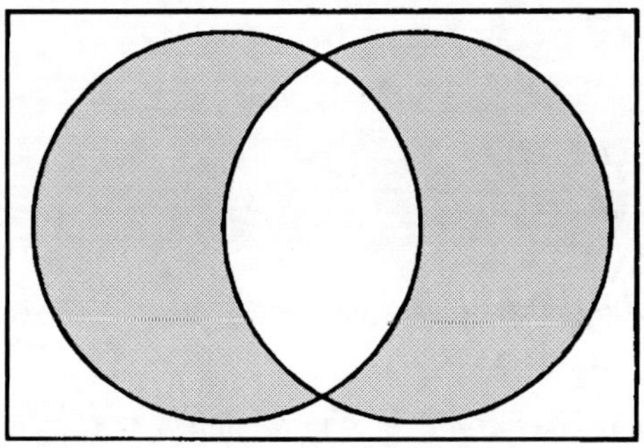

Fig. 9.7. The symmetric difference of *A* and *B*

Cartesian Product

A new set can be constructed by associating every element of one set with every element of another set. The *Cartesian product* of two sets A and B, denoted by $A \times B$ is the set of all ordered pairs (a, b) such that a is a member of A and b is a member of B.

Examples:

- {1, 2} × {red, white} = {(1, red), (1, white), (2, red), (2, white)}.
- {1, 2, green} × {red, white, green} = {(1, red), (1, white), (1, green), (2, red), (2, white), (2, green), (green, red), (green, white), (green, green)}.
- {1, 2} × {1, 2} = {(1, 1), (1, 2), (2, 1), (2, 2)}.

Some basic properties of cartesian products:

- $A \times \emptyset = \emptyset$.
- $A \times (B \cup C) = (A \times B) \cup (A \times C)$.
- $(A \cup B) \times C = (A \times C) \cup (B \times C)$.

Let A and B be finite sets. Then

- $|A \times B| = |B \times A| = |A| \times |B|$.

9.6. PROPERTIES OF CRISP SETS

Basic properties of set operations are discussed here. **1 – 6** directly correspond to identities and implications of propositional logic, and **7 – 11** also follow immediately from them as illustrated below.

1. $A \cup \emptyset = A$
   $A \cap U = A$ ...Identity Laws
2. $A \cup U = U$
   $A \cap \emptyset = \emptyset$ ...Domination Laws
3. $A \cup A = A$
   $A \cap A = A$ ...Idempotent Laws
4. $A \cup B = B \cup A$
   $A \cap B = B \cap A$ ...Commutative Laws
5. $(A \cup B) \cup C = A \cup (B \cup C)$
   $(A \cap B) \cap C = A \cap (B \cap C)$ ...Associative Laws
6. $A \cup (B \cap C) = (A \cup B) \cap (A \cup C)$
   $A \cap (B \cup C) = (A \cap B) \cup (A \cap C)$ ...Distributive Laws
7. If $A \subseteq B$ and $C \subseteq D$, then $A \cup C \subseteq B \cup D$, and $A \cap C \subseteq B \cap D$.
8. If $A \subseteq B$, then $A \cup B = B$ and $A \cap B = A$.
9. $A \cup (B - A) = A \cup B$.
10. $A \cap (B - A) = \emptyset$.
11. $A - (B \cup C) = (A - B) \cap (A - C)$ (cf. $\overline{A \cup B} = \overline{A} \cap \overline{B}$)
    $A - (B \cap C) = (A - B) \cup (A - C)$ (cf. $\overline{A \cap B} = \overline{A} \cup \overline{B}$) ...De Morgan's Laws
12. $B = \overline{A}$ if and only if $A \cup B = U$ and $A \cap B = \emptyset$
13. $\overline{\overline{A}} = A$.

Additional properties:

14. $A \subseteq A \cup B$.
15. $A \cap B \subseteq A$.

The properties 1 ~ 6, and 11 can be proven using equivalences of propositional logic. The others can also be proven similarly by going to logic, though they can be proven also using some of these properties (after those properties are proven, needless to say). Let us prove some of these properties.

Proof for 4: $A \cup B = B \cup A$

We are going to prove this by showing that every element that is in $A \cup B$ is also in $B \cup A$ and vice-versa.

Consider an arbitrary element x. Then by the definition of set union

$x \in A \cup B \Leftrightarrow x \in A \vee x \in B$

$\Leftrightarrow x \in A \vee x \in B$ by the commutativity of $\vee$

$\Leftrightarrow x \in B \cup A$ by the definition of set union.

Note here the correspondence of the commutativity of $\cap$ and that of $\vee$. This correspondence holds not just for the commutativity but also for others.

Furthermore a similar correspondence exists between $\cap$ and $\wedge$, and between $\subseteq$ and $\rightarrow$.

Proof for 6: By the definition of the equality of sets, we need to prove that

$$\forall\ x \mid x \in A \cup (B \cap C) \text{ if and only if } x \in (A \cup B) \cap (A \cup C)$$

For that we need to show that for an arbitrary element in the universe x,

$$x \in A \cup (B \cap C) \text{ if and only if } x \in (A \cup B) \cap (A \cup C)$$

Here the only if part is going to be proven. The *if* part can be proven similarly.

$x \in A \cup (B \cap C) \Leftrightarrow x \in A \vee x \in (B \cap C)$ by the definition of $\cup$.

$\Leftrightarrow\ x \in A \vee (x \in B \wedge x \in C)$ by the definition of $\cap$.

$\Leftrightarrow\ (x \in A \vee x \in B) \wedge (x \in A \vee x \in C)$ by the distribution from the equivalences of propositional logic.

$\Leftrightarrow\ (x \in A \cup B) \wedge (x \in A \cup C)$ by the definition of $\cup$.

$\Leftrightarrow\ x \in (A \cup B) \cap (A \cup C)$ by the definition of $\cap$.

Proof for 8: (*a*) If $A \subseteq B$ then $A \cup B = B$.

Let x be an arbitrary element in the universe.

Then, $x \in A \cup B \Leftrightarrow x \in A \vee x \in B$

Since $A \subseteq B\ x \in A \rightarrow x \in B$

Hence, $x \in B \rightarrow x \in B$

Hence, $A \cup B \subseteq B$

Since, $B \subseteq A \cup B$ (use "addition" rule), $A \cup B = B$ follows.

(*b*) Similarly for $A \cap B = A$.

Alternative proof :

These can also be proven using 8, 14 and 15. For example, (*b*) can be proven as follows :

First by 15 $A \cap B \subseteq A$.

Then since $A \notin A$, and $A \subseteq B$, by 7 $A \cap A \subseteq A \cap B$.

Since $A \cap A = A$ by 3, $A \subseteq A \cap B$.

Proof for 9 : Let x be an arbitrary element in the universe.

Then $[x \in A \cup (B - A)]\ \Leftrightarrow [x \in A \vee (x \in B \wedge x \notin A)]$

$\Leftrightarrow\ [(x \in A \vee x \in B) \wedge (x \in A \vee x \notin A)]$

$\Leftrightarrow [(x \in A \vee x \in B) \wedge \text{True}]$

$\Leftrightarrow\ [x \in A \vee x \in B]$

Hence, $A \cup (B - A) = A \cup B$.

Alternative proof:

This can also proven using set properties as follows :

$A \cup (B - A) = A \cup (B \cap \overline{A})$ by the definition of $(B - A)$.

$= (A \cup B) \cap (A \cup \overline{A})$ by the distribution

$= (A \cup B) \cap U$

$= (A \cup B)$ by 1.

Proof for 10 : Suppose $A \cap (B - A) \neq \emptyset$

Then there is an element x that is in $A \cap (B - A)$, *i.e.*,

$x \in A \cap (B - A) \Leftrightarrow x \in A \wedge x \in B - A$

$\Leftrightarrow\ x \in A \wedge (x \in B \wedge x \notin A)$

$\Leftrightarrow\ (x \in A \wedge \notin A) \wedge x \in B$

$\Leftrightarrow\ F$

Hence, $A \cap (B - A) \neq \emptyset$ does not hold.

Hence, $A \cap (B - A) = \emptyset$.

This can also be proven in the similar manner to 9 above.

Proof for 11: Let x be an arbitrary element in the universe.

Then $x \in A - (B \cup C) \Leftrightarrow x \in A \wedge x \notin B \cup C \Leftrightarrow x \in A \wedge \neg(x \in B \vee x \in C)$

$\Leftrightarrow\ x \in A \wedge (x \notin B \wedge x \notin C)$

$\Leftrightarrow\ x \in A - B \wedge x \in A - C$

$\Leftrightarrow\ x \in (A - B) \cap (A - C)$

Hence, $A - (B \cup C) = (A - B) \cap (A - C)$

$$A \cup B = U \wedge A \cap B = \emptyset \Rightarrow B = \overline{A}$$

Proof for 12 : (*a*)

Try to prove $B \subseteq \overline{A}$ and $\overline{A} \subseteq B$

Let x be an arbitrary element in the universe.

Then if $x \in B$, then $x \notin A$ since $A \cap B = \emptyset$. Hence $x \notin \overline{A}$

Hence $B \subseteq \overline{A}$

If $x \in \overline{A}$, then $x \notin A$. Since $x \in A \vee x \in B$ (from $A \cup B = U$), $x \in B$ must hold. Hence $\overline{A} \subseteq B$

Hence $B = \overline{A}$.

(*b*) $B = \overline{A} \Rightarrow (A \cup B = U \wedge A \cap B = \emptyset)$

Since $B = \overline{A}$, $A \cup B = A \cup (U - A) = A \cup U = U$ since $A \subseteq U$.

Also, $A \cap B = A \cap (U - A) = \emptyset$ by 10 above.

Proof for 13 : Since $A \cup \overline{A} = \overline{A} \cup A$, $\overline{A} \cup A = U$.

Also since $A \cap \overline{A} = \overline{A} \cap A$, $\overline{A} \cap A = \emptyset$

Hence A satisfies the conditions for the complement of $\overline{A}$.

Hence $A = \overline{\overline{A}}$.

9.7. APPLICATIONS OF CRISP SETS

Set theory is seen as the foundation from which virtually all of mathematics can be derived. For example, structures in abstract algebra, such as groups, fields and rings, are sets closed under one or more operations.

One of the main applications of set theory is constructing relations. A relation from a domain A to a codomain B is a subset of the Cartesian product $A \times B$. Given this concept, we are quick to see that the set F of all ordered pairs (x, x^2), where x is real, is quite familiar. It has a domain set **R** and a codomain set that is also **R**, because the set of all squares is subset of the set of all reals. If placed in functional notation, this relation becomes $f(x) = x^2$. The reason these two are equivalent is for any given value, y that the function is defined for, its corresponding ordered pair, (y, y^2) is a member of the set F

9.8. FUZZY SETS

In crisp sets an element is either belongs to or does not belongs to a set, while in fuzzy set theory many degrees of membership (between 0 and 1) are allowed. Thus a membership function μ_A is associated with fuzzy set A such that the function maps every element of the universe of discourse X to the interval [0,1].

A fuzzy set is always written as a collection of ordered pairs (member with its membership function) μ_A.

$$A = \{(x, \mu_A(x)), x \in X\}$$

But in crisp sets only the member is written but not the membership function.

9.8.1. Membership Functions

The membership function values need not always be described by discrete values, but they can be described by a continuous function. Different types of membership functions are : Triangular, Trapezoidal, S Shaped and n shaped

9.8.2. Triangular Membership Function

It is defined by parameters $[a, b, c]$, where is the membership function left intercept at grade equal to 0, b is the center peak where grade equals 1 and c is the right intercept at grade equal to 0.

For example, A triangular membership function "x close to 33' defined over x [0 : 1 : 50] with $[a\ b\ c]$ = [23 33 43] would be created with

$$\frac{A}{x} = \begin{cases} 0 & \text{if } x \le a \\ \dfrac{(x-a)}{(m-a)} & \text{if } x \in (a, m) \\ \dfrac{(b-x)}{(b-m)} & \text{if } x \in (m, b) \\ 1 & \text{if } x \ge b \end{cases}$$

$a < m < b$ m is the modal value.

9.8.3. Trapezoidal Membership Function

A temperature value of 0 would have a 0 membership to all fuzzy sets. Therefore we should use trapezoidal; membership function to define the cool and hot fuzzy sets.

$$x = [0 : 100]$$
$$\text{cool} = [0\ 0\ 25\ 50]$$
$$\text{medium} = [15\ 75\ 100\ 100]$$
$$\text{hot} = [50\ 75\ 100\ 100]$$

9.8.4. S Shaped Membership Function

It is defined by three parameters [α β γ] where

$$\alpha = \text{the point where } \mu(x) = 0$$
$$\beta = \text{the point where } \mu(x) = 0.5$$
$$\gamma = \text{the point where } \mu(x) = 1.0$$

9.8.5. n Shaped Membership Function

It is defined by two parameters [γ, β] where

γ = center of membership function

β = width of membership at grade = 0.5

$$x = [0 :100]$$
$$\text{cool} = [25\ 20]$$
$$\text{medium} = [50\ 20]$$
$$\text{hot} = [75\ 20]$$

9.9. BASIC FUZZY SET OPERATIONS

Fuzzy Set Operations

1. Union

The membership function of the Union of two fuzzy sets A and B with membership functions μ_A and μ_B respectively is defined as the maximum of the two individual membership functions. This is called the *maximum* criterion.

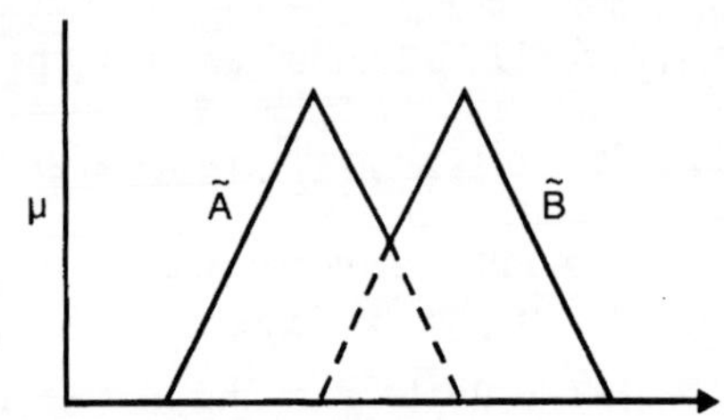

Fig. 9.8. Membership Function for Union

$$\mu_{A \cup B} = \max(\mu_A, \mu_B)$$

The Union operation in Fuzzy set theory is the equivalent of the OR operation in Boolean algebra.

Example 9.1. Let A and B are two fuzzy sets. Where $A = \{(x_1, 0.5), (x_2, 0.7), (x_3, 1)\}$ and $B = \{(x_1, 0.8), (x_2, 0.2), (x_3, 0)\}$.

Then, find the union of A and B.

Solution. $\mu_{A \cup B} = \{(x_1, 0.8), (x_2, 0.7), (x_3, 1)\}$

Since union is maximum of the membership functions. If we compare the membership values of x_1 from A and B then maximum of 0.5 and 0.8 is 0.8. Similarly max of 0.7 and 0.2 is 0.7 and max of 0 and 1 for x_3 is 1.

2. Intersection

The membership function of the Intersection of two fuzzy sets A and B with membership functions μ_A and μ_B respectively is defined as the minimum of the two individual membership functions. This is called the *minimum* criterion.

$$\mu_{A \cap B} = \min (\mu_A, \mu_B)$$

The Intersection operation in Fuzzy set theory is the equivalent of the AND operation in Boolean algebra.

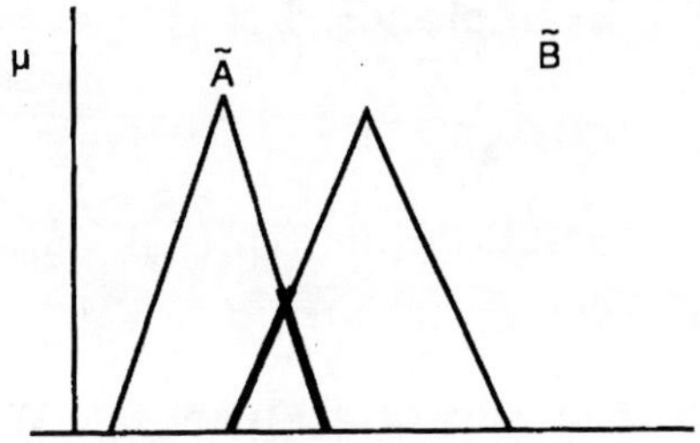

Fig. 9.9. Membership Function for Intersection

Example 9.2. Let A and B are two fuzzy sets. Where $A = \{(x_1, 0.5), (x_2, 0.7), (x_3, 1)\}$ and $B = \{(x_1, 0.8), (x_2, 0.2), (x_3, 0)\}$. Then find the intersection of A and B.

Solution. $\mu_{A \cap B} = \{(x_1, 0.5), (x_2, 0.2), (x_3, 0)\}$

Since intersection is minimum of the membership functions. If we compare the membership values of x_1 from A and B then minimum of 0.5 and 0.8 is 0.5. Similarly min of 0.7 and 0.2 is 0.2 and min of 0 and 1 for x_3 is 0.

3. Complement

The membership function of the Complement of a Fuzzy set A with membership function μ_A is defined as the negation of the specified membership function. This is called the *negation* criterion.

$$\mu_{\bar{A}} = 1 - \mu_A$$

The Complement operation in Fuzzy set theory is the equivalent of the NOT operation in Boolean algebra.

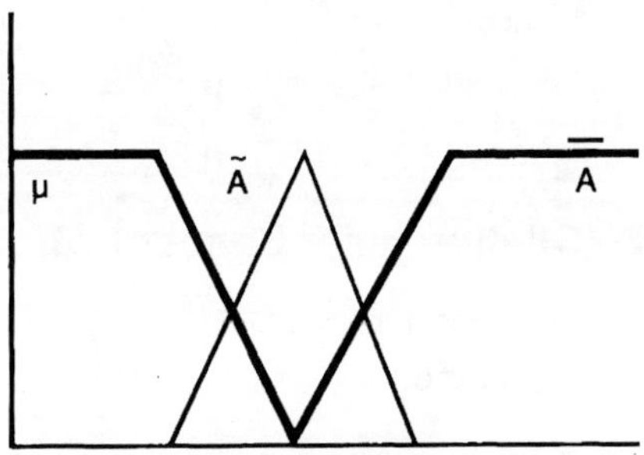

Fig. 9.10. Membership Function for Complement

Example 9.3. Let $A = \{(x_1, 0.5), (x_2, 0.7), (x_3, 0)\}$. Find the complement of A.

Solution. Since $\mu_{\bar{A}} = 1 - \mu_A$ therefore

complement of fuzzy set A is $= \{(x_1, 0.5), (x_2, 0.3), (x_3, 1)\}$.

[by subtracting each of the membership value from 1 *i.e.*, $1 - 0.5 = 0.5$, $1 - 0.7 = 0.3$, $1 - 0 = 1$].

4. Product of Two Fuzzy Sets

The product of two fuzzy sets A and B is a new fuzzy set $A.B$ whose membership function is defined as $\mu_{A.B}(x) = \mu_A(x).\, \mu_B(x)$.

Example 9.4. Let A and B are two fuzzy sets. Where $A = \{(x_1, 0.5), (x_2, 0.7), (x_3, 1)\}$ and $B = \{(x_1, 0.8), (x_2, 0.2), (x_3, 0)\}$. Then find the product of A and B.

Solution. $A.B = \{(x_1, 0.4), (x_2, 0.14), (x_3, 0)\}$

Since
$$\mu_{A.B}(x_1) = \mu_A(x_1).\, \mu_B(x_1) = 0.5*0.8 = 0.4$$
$$\mu_{A.B}(x_2) = \mu_A(x_2).\, \mu_B(x_2) = 0.7*0.2 = 0.14$$
$$\mu_{A.B}(x_3) = \mu_A(x_3).\, \mu_B(x_3) = 0*1 = 0$$

Equality

Two fuzzy sets A and B are said to be equal if $\mu_A(x) = \mu_B(x)$

Example 9.5. Let $A = \{(x_1, 0.2), (x_2, 0.8)\}$
$B = \{(x_1, 0.6), (x_2, 0.8)\}$
$C = \{(x_1, 0.2), (x_2, 0.8)\}$

Solution. Since $\mu_A(x_1)! = \mu_B(x_1)$ therefore $A \neq B$ although $\mu_A(x_2) = \mu_B(x_2)$
But $A = C$ since $\mu_A(x_1) = \mu_C(x_1) = 0.2$ and $\mu_A(x_2) = \mu_C(x_2) = 0.8$

5. Product of a Fuzzy Set With a Crisp Number

Multiplying a Fuzzy Set A by a crisp number 'a' results in a new fuzzy set product $a.A$ with the membership function.

$$\mu_{A.a}(x) = a.\ \mu_A(x)$$

Example 9.6. Consider $A= \{(x_1, 0.2), (x_2, 0.8), (x_3, 0.6)\}$ and $a = 0.3$. Find the product of a fuzzy set with a crisp number.

Solution. $a.\ A = \{(x_1, 0.05), (x_2, 0.24), (x_3, 0.18)\}$

Since
$$\mu_{A.a}(x_1) = a.\ \mu_A(x_1) = 0.2 * 0.3 = 0.05$$
$$\mu_{A.a}(x_2) = a.\ \mu_A(x_2) = 0.8 * 0.3 = 0.24$$
$$\mu_{A.a}(x_1) = a.\ \mu_A(x_1) = 0.6 * 0.3 = 0.18$$

6. Dilation and Concentration

Raising a fuzzy set to its second power is called concentration and taking the square root is called dilation.

7. Difference

The difference of 2 fuzzy sets A and B is a new fuzzy set $A - B$ defined as $A - B = (A \cap (\neg B)\ )$.

Example 9.7. Let A and B are two fuzzy sets. Where $A = \{(x_1, 0.2), (x_2, 0.5), (x_3, 0.6)\}$ and $B = \{(x_1, 1), (x_2, 0.4), (x_3, 0.5)\}$. Then find the difference of A and B.

Solution. $A - B = \{(x_1, 0.2), (x_2, 0.5), (x_3, 0.5)\}$

Since $(\neg B) = 1 - \mu_B(x) = \{(x_1, 0), (x_2, 0.6), (x_3, 0.5)\}$

And $(A \cap (\neg B)) = \min\ (\mu_A(x), \min\ (\mu(\neg B)(x))$
$= \min\ (0.2, 1) = 0$ and so on.

Disjunction Sum: The disjunctive sum of 2 fuzzy sets A and B is a new fuzzy set A exor $B = (\bar{A} \cap B) \cup (A \cap \bar{B})$

Example 9.8. Consider 2 fuzzy sets A and B where $A = \{(x_1, 0.4), (x_2, 0.8), (x_3, 0.6)\}$ and $B = \{(x_1, 0.2), (x_2, 0.6), (x_3, 0.9)\}$. Then find the disjunctive sum of A and B.

Solution. Since A exor $B = (\bar{A} \cap B) \cup (A \cap \bar{B})$

$$(\bar{A} \cap B) = \{(x_1, 0.2), (x_2, 0.2), (x_3, 0.4)\}$$

$$(A \cap \bar{B}) = \{(x_1, 0.4), (x_2, 0.4), (x_3, 0.1)\}$$

Therefore A exor $B = \{(x_1, 0.4), (x_2, 0.4), (x_3, 0.4)\}$

9.10. OPERATIONS ON FUZZY SET THEORY

De Morgans Law:

$$\overline{(A \cap B)} = \overline{A} \cap \overline{B}, \overline{(A \cup B)} = \overline{A} \cap \overline{B}$$

Associativity:

$$(A \cap B) \cap C = A \cap (B \cap C)$$

$$(A \cup B) \cup C = A \cup (B \cup C)$$

Commutativity:

$$A \cap B = B \cap A, A \cup B = B \cup A$$

Distributivity:

$$A \cap (B \cup C) = (A \cap B) \cup (A \cap C)$$

$$A \cup (B \cap C) = (A \cup B) \cap (A \cup C)$$

Idempotency:

$$A \cup A = A \text{ and } A \cap A = A$$

Identity : $A \uplus \emptyset = A$ and $A \cap X = A$ and $A \cap \emptyset = \emptyset$ and $A \cup X = X$

Transitivity : If $A \cup B \cup C$, then $A \cup C$

Involution : $\overline{A} = A$

Other Important Properties and Characteristics of Fuzzy Sets

1. Having two fuzzy sets A and B based on X, then both are *equal* if their membership functions are equal, *i.e.*,

$$A = B \Leftrightarrow \mu_A(x) = \mu_B(x),\ x \in X$$

2. The universal set μ is defined as

$$\mu_V(x) = 1,\ x \in X$$

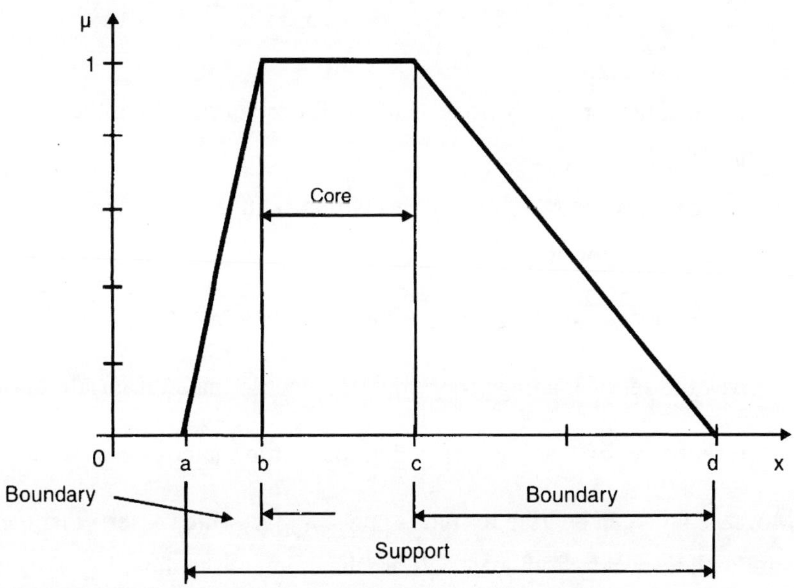

Fig. 9.11. Some Characteristics of a Membership Function

3. The height of a fuzzy set A is the largest membership grade obtained by any element in that set, *i.e.*,

$$\text{hgt}(A) = \sup \mu_A(x),\ x \in X$$

4. A fuzzy set A is called normal when hgt(A) = 1, and it is subnormal when hgt(A) < 1.

5. The *support* of a fuzzy set A is the crisp set that contains all the elements of X that have nonzero membership grades in A, *i.e.*,

$$\text{supp}(A) = \{x \in X \mid \mu_A(x) > 0\}$$

6. The *core* of a normal fuzzy set A is the crisp set that contains all the elements of X that have the membership grades of one in A, *i.e.*,

$$\text{core}\ (A) = \{x \in X \mid \mu_A(x) = 1\}$$

7. The *boundary* is the crisp set that contains all the elements of X that have the membership grades of $0 < \mu_A(x) < 1$ in A, *i.e.*,

$$\text{bnd}(A) = \{x \in X \mid q < \mu_A(x) < 1\}$$

8. Having two fuzzy sets A and B based on X, then both are *similar* if

$$\text{core}(A) = \text{core}(B) \quad \text{and} \quad \text{supp}(A) = \text{supp}(B)$$

9. If the support of a normal fuzzy set consists of a single element x_0 of X, which has the property

$$\text{supp}(A) = \text{core}(A) = \{x_0\},$$

Example 9.9. A simple hollow shaft is 1-m radius and has a wall thickness of (1/2p) m. The shaft is built up stacking a ductile section and a brittle section. A downward force P and a torque T are simultaneously applied to the shaft. The failure properties of the two sections can be described by the following fuzzy sets A and B for the ductile and brittle sections as follows:

$$\tilde{A} = \left\{\frac{1}{2} + \frac{0.5}{3} + \frac{0.3}{4} + \frac{0.2}{5}\right\} \text{ and } \tilde{B} = \left\{\frac{0.5}{2} + \frac{0.7}{3} + \frac{0.2}{4} + \frac{0.4}{5}\right\}$$

We can see the following:

1. The set of loadings for which either material B or material D will be "safe" can be obtained by getting $\tilde{A} \cup \tilde{B}$.
2. The set of loadings for which one expects that both material B and material D are "safe" can be obtained by forming $\tilde{A} \cap \tilde{B}$.
3. The complements $\overline{\tilde{A}}$ and $\overline{\tilde{B}}$ represents the set of loadings for material D and B are unsafe.
4. $\tilde{A} \mid \tilde{B}$ gives the set of loadings for which the ductile material is safe but the brittle is not.
5. $\overline{\tilde{B}} \mid \overline{\tilde{A}}$ gives the set of loadings for which the brittle material is safe but the ductile not.
6. De Morgans laws can be used to find $\tilde{A} \cap \tilde{B} = \overline{\tilde{A}} \cup \overline{\tilde{B}}$ which asserts that the loadings that are not safe with respect to both materials are the union of the those that are unsafe with respect to the brittle material with those that are unsafe for with respect to the ductile material.

7. De Morgans law $\overline{\tilde{A} \cup \tilde{B}} = \overline{\tilde{A}} \cap \overline{\tilde{B}}$ asserts that the loads that are safe for neither material D nor material B are the intersection of those that are unsafe for material D with those that are unsafe for material B.

Consequently, we can find the following:

$$\tilde{A} = \left\{\frac{1}{2} + \frac{0.5}{3} + \frac{0.3}{4} + \frac{0.2}{5}\right\} \text{ and } \tilde{B} = \left\{\frac{0.5}{2} + \frac{0.7}{3} + \frac{0.2}{4} + \frac{0.4}{5}\right\}$$

Complements:
$$\overline{\tilde{A}} = \left\{\frac{1}{1} + \frac{0}{2} + \frac{0.5}{3} + \frac{0.7}{4} + \frac{0.8}{5}\right\} \text{ and } \overline{\tilde{B}} = \left\{\frac{0.5}{2} + \frac{0.3}{3} + \frac{0.8}{4} + \frac{0.6}{5}\right\}$$

Union:
$$\tilde{A} \cup \tilde{B} = \left\{\frac{1}{2} + \frac{0.7}{3} + \frac{0.3}{4} + \frac{0.4}{5}\right\}$$

Intersection :
$$\tilde{A} \cap \tilde{B} = \left\{\frac{0.5}{2} + \frac{0.3}{3} + \frac{0.3}{4} + \frac{0.2}{5}\right\} \text{ etc.}$$

Alpha Cuts

One of the most important concepts of fuzzy sets is the concept of an α-cut and its variant, a *strong* α*-cut.* Given a fuzzy set A defined on X and any number $\alpha \in [0, 1]$, the α-cut, ${}^{\alpha}A$, and the strong α-cut, ${}^{\alpha+}A$, are the crisp sets

$$ {}^{\alpha}A = \{x \mid A(x) \geq \alpha\}$$
$$ {}^{\alpha+}A = \{x \mid A(x) > \alpha\}.$$

That is, the α-cut (or the strong α-cut) of a fuzzy set A is the crisp ${}^{\alpha}A$ (or the crisp set ${}^{\alpha+}A$) that contains all the elements of the universal set X whose membership grades in A are greater than or equal to (or only greater than) the specified value of α.

There are at least two roles that alpha-cuts play in fuzzy logic.

1. Alpha cuts are used to decompose a fuzzy set into a weighted combination of classical sets using the resolution identity principle.

   The principle is important in fuzzy set theory because it establishes a bridge between fuzzy sets and crisp sets. Hence, it has been used as the foundation for generalizing concepts and methods based on crisp sets into those based on fuzzy sets.
2. Alpha cuts can be used to describe intermediate fuzzy conclusions generated by fuzzy inference.

An alpha cut is a crisp set that contains the elements that have a support (also called membership grade) greater than a certain value.

The function α is written to return the minimum and the maximum value where an alpha cut is one.

Additional Properties of Alpha Cut

Let $A, B \in F(X)$. Then, the following properties hold for all $\alpha, \beta \in [0, 1]$:

(*i*) ${}^{\alpha+}A \subseteq {}^{\alpha}A$;

(*ii*) $\alpha \leq \beta$ implies ${}^{\alpha}A \supseteq {}^{\beta}A$ and ${}^{\alpha+}A \supseteq {}^{\beta+}A$;

(*iii*) ${}^{\alpha}(A \cap B) = {}^{\alpha}A \cap {}^{\alpha}B$ and ${}^{\alpha}(A \cup B) = {}^{\alpha}A \cup {}^{\alpha}B$;

(*iv*) $^{\alpha+}(A \cap B) = {}^{\alpha+}A \cap {}^{\alpha+}B$ and $^{\alpha+}(A \cup B) = {}^{\alpha+}A \cup {}^{\alpha+}B$;

(*v*) $^{\alpha}(\bar{A}) = {}^{(1-\alpha)+}\bar{A}$

Let $A, B \in F(X)$. Then, for all $\alpha \in [0, 1]$

$A \subseteq B$ iff $^{\alpha}A \subseteq {}^{\alpha}B$;

$A \subseteq B$ iff $^{\alpha+}A \subseteq {}^{\alpha+}B$;

$A = B$ iff $^{\alpha}A = {}^{\alpha}B$;

$A = B$ iff $^{\alpha+}A = {}^{\alpha+}B$.

Each fuzzy set can uniquely be represented by either the family of all its α-cuts or the family of all its string α-cuts.

Consider the fuzzy set

$$A = \frac{0.2}{x_1} + \frac{0.4}{x_2} + \frac{0.6}{x_3} + \frac{0.8}{x_4} + \frac{1}{x_5}$$

This can be represented by its α-cuts:

$$^{0.2}A = \frac{1}{x_1} + \frac{1}{x_2} + \frac{1}{x_3} + \frac{1}{x_4} + \frac{1}{x_5},$$

$$^{0.4}A = \frac{0}{x_1} + \frac{1}{x_2} + \frac{1}{x_3} + \frac{1}{x_4} + \frac{1}{x_5},$$

$$^{0.6}A = \frac{0}{x_1} + \frac{0}{x_2} + \frac{1}{x_3} + \frac{1}{x_4} + \frac{1}{x_5},$$

$$^{0.8}A = \frac{0}{x_1} + \frac{0}{x_2} + \frac{0}{x_3} + \frac{1}{x_4} + \frac{1}{x_5},$$

$$^{1}A = \frac{0}{x_1} + \frac{0}{x_2} + \frac{0}{x_3} + \frac{0}{x_4} + \frac{1}{x_5},$$

Define a fuzzy set

$$_{\alpha}A(x) = \alpha . {}^{\alpha}A(x)$$

We obtain

$$^{0.2}A = \boxed{\frac{0.2}{x_1}} + \frac{0.2}{x_2} + \frac{0.2}{x_3} + \frac{0.2}{x_4} + \frac{0.2}{x_5},$$

$$^{0.4}A = \frac{0}{x_1} + \boxed{\frac{0.4}{x_2}} + \frac{0.4}{x_3} + \frac{0.4}{x_4} + \frac{0.4}{x_5},$$

$$^{0.6}A = \frac{0}{x_1} + \frac{0}{x_2} + \boxed{\frac{0.6}{x_3}} + \frac{0.6}{x_4} + \frac{0.6}{x_5},$$

$$^{0.8}A = \frac{0}{x_1} + \frac{0}{x_2} + \frac{0}{x_3} + \boxed{\frac{0.8}{x_4}} + \frac{0.8}{x_5},$$

$$^{1}A = \frac{0}{x_1} + \frac{0}{x_2} + \frac{0}{x_3} + \frac{0}{x_4} + \boxed{\frac{1}{x_5}}.$$

Now, it is easy to see that

$$A = {}_{0.2}A \cup {}_{0.4}A \cup {}_{0.6}A \cup {}_{0.8}A \cup {}_{1}A.$$

$$A = \frac{0.2}{x_1} + \frac{0.4}{x_2} + \frac{0.6}{x_3} + \frac{0.8}{x_4} + \frac{1}{x_5}$$

$$^{0.2}A = \frac{1}{x_1} + \frac{1}{x_2} + \frac{1}{x_3} + \frac{1}{x_4} + \frac{1}{x_5},$$

$$^{0.4}A = \frac{0}{x_1} + \frac{1}{x_2} + \frac{1}{x_3} + \frac{1}{x_4} + \frac{1}{x_5},$$

$$^{0.6}A = \frac{0}{x_1} + \frac{0}{x_2} + \frac{1}{x_3} + \frac{1}{x_4} + \frac{1}{x_5},$$

$$^{0.8}A = \frac{0}{x_1} + \frac{0}{x_2} + \frac{1}{x_3} + \frac{1}{x_4} + \frac{1}{x_5},$$

$$^{1}A = \frac{0}{x_1} + \frac{0}{x_2} + \frac{0}{x_3} + \frac{0}{x_4} + \frac{1}{x_5},$$

9.11. CRISP RELATIONS

The concept of relations between sets is built on the Cartesian product operator of sets.

9.11.1. Cartesian Product

Cartesian product of 2 sets X and Y denoted by $X \times Y$ is the set of all ordered pairs such that first element in the pair belongs to X and second element belongs to Y thus producing ordered relationships among sets

$X \times Y = \{(x,y) \mid x \in X,\ y \in Y\}$

If $X \neq Y$ and X and Y are non-empty then $X \times Y \neq Y \times X$

All the $Ar = A$

$A_1 \times A_2 \times \ldots \times A_r = A_r$

Crisp relations are measured by characteristic function: χ

$$X \times Y = \{(x, y) \mid x \in X,\ y \in Y\}$$

For $r = 2$ the relation $R(X_1, X_2)$ is termed as, Binary relation

$$\chi_{X \times Y}(x,y) = 1,\ (x, y) \in X \times Y$$
$$0,\ (x, y) \quad X \times Y$$

$$\chi_R(x, y) = 1,\ (x, y) \in X \times Y$$
$$0,\ (x, y) \quad X \times Y$$

Example :

Set $A = \{0, 1\}$

Set $B = \{a, b, c\}$

$$A \times B = \{(0, a), (0, b), (0, c), (1, a), (1, b), (1, c)\}$$

$$B \times A = \{(a, 0), (a, 1), (b, 0), (b, 1), (c, 0), (c, 1)\}$$

$$A \times A = A^2 = \{(0,0),(0,1),(1,0),(1,1)\}$$

$$B \times B = B^2 = \{(a, a), (a, b), (a, c), (b, a), (b, b), (b, c), (c, a), (c, b), (c, c)\}$$

Example:

$X = \{1, 2, 3\}$ $Y = \{a, b, c\}$ make the sagittal diagram and the relation matrix.

Sagittal graph:

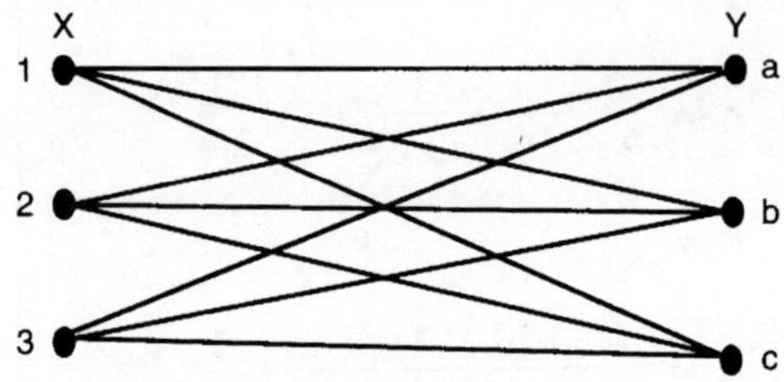

Fig. 9.12

Relation matrix

$$R = \begin{array}{c} \\ 1 \\ 2 \\ 3 \end{array} \begin{array}{c} \begin{array}{ccc} a & b & c \end{array} \\ \begin{bmatrix} 1 & 1 & 1 \\ 1 & 1 & 1 \\ 1 & 1 & 1 \end{bmatrix} \end{array}$$

9.11.2. Cardinality of Crisp Relations

Let X consists of n elements X and Y consists of m elements

n_X : the cardinality of X

n_Y : the cardinality of Y

Then Cardinality of the relation

$$n_{X \times Y} = n_X * n_Y$$

Power set:

The cardinality: $P(X \times Y)$

$$n_{P(X \times Y)} = 2^{(nXnY)}$$

Operations on Crisp Relations

Union:

$$R \cup S \rightarrow \chi_{R \cup S}(x, y) : \chi_{R \cup S}(x, y) = \max [\chi_R(x, y), \chi_S(x, y)]$$

Intersection:

$$R \cap S \rightarrow \chi_{R \cap S}(x, y) : \chi_{R \cap S}(x, y) = \min [\chi_R(x, y), \chi_S(x, y)]$$

Complement:

$$\overline{R} \rightarrow \chi_R(x, y) : \chi_R(x, y) = 1 - \chi_R(x, y)$$

Containment:

$$R \subset S \rightarrow \chi_R(x, y) : \chi_R(x, y) \leq \chi_S(x, y)$$

Identity:

$(\emptyset \rightarrow 0$ and $X \rightarrow E)$

Properties of Crisp Relations

Commutative:

$$A \cup B = B \cup A, A \cap B = B \cap A$$

Associative:

$$(A \cup B) \cup C = A \cup (B \cup C) = A \cup B \cup C$$

Distributive:

$$A \cap (B \cup C) = (A \cap B) \cup (A \cap C)$$

Idempotence:

$$A \cup A = A$$

$$A \cap A = A$$

DeMorgans Law:

$$\overline{A \cup B} = \bar{A} \cap \bar{B}, \overline{A \cap B} = \bar{A} \cup \bar{B},$$

Law of excluded middle:

$$A \cup \bar{A} = U$$

Composition of relation is denoted by R o S

$$R = \{(X_1, Y_1), (X_1, Y_3), (X_2, Y_4)\}$$

$$S = \{(Y_1, Z_2), (Y_3, Z_2)\}$$

9.12. FUZZY RELATIONS

Fuzzy relations play an important role in fuzzy modeling, fuzzy diagnosis, and fuzzy control. They also have applications in fields such as psychology, medicine, economics, and standard *Mathematical* functions etc.

A fuzzy relation is characterized by the same two items as a fuzzy set. First is a list containing element and membership grade pairs, $\{\{v_1, w_1\}, R_{11}\}, \{\{v_1, w_2\}, R_{12}\}, ..., \{\{v_n, w_m\}, R_{nm}\}\}$. Note that the elements of the relation are defined as ordered pairs, $\{v_1, w_1\}, \{v_1, w_2\}, ..., \{v_n, w_m\}$. These elements are again grouped with their membership grades, $\{R_{11}, R_{12}, ..., R_{nm}\}$, which are values that range from 0 to 1 inclusive.

The second item characterizing fuzzy relations is the universal space. For relations, the universal space consists of a pair of ordered pairs, $\{\{V_{min}, V_{max}, c_1\}, \{W_{min}, W_{max}, c_2\}\}$.

The first pair defines the universal space to be used for the first set under consideration in the relation, and the second pair defines the universal space for the second set. The following is an example showing how fuzzy relations are represented in this package.

Universal spaces for fuzzy sets and fuzzy relations are defined with three numbers in this package. The first two numbers specify the start and end of the universal space, and the third argument specifies the increment between discrete elements. Here is an example.

Fuzzy Relation [{{{1, 1}, 0.1}, {{1, 2}, 0.5}, {{2, 1}, 0.9}, {{2, 2}, 0.7}, {{3, 1}, 0.4}, {{3, 2}, 1}},

Universal Space → {{0, 3, 1}, {0, 2, 1}}]

Assuming that V and W are two collections of objects, an arbitrary fuzzy set R, defined in the Cartesian product $V \times W$, will be called a fuzzy relation in the space $V \times W$.

R is thus a function defined in the space $V \times W$, which takes values from the interval [0, 1].

$$R : V \times W \rightarrow [0, 1]$$

In the case where $V = W$, we have a binary fuzzy relation on a single set V.

9.12.1. Union of Fuzzy Relations

Let R and S be fuzzy relations in $V \times W$. The union of two fuzzy relations R and S, denoted by $R \cup S$, is a fuzzy relation in $V \times W$, such that for all (v, w) in $V \times W$. Then $(R \cup S)(v, w) =$ Max $(R(v, w), S(v, w))$.

9.12.2. Intersection of Fuzzy Relations

The intersection of two fuzzy relations, R and S, in $V \times W$ is the fuzzy relation in $V \times W$ such that for all (v, w) in $V \times W$, $(R \cap S)(v, w) = \text{Min}(R(v, w), S(v, w))$.

9.12.3. Algebraic Product of Fuzzy Relations

The algebraic product of the two fuzzy relations R and S in the space $V \times W$ is defined as a fuzzy set in $V \times W$, whose elements satisfy the relation, for all (v, w) in $V \times W$, AlgProduct$(R, S)(v, w)$ $= R(v, w) * S(v, w)$.

Summary of operations on fuzzy relations

Union:

$$\mu_{R \cup S}(x, y) = \max(\mu_R(x, y), \mu_S(x, y))$$

Intersection:

$$\mu_{R \cap S}(x, y) = \min(\mu_R(x, y), \mu_S(x, y))$$

Complement:

$$\mu_R(x, y) = 1 - \mu_R(x, y)$$

Containment:

$$R \subset S \Rightarrow \mu_R(x, y) \leq \mu_S(x, y)$$

Properties of fuzzy relations:

1. Law of excluded middle :

$$\underset{\sim}{R} \cup \overline{\underset{\sim}{R}} \neq E$$

$$\underset{\sim}{R} \cap \overline{\underset{\sim}{R}} \neq O$$

Cartesian product space:

$$\underset{\sim}{A} \times \underset{\sim}{B} = \underset{\sim}{R} \subset X \times Y$$

Fuzzy relation R has membership function

$$\mu_R(x, y) = \mu_{A \times B}(x, y) = \min(\mu_A(x), \mu_B(y))$$

Fuzzy max-min composition

$$T = R \circ S$$

$$\mu_T(x, z) = \underset{x=Y}{\vee} (\mu_R(x, y) \wedge \mu_S(y, z))$$

Fuzzy max-product composition

$$T = R \bullet S$$

$$\mu_T(x, z) = \bigvee_{x=Y} (\mu_R(x, y) \wedge \mu_S(y, z))$$

Example :

$$A = \frac{0.2}{x_1} + \frac{0.5}{x_2} + \frac{1}{x_3} \qquad B = \frac{0.3}{y_1} + \frac{0.9}{y_2}$$

$$A \times B = R = \begin{matrix} & y_1 \quad y_2 \\ \begin{matrix} x_1 \\ x_2 \\ x_3 \end{matrix} & \begin{bmatrix} 0 & 2 & 0 & 2 \\ 0 & 3 & 0 & 5 \\ 0 & 3 & 0 & 9 \end{bmatrix} \end{matrix}$$

Example :

$$X = \{x_1, x_2\}, \quad Y = \{y_1, y_2\} \quad Z = \{z_1, z_2, z_3\}$$

$$R = \begin{matrix} & y_1 \quad y_2 \\ \begin{matrix} x_1 \\ x_2 \end{matrix} & \begin{bmatrix} 0 & 70 & 5 \\ 0 & 80 & 4 \end{bmatrix} \end{matrix} \qquad S = \begin{matrix} & z_1 \quad z_2 \quad z_3 \\ \begin{matrix} y_1 \\ y_2 \end{matrix} & \begin{bmatrix} 0 & 90 & 60 & 2 \\ 0 & 10 & 70 & 5 \end{bmatrix} \end{matrix}$$

Max-min composition

$$T = \begin{matrix} & x_1 \quad x_2 \quad x_3 \\ \begin{matrix} x_1 \\ x_2 \end{matrix} & \begin{bmatrix} 0.7 & 0.6 & 0.5 \\ 0.8 & 0.6 & 0.4 \end{bmatrix} \end{matrix}$$

Max-product composition

$$T = \begin{matrix} & z_1 \quad z_2 \quad z_3 \\ \begin{matrix} x_1 \\ x_2 \end{matrix} & \begin{bmatrix} 0.63 & 0.42 & 0.25 \\ 0.72 & 0.48 & 0.20 \end{bmatrix} \end{matrix}$$

Tolerance and Equivalence relations: there are three types of relations : reflexive, transitive, symmetrical relations.

1. Reflexive

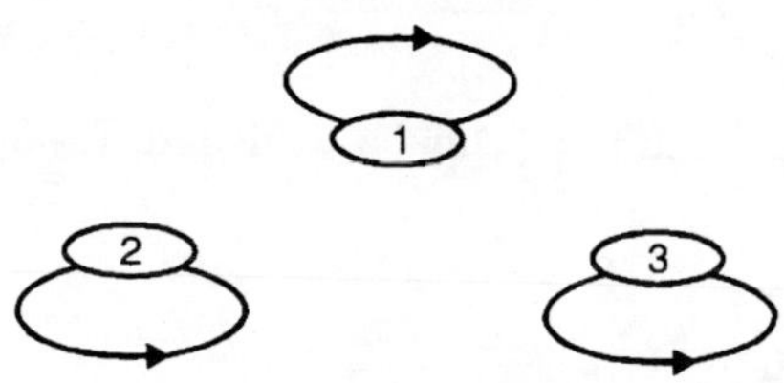

2. Symmetry

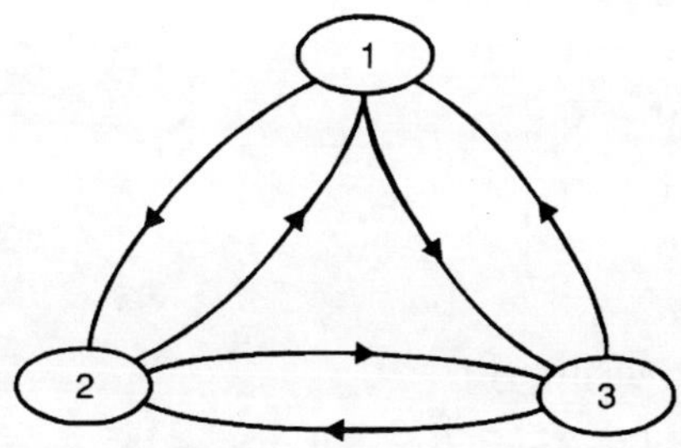

3. Transtivity

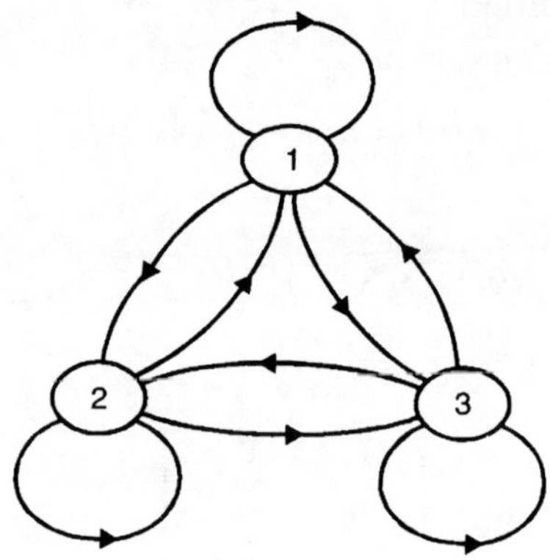

Crisp Equivalence Relation: Crisp equivalence relations are those which follow transitivity, symmetry and reflexivity properties *i.e.,*

Transitivity:

$(x_i, x_j) \in R$ and $(x_j, x_k) \in R \rightarrow (x_j, x_k) \in R$

or $\chi_R(x_i, xj)$ and $\chi_R(x_j, x_k) = 1 \rightarrow \chi_R(x_i, x_k) = 1$

Reflexivity:

$(x_i, x_j) \in R$ or $\chi_R(x_i, x_i) = 1$

Symmetry:

$(x_i, x_j) \in R \rightarrow (x_j, x_i) \in R$

or $\chi_R(x_i, x_j) = \chi_R(x_i, x_i)$

Crisp Tolerance Relation

- Crisp tolerance relations are also called proximity relation.
- Only the reflexivity and symmetry are followed.
- Crisp tolerance relations can be reformed into an equivalence relation.

By at most $(n - 1)$ compositions with itself

$$R_1^{n-1} = R_1 \circ R_1 \circ \ldots \circ R_1 = R$$

Example : $X = \{x_1, x_2, x_3, x_4, x_5\}$ = {Omaha, Chicago, Rome, London, Detroit}

$$R_1 = \begin{array}{c} \\ x_1 \\ x_2 \\ x_3 \\ x_4 \\ x_5 \end{array} \begin{array}{c} \begin{array}{ccccc} x_1 & x_2 & x_3 & x_4 & x_5 \end{array} \\ \begin{bmatrix} 1 & 1 & 0 & 0 & 0 \\ 1 & 1 & 0 & 0 & 1 \\ 0 & 0 & 1 & 0 & 0 \\ 0 & 0 & 0 & 1 & 0 \\ 0 & 1 & 0 & 0 & 1 \end{bmatrix} \end{array}$$

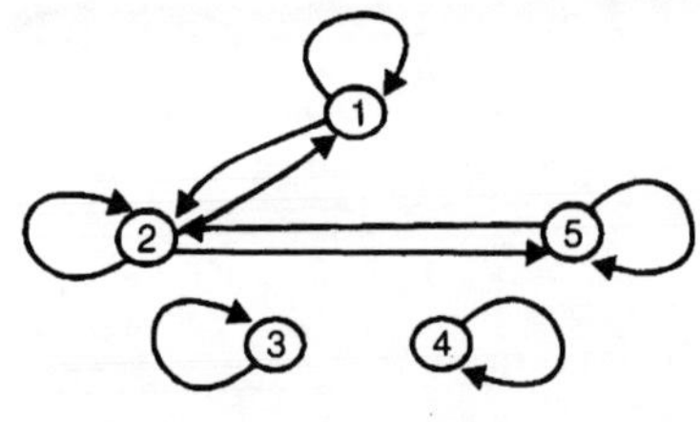

R_1 does not properties of transitivity

For example, $(x_1, x_2) \in R_1$ $(x_2, x_5) \in R_1$ but $(x_1, x_5) \notin R_1$

Example : R_1 can become an equivalence relation through two compositions

$$R_1 \text{ o } R_1 = \begin{bmatrix} 1 & 1 & 0 & 0 & 1 \\ 1 & 1 & 0 & 0 & 1 \\ 0 & 0 & 1 & 0 & 0 \\ 0 & 0 & 0 & 1 & 0 \\ 1 & 1 & 0 & 0 & 1 \end{bmatrix} = R$$

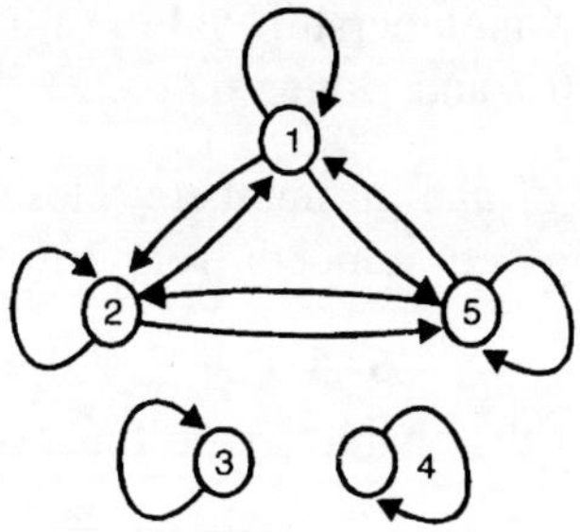

Fuzzy Tolerance and Equivalence Relations

Reflexivity:

$$\mu_R(X_i, X_i) = 1$$

Symmetry:

$$\mu_R(X_i, X_j) = \mu_R(X_j, X_i)$$

Transitivity:

$$\mu_R(X_i, X_j) = \lambda_1 \text{ and } \mu_R(X_j, X_k) = \mu_2 \rightarrow \mu_R(X_i, X_k) = \lambda$$

where $\lambda \geq \min [\lambda_1, \lambda_2]$

Equivalence relations

Fuzzy tolerance relation R_1 can be reformed into an equivalence relation

By at most $(n - 1)$ compositions with itself

$$R_1^{n-1} = R_1 \text{ o } R_1 \ldots \text{o } R_1 = R$$

Example :

$$R_1 = \begin{bmatrix} 1 & 0.8 & 0 & 0.1 & 0.2 \\ 0.8 & 1 & 0.4 & 0 & 0.9 \\ 0 & 0.4 & 1 & 0 & 0 \\ 0.1 & 0 & 0 & 1 & 0.5 \\ 0.2 & 0.9 & 0 & 0.5 & 1 \end{bmatrix}$$

It is not transitive

$\mu_R(x_1, x_2) = 0.8$ $\mu_R(x_2, x_5) = 0.9$ but $\mu_R(x_1, x_5) = 0.2 \leq \min (0.8, 0.9)$

$$R_1^2 = R_1 \text{ o } R_1 = \begin{bmatrix} 1 & 0.8 & 0.4 & 0.2 & 0.8 \\ 0.8 & 1 & 0.4 & 0.5 & 0.9 \\ 0.4 & 0.4 & 1 & 0 & 0.4 \\ 0.2 & 0.5 & 0 & 1 & 0.5 \\ 0.8 & 0.9 & 0.4 & 0.5 & 1 \end{bmatrix}$$

Example : Suppose some one wants to describe the class of cars having the property of being expensive by considering BMW, Rolls Royce, Mercedes, Ferrari, Fiat, Honda and Renault. Some cars like Ferrari and Rolls Royce are definitely expensive and some like Fiat and Renault are not expensive in comparison and do not belong to the set. Using a fuzzy set, the fuzzy set of expensive cars can be described as:

{(Ferrari,1), (Rolls Royce,1), (Mercedes, 0.8), (BMW, 0.7), (Honda,0.4)}. Obviously, Ferrari and Rolls Royce have membership value of 1 whereas BMW, which is less expensive, has a membership value of 0.7 and Honda 0.4.

Example : A set of natural numbers 'close to 6' can be defined as a fuzzy set. This can be done, say, buy including all numbers from 3 to 9 as follows:

$$\tilde{6} = \{(3, 0.1), (4, 0.2), (5, 0.5), (6, 1), (7, 0.5), (8, 0.2), (9, 0.1)\}$$

The expression for the characteristic function or membership function can be written as :

$$\mu_{\tilde{6}}(x) = \frac{1}{(1+(x-6)^2}$$

The fuzzy set $\tilde{6}$ contains the elements like (6, 1), (5.5, 0.8), and (100, 0.000113162) etc.

Zadeh proposed an alternate representation for fuzzy sets, which is more convenient. Suppose A is a finite crisp set with elements $\{x_1, x_2, ..., x_n\}$, then an alternative representation for C is :

$$C = \{x_1, x_2 + x_3 + ... + x_n\}$$

Here + denotes an enumeration or listing rather than addition. Yet another way is to include the characteristic function also into its fold is : $\frac{\mu(x)}{x}$ *where the line between the top and bottom entries is just a delimiter or separator.*

The fuzzy set of expensive cars can be now written using this notation as :

$$\left\{\frac{1}{\text{Ferrari}} + \frac{1}{\text{Rollsroyee}} + \frac{0.8}{\text{Mercedes}} + \frac{0.7}{\text{BMW}} + \frac{0.4}{\text{Honda}} + \frac{0}{\text{Fiat}} + \frac{0}{\text{Renault}}\right\}$$

The set $F = \{(x, \mu_F(x)) \mid x \in X\}$ can then be written as :

$$F = \frac{\mu_F(x_1)}{x_1} + ... + \frac{\mu_F(x_n)}{x_n} = \sum_{i=1}^{n} \frac{\mu_F(x_i)}{x_i}$$

Where + satisfies the condition :

$$\frac{a}{x} + \frac{b}{x} = \max\left(\frac{a, b}{x}\right)$$

i.e., if the same element has two membership values, say 0.8 and 0.6, then its membership degree becomes 0.8, the larger of the two. Any countable or discrete universe U allows such a notation.

$$A = \sum_{x \in X} \frac{\mu_F(x)}{x}$$

but when the set is uncountable, or continuous, it can be written as :

$$A = \int \frac{\mu_A(x)}{x}$$

Here the symbol $\int_0^x$ denotes a listing or collection rather than integration.

The set 'close to 6' can now be re-written as :

$$\tilde{6} = \left\{\frac{0.1}{3}, \frac{0.2}{4}, \frac{0.5}{5}, \frac{1}{6}, \frac{0.5}{7}, \frac{0.2}{8}, \frac{0.1}{9}\right\}$$

Example : X can be the set of integers described by the membership function :

$$\mu_F(x) = \begin{cases} 1 - \sqrt{\dfrac{x-6}{3}} & 3 \le x \le 6 \\ 0 & \text{otherwise} \end{cases}$$

Then, F can be expressed as : $A = \int \frac{\mu_F(x)}{x}$ where the operation is on the set of all real integers.

Example : Consider the set of old people belonging to the universe of people in the age of 0 to 120. X = [0, 120]

We can define the membership function as :

$$\mu_{\text{old}} = \begin{cases} 0 & ;\ 0 \le x \le 60 \\ (x-60)/20 & ;\ 60 \le x \le 80 \\ 1 & ;\ x \ge 80 \end{cases}$$

then we can call the set as :

$$\int_0^{120} \frac{\mu_{old}(x)}{x}$$

9.13. THE COMPONENTS OF FUZZY SYSTEM

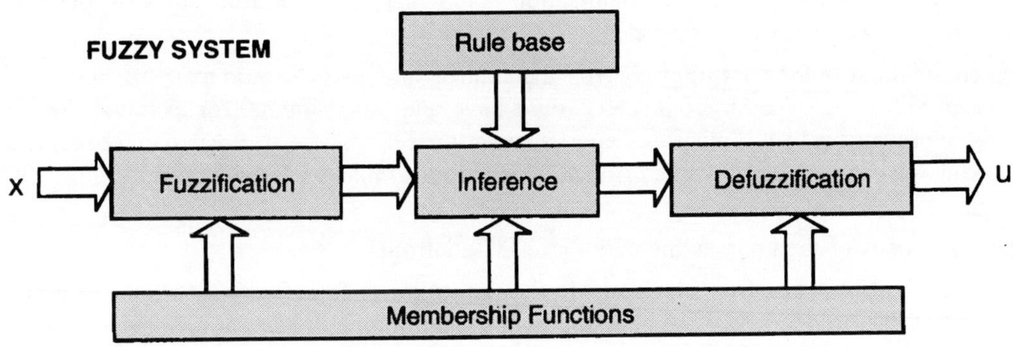

Fig. 9.13. Components of a Fuzzy System.

The input signals combined to the vector $x = [x_1, x_2, ..., x_q]^T$ are crisp values, which are transformed into fuzzy sets in the fuzzification block. The output u comes out directly from the defuzzification block, which transforms an output fuzzy set back to a crisp value. The set of membership functions responsible for the transforming part and the rule base as the relational part contain as a whole the modelling information about the system, which is processed by the inference machine. This rule-based fuzzy system is the basis of a fuzzy controller.

Fuzzy Logic Controller

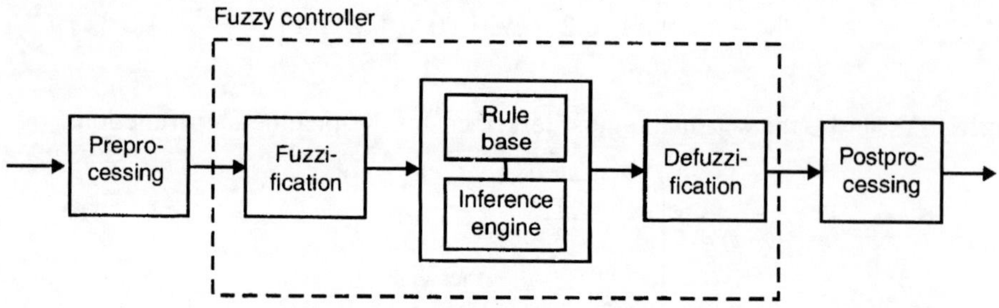

Fig. 9.14. Blocks of a Fuzzy Controller

Preprocessing

The inputs are most often hard or *crisp* measurements from some measuring equipment, rather than linguistic. A preprocessor, the first block in above fig , conditions the measurements before they enter the controller. Examples of preprocessing are:

- Quantisation in connection with sampling or rounding to integers;
- normalisation or scaling onto a particular, standard range;
- filtering in order to remove noise;
- averaging to obtain long term or short term tendencies;
- a combination of several measurements to obtain key indicators; and
- differentiation and integration or their discrete equivalences.

A quantiser is necessary to convert the incoming values in order to find the best level in a discrete universe. Assume, for instance, that the variable error has the value 4.5, but the universe is u = (–5, –4, ..., 0, 0, ..., 4, 5). The quantiser rounds to 5 to fit it to the nearest level. Quantisation is a means to reduce data, but if the quantisation is too coarse the controller may oscillate around the reference or even become unstable.

When the input to the controller is error, the control strategy is a static mapping between input and control signal. A dynamic controller would have additional inputs, for example derivatives, integrals, or previous values of measurements backwards in time. These are created in the preprocessor thus making the controller multi-dimensional, which requires many rules and makes it more difficult to design.

The preprocessor then passes the data on to the controller.

1. Fuzzification: The first block inside the controller is fuzzification, which converts each piece of input data to degrees of membership by a lookup in one or several membership functions. The fuzzification block thus matches the input data with the conditions of the rules to determine how well the condition of each rule matches that particular input instance. There is a degree of membership for each linguistic term that applies to that input variable.

2. Rule Base: The rules may use several variables both in the condition and the conclusion of the rules. The controllers can therefore be applied to both multi-input-multi-output (MIMO) problems and single-input-single-output (SISO) problems. The typical SISO problem is to regulate a control signal based on an error signal. The controller may actually need both the *error,* the *change in error,* and the *accumulated error* as inputs, but we will call it single-loop control, because in principle all three are formed from the error measurement. To simplify.

Rule formats: Basically a linguistic controller contains rules in the *if-then* format, but they can be presented in different formats. In many systems, the rules are presented to the end-user in a format similar to the one below:

1. If error is Neg and change in error is Neg then output is NB
2. If error is Neg and change in error is Zero then output is NM
3. If error is Neg and change in error is Pos then output is Zero
4. If error is Zero and change in error is Neg then output is NM
5. If error is Zero and change in error is Zero then output is Zero
6. If error is Zero and change in error is Pos then output is PM
7. If error is Pos and change in error is Neg then output is Zero
8. If error is Pos and change in error is Zero then output is PM
9. If error is Pos and change in error is Pos then output is PB.

The names Zero, Pos, Neg are labels of fuzzy sets as well as NM, NM, PB and PM (negative big, negative medium, positive big, and positive medium respectively). The same set of rules could be presented in a relational format, a more compact representation.

Error	*Change in error*	*Output*
Neg	Pos	Zero
Neg	Zero	NM
Neg	Neg	NB
Zero	Pos	PM
Zero	Zero	Zero
Zero	Neg	NM
Pos	Pos	PB
Pos	Zero	PM
Pos	Neg	Zero

The top row is the heading, with the names of the variables. It is understood that the two leftmost columns are inputs, the rightmost is the output, and each row represents a rule. This format is perhaps better suited for an experienced user who wants to get an overview of the rule base quickly. The relational format is certainly suited for storing in a relational database. It should be emphasized, though, that the relational format implicitly assumed that the connective between the inputs is always logical and – or logical or for that matter as long as it is the same operation for all rules — and not a mixture of connectives. Incidentally, a fuzzy rule with an or combination of terms can be converted into an equivalent and combination of terms using laws of logic (DeMorgan's laws among others). A third format is the tabular linguistic format.

		Change in error		
		Neg	Zero	Pos
	Neg	NB	NM	Zero
Error	Zero	NM	Zero	PM
	Pos	Zero	PM	PB

This is even more compact. The input variables are laid out along the axes, and the output variable is inside the table. In case the table has an empty cell, it is an indication of a missing rule, and this format is useful for checking completeness. When the input variables are error and change in error, as they are here, that format is also called a linguistic phase plane. In case there are $n > 2$ input variables involved, the table grows to an n-dimensional array; rather user -unfriendly.

To accommodate several outputs, a nested arrangement is conceivable. A rule with several outputs could also be broken down into several rules with one output.

Lastly, a graphical format which shows the fuzzy membership curves is also possible.

Defuzzification : The resulting fuzzy set must be converted to a number that can be sent to the process as control signal. The operation is called DEFUZZIFICATION. There are various methods of defuzzification which are described in previous section.

Postprocessing

Output scaling is also relevant. In case the output is defined on a standard universe this must be scaled to *engineering units,* for instance, volts, meters, or tons per hour. An example is the scaling from the standard universe [–1, 1] to the physical units [–10, 10] volts.

The postprocessing block often contains an output gain that can be tuned, and sometimes also an integrator.

Applications

Applications which have benefited through the use of fuzzy systems theory have been information retrieval systems, a navigation system for automatic cars, a predictive fuzzy-logic controller for automatic operation of trains, laboratory water level controllers, controllers for robot arc-welders, feature-definition controllers for robot vision, graphics controllers for automated police sketchers, and more.

Expert systems have been the most obvious recipients of the benefits of fuzzy logic, since their domain is often inherently fuzzy. Examples of expert systems with fuzzy logic central to their control are decision-support systems, financial planners, diagnostic systems for determining soybean pathology, and a meteorological expert system in China for determining areas in which to establish rubber tree orchards.

Other applications are:

- Automobile and other vehicle subsystems, such as automatic transmissions,
- Air conditioners
- The Massive engine used in the Lord of the Rings films, which helped huge scale armies create random, yet orderly movements
- Cameras
- Digital image processing, such as edge detection
- Rice cookers
- Dishwashers
- Elevators
- Washing machines and other home appliances
- Video game artificial intelligence
- Language filters on message boards and chat rooms for filtering out offensive text

- Pattern recognition in Remote Sensing
- Hydrometeor classification algorithms for polar metric weather radar
- Fuzzy logic has also been incorporated into some microcontrollers and microprocessors.
- Mineral Deposit estimation

Summary

- To deal with uncertainty, Fuzzy logic is the most effective tool.
- In fuzzy logic, the event can take any value between 0 and 1 while in crisp logic the event can take only 2 values either true or false.
- Fuzzy sets always a combination of the member and its membership function.
- The membership function need not be discrete always, it can be continuous also.
- The properties like commutatively, distributivity, De Morgans law are same for both Fuzzy and Crisp sets.
- List of fuzzy set operations.

  $\mu_{A \cup B} = \max(\mu_A, \mu_B)$

  $\mu_{A, B}(x) = \mu_A(x) \cdot \mu_B(x)$

  A exor $B = (\bar{A} \cap B) \cup (A \cap \bar{B})$

  $\mu_{A \cap B} = \min(\mu_A, \mu_B)$

Review Questions

1. Why do we need Fuzzy Set theory ?
2. Write short note on :
   (*a*) Randomness vs. Fuzziness
   (*b*) Fuzzy sets vs. Crisp sets
   (*c*) Membership Functions
   (*d*) Linguistic variable
   (*e*) Fuzzy logic vs Probability
3. Define the following standard operations on fuzzy sets: intersection, union and complement.
4. When is a fuzzy set normal ?
5. Define core and support of a fuzzy set ?
6. Show that $\overline{A \cup B} = \bar{A} \cap \bar{B}$. Consider fuzzy sets defined on the interval $X = [0, 10]$ of natural by the membership grade functions

$$A(x) = \frac{x}{(x+2)}$$

$$B(x) = 2 - x$$

$$C(x) = \frac{1}{(1+10(x-2)^2)}$$

7. Find and plot membership grade function for the following
   (*a*) $\bar{A}, \bar{B}, \bar{C}$ (*b*) $A \cup B$, $A \cup C$, $C \cup B$

❑❑❑

CHAPTER

10

GENETIC ALGORITHMS

INSIDE THIS CHAPTER

10.1. INTRODUCTION TO GENETIC ALGORITHMS (GA)

Genetic algorithms (GA) are an evolutionary optimization approach, which are an alternative to traditional optimization methods. GA is one of the most appropriate methods for complex non-linear models where location of the global optimum is a difficult task. GA follows the concept of solution evolution by stochastically developing generations of solution populations using a given fitness. They are particularly applicable to problems, which are large, non-linear and possibly discrete in nature.

10.2. DIFFERENCE BETWEEN GAS AND OTHER NORMAL OPTIMIZATION AND SEARCH PROCEDURES

Advantages of Genetic Algorithm

- GAs work with a coding for the parameter set, not the parameters themselves.
- GAs search from a population of points, not a single point.
- GAs use probabilistic transition rules, not deterministic rules.
- The Genetic Algorithms are first suited for the optimization problem.
- GAs are robust with respect to local minima, maxima.
- GAs work well on mixed discrete/continuous problems.
- GAs can operate on various representations.
- GAs are stochastic.

- GAs are easily parallelised.
- GAs are easy to implement and lend themselves well to hybridization.

Limitations of Genetic Algorithm

- GA implementation is still an art. They often require a lot of tweaking, then again sometimes you can tweak all you want and they will still find the same result.
- GAs are computationally expensive.
- GAs typically require less information about the problem (*e.g.*, no gradients), but designing an objective function and getting the representation and operators right can be difficult.

Summarizing: It is suited for the dynamic program, because the genetic programming paradigm parallels nature in that it is a never-ending process. However, as a practical matter, a run of the genetic programming paradigm terminates when the 'termination criterion is satisfied'.

10.3. FROM BIOLOGY TO GENETIC ALGORITHMS

We can borrow some terms (and ideas) from biology ...

- **Chromosome:** The coding of a possible solution for a given problem, usually represented with an array of bits or characters.
- **Gene:** A single bit or a set of bits coding part of the solution.
- **Allele:** One of the elements used to code the genes.
- **Fitness:** Evaluation of the actual solution ... to model a learning process as evolution.
- **Crossover:** Generate new solution by "mixing" two existing solutions.
- **Mutation:** Random change in the solution.

10.4. WORKING PRINCIPLE OF GENETIC ALGORITHMS

The GA procedure is based on the Darwinian principle of survival of the fittest.

- An initial population is created containing a predefined number of individuals, each represented by a genetic string.
- Each individual has an associated fitness measure. The concept that the fittest (or best) individuals in a population will produce a fitter offspring is then implemented in order to reproduce the next population.
- Selected individuals are chosen for reproduction (or crossover) at each generation, with an appropriate mutation factor to randomly modify the genes of an individual, in order to develop the new population.
- The result is another set of individuals based on the original subjects leading to subsequent populations with better fitness and those with lower fitness will naturally get discarded from the population.

The GA consists of four main stages: *evaluation*, *selection*, *crossover* and *mutation*.

- The ***evaluation*** procedure measures the fitness of each individual solution in the population and assigns to it a score.

- The ***selection*** procedure randomly selects individuals of the current population for development of the next generation.
- The ***crossover*** procedure takes two selected individuals and combines them about a crossover point thereby creating two new individuals.
- The ***mutation*** procedure randomly modifies the genes of an individual subject to a small mutation factor, introducing further randomness into the population. This iterative process continues until one of the possible termination criteria is met: if a known optimal or acceptable solution level is attained; or if a maximum number of generations have been performed; or if a given number of generations without fitness improvement occur, (Goldberg 1989; Lewin 1994, 1996).

10.5. STRUCTURE OF GAS

GAs are categorised under an umbrella term Evolutionary Algorithms, which are used to describe computer-based problem solving systems which use computational models of evolutionary processes as key elements in their design and implementation. A variety of evolutionary algorithms have been proposed, of which the major ones are: GAs, evolutionary programming, evolution strategies, classifier systems, and genetic programming (check out www.google.com on these keywords). They all share a common conceptual base of simulating the 'evolution' of individual structures via processes of selection, mutation and reproduction.

The existing GAs are founded upon the following main principles:

1. Reproduction
2. Fitness
3. Crossover
4. Mutation

There are various flavours of GAs in circulation, varying in implementation of these three parameters, but in essence the algorithms all follow a standard procedure:

1. [Start] Generate random population of n chromosomes (suitable solutions).
2. [Fitness] Evaluate the fitness $f(x)$ of each chromosome x in the population.
3. [New population] Create a new population by repeating following steps until the new population is complete:
   (*a*) [Selection] Select two parent chromosomes from a population according to their fitness (the better fitness, the bigger chance to be selected).
   (*b*) [Crossover] With a crossover probability cross over the parents to form new offspring. If no crossover was performed, offspring is the exact copy of parents.
   (*c*) [Mutation] With a mutation probability mutate new offspring at each locus.
   (*d*) [Accepting] Place new offspring in the new population.
4. [Replace] Use new generated population for a further run of the algorithm
5. [Test] If the end condition is satisfied, return the best solution in current population
6. [Loop] Go to step 2

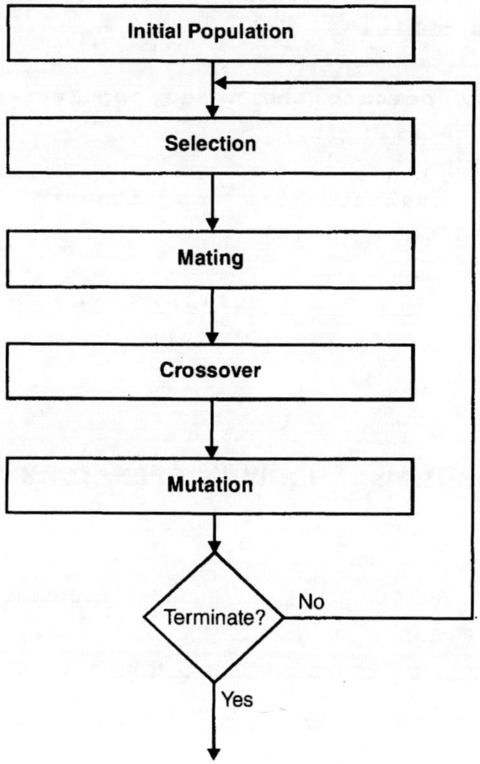

Fig. 10.1. Simplified Flow Chart of a Genetic Algorithm

PSEUDOCODE

Algorithm GA is

```
// start with an initial time
t := 0;

// initialise a usually random population of individuals
   initpopulation P (t);

// evaluate fitness of all initial individuals in population
   evaluate P (t);

// test for termination criterion (time, fitness, etc.)
   while not done do

        // increase the time counter
        t := t + 1;

         // select sub-population for offspring production
        P' := selectparents P (t);
```

```
        // recombine the "genes" of selected parents
        recombine P' (t);

        // perturb the mated population stochastically
        mutate P' (t);

        // evaluate it's new fitness
        evaluate P' (t);

        // select the survivors from actual fitness
        P := survive P,P' (t);
    od
    end GA.
```

10.6. GENETIC ALGORITHMS: ENCODING OPERATORS

10.6.1. Encoding (I)

Binary encoding is the most common one (mainly because the first research of GA used this type of encoding)

- In binary encoding, every chromosome is a string of bits (0 or 1)
- Simple Implementation of the genetic operators
- Not always natural for many problems

Cromosome 1	1 1 0 1 1 0 0 1 0 0 1 1 0 1 1 0
Cromosome 2	1 1 1 0 1 1 1 0 0 0 0 1 1 1 1 0

Example of Problem: Knapsack problem

- There are things with given value and size. The knapsack has given capacity. Select things to maximize the value of things in knapsack, but do not extend knapsack capacity.
- Each bit says, whether the corresponding thing is in knapsack.

10.6.2. Encoding (II)

Permutation encoding can be used in ordering problems

- Every chromosome is a string of numbers that represent a position in a sequence
- Crossover and mutation must be designed to leave the chromosome consistent (i.e. have real sequence in it).

Cromosome 1:	15 7 8 3 5 1310111612 1 14 2 4 6 9
Cromosome 2:	2 9 14 1 11 5 8 1513 6 1216 7 3 10 4

Example of Problem: Travelling salesman problem (TSP)

- There are cities and given distances between them. Travelling salesman has to visit all of them, but he does not want to travel more than necessary. Find a sequence of cities with a minimal travelled distance.
- Chromosome describes the order of cities.

10.6.3. Encoding (III)

Direct value encoding can be used in problems where some more complicated values are required

- Every chromosome is a sequence of some values connected to the problem, such as (real) numbers, chars or any objects
- Good choice for some special problems, but necessary to develop some specific crossover and mutation

Cromosome:	D S A B H Y V V
Cromosome:	2.5678 1.4361 3.3426 7.8761
Cromosome:	open walk back close

Example of Problem: Finding weights for a neural network

- A neural network is given with defined architecture. Find weights between neurons to get the desired output from the network
- Real values in chromosomes represent weights in the neural network

10.6.4. Encoding (IV)

Tree encoding is used mainly for evolving programs or expressions (*i.e.*, genetic programming)

- Every chromosome is a tree of some objects, such as functions or commands in programming language.
- Programming language LISP is often used for this purpose, so crossover and mutation can be done relatively easily.

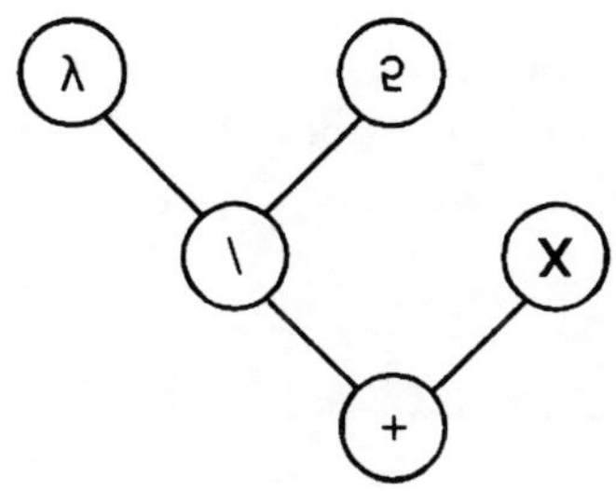

Fig. 10.2. Cromosome (+X(/5y))

Example of Problem: Finding a function that would approximate given pairs of values

- Input and output values are given. The task is to find a function that will give the best outputs for all inputs.
- Chromosome are functions represented in a tree

10.6.5. Genetic Algorithms: Selection Operators

According to Darwin's theory of evolution the best chromosome survive to create new offspring. There are many methods in selecting the best chromosomes:

- Roulette wheel selection,
- Rank selection

- Tournament selection
- Boltzmann selection

10.6.5.1. Roulette Wheel Selection

The simplest selection scheme is roulette-wheel selection, also called stochastic sampling with replacement. This is a stochastic algorithm and involves the following technique:

The individuals are mapped to contiguous segments of a line, such that each individual's segment is equal in size to its fitness. A random number is generated and the individual whose segment spans the random number is selected. The process is repeated until the desired number of individuals is obtained (called mating population). This technique is analogous to a roulette wheel with each slice proportional in size to the fitness.

This can be simulated by the following algorithm:

1. [Sum] Calculate sum S of all chromosome fitness values in population.
2. [Select] Generate random number r from interval $(0, S)$.
3. [Loop] Go through the population and sum fitness values from $0 - S$. When the sum is greater then r, stop and return the chromosome where you are.

PSEUDOCODE

```
for all members of population
     sum += fitness of this individual
end for

for all members of population
     probability = sum of probabilities + (fitness / sum)
     sum of probabilities += probability
end for

loop until new population is full
      do this twice
           number = Random between 0 and 1
         for all members of population
            if number > probability but less than next probability
                   then you have been selected
        end for
      end
      create offspring
end loop
```

Table shows the fitness value and the selection probability for 8 individuals. Individual 5 is the most fit individual and occupies the largest interval, and individual 1 is the least fit individual having the smallest interval on the line. For selecting the mating population the appropriate number of uniformly distributed random numbers (uniform distributed between 0.0 and 1.0) is independently generated.

No.	Initial Population	Fitness value F_i	Probability $P_i = F_i/sum$	Expected Count= (n=8)* P_i	Cummu-lative P_i	Random No.	Chromo-some No.	Selected Population
1	0000 0000	1	0.0429	0.33	0.0429	0.259	3	0001 0101
2	0010 0001	2.1	0.09	0.72	0.1329	0.038	1	0000 0000
3	0001 0101	3.11	0.1336	1.064	0.266	0.486	5	0110 1010
4	0010 1000	4.01	0.1723	1.368	0.438	0.428	4	0010 1000
5	0110 1010	4.66	0.2	1.6	0.638	0.095	2	0010 0001
6	1110 1000	1.91	0.082	0.656	0.72	0.3	4	0010 1000
7	1110 1101	1.93	0.0829	0.664	0.809	0.616	5	0110 1010
8	0111 1100	4.55	0.1955	1.56	1	0.897	8	0111 1100
Sum of all fitness values: sum = 23.264 Average fitness = 23.264/8 = 2.908								

After one iteration, instead of the least fit individual 1 the next least fit individual 6 has been discarded from the pool. This shows the noisiness in roulette-wheel selection. After repeating the selection procedure n = 8 times the count of each chromosome is shown below.

No.	1	2	3	4	5	6	7	8
Initial Population	0000 0000	0010 0001	0001 0101	00101000	0110 1010	1110 1000	1110 1101	0111 1100
Count in the pool	1	1	1	2	2	0	0	1
Selected Population	0000 0000	0010 0001	0001 0101	00101000	00101000	0110 1010	0110 1010	0111 1100

Referring the expected counts 5 and 8 gets two copies and 1 and 6 get no copies. After 8 iterations the pool we get is also very similar to the expected count.

10.6.5.2. Rank Selection

The previous selection will have problems when the fitness values differ very much. For example, if the best chromosome fitness is 90% of all the roulette wheel then the other chromosomes will have very few chances to be selected.

Rank selection first ranks the population and then every chromosome receives fitness from this ranking. The worst will have fitness 1, second worst 2 etc. and the best will have fitness N (number of chromosomes in population).

After this all the chromosomes have a chance to be selected. But this method can lead to slower convergence, because the best chromosomes do not differ so much from other ones.

10.6.5.3. Tournament Selection

All methods above rely on global population statistics.

➡ Could be a bottleneck esp. on parallel machines.

- Relies on presence of external fitness function which might not exist: e.g. evolving game players.

Tournament selection is an informal procedure:

- Pick *k* members at random then select the best of these
- Repeat to select more individuals

Probability of selecting i will depend on:

- Rank of *i*
- Size of sample *k*

Higher k increases selection pressure

- Whether contestants are picked with replacement.

Picking without replacement increases selection pressure

- Whether fittest contestant always wins (deterministic) or this happens with probability *p*

For k = 2, time for fittest individual to take over population is the same as linear ranking with s = 2 • p

10.6.5.4. Steady-State Selection

This is not particular method of selecting parents. Main idea of this selection is that big part of chromosomes should survive to next generation.

GA then works in a following way. In every generation are selected a few (good - with high fitness) chromosomes for creating a new offspring. Then some (bad - with low fitness) chromosomes are removed and the new offspring is placed in their place. The rest of population survives to new generation.

10.6.5.5. Elitism

Idea of elitism has been already introduced. When creating new population by crossover and mutation, we have a big chance, that we will loose the best chromosome.

Elitism is name of method, which first copies the best chromosome (or a few best chromosomes) to new population. The rest is done in classical way. Elitism can very rapidly increase performance of GA, because it prevents losing the best found solution.

10.7. GENETIC ALGORITHM: CROSSOVER AND MUTATION OPERATORS

There are many methods for crossovers and mutations depending on the kind of encoding applied to represent the chromosomes.

10.7.1. Crossover/Mutation (I)

For **binary encoding** we have many Crossover operators:

- **Single point crossover:** One crossover point is selected, binary string from the beginning of the chromosome to the crossover point is copied from the first parent, the rest is copied from the other parent

  11001011+11011**111** = **11001111**

- **Two point crossover:** Two crossover points are selected, binary string from the beginning of the chromosome to the first crossover point is copied from the first parent, the part from the first to the second crossover point is copied from the other parent and the rest is copied from the first parent again.

11001011 + 11011111 = **11011111**

- *Uniform crossover:* Bits are randomly copied from the first or from the second parent

  11001011 + 11011101 = 11011111

- *Arithmetic crossover:* Some arithmetic operation is performed to make a new offspring (*e.g.*, logic AND)

  11001011 + 11011111 = 11001001 (AND)

Mutation: Inversion of selected bits

11001001 => 10001001

10.7.2. Crossover/Mutation (II)

For **permutation encoding** we have to preserve consistency:

- *Single point crossover:* One crossover point is selected, the permutation is copied from the first parent till the crossover point, then the other parent is scanned looking the other numbers

  (1 2 3 4 5 6 7 8 9) + (4 5 3 6 8 9 7 2 1) = (1 2 3 4 5 6 8 9 7)

Order changing mutation: Two numbers are selected and exchanged

(1 2 3 4 5 6 8 9 7) => (1 8 3 4 5 6 2 9 7)

10.7.3. Crossover/Mutation (III)

For **real value encoding** we can reuse crossover from binary encoding:

Mutation: A small number is added (or subtracted) to selected values

(1.29 5.68 2.86 4.11 5.55) => (1.29 5.68 2.73 4.22 5.55)

10.7.4. Crossover/Mutation (IV)

Specific operators have to be selected also for tree encoding:

- **Tree crossover:** One crossover point is selected in both parents, parents are divided in that point and the parts below crossover points are exchanged to produce new offspring.

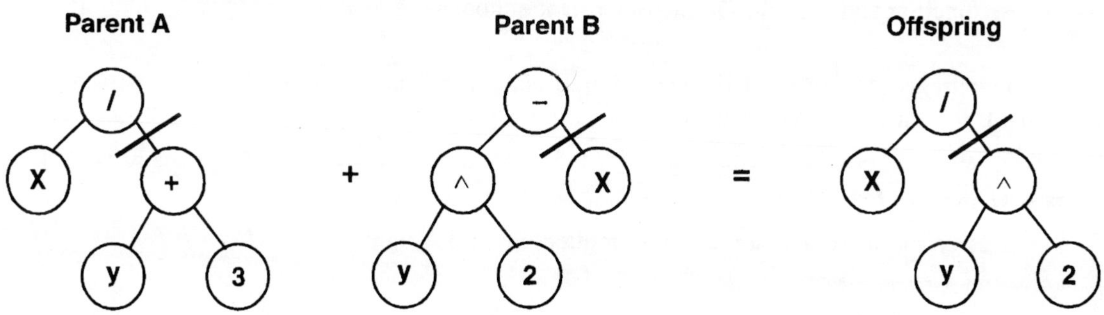

Fig. 10.3

Changing mutation: The operator, number, or variable in a randomly selected node is changed.

Parameters of GA

Today there is no general theory which would describe parameters of GA for any problem. Recommendations are often results of some empiric studies of GAs, which were often performed only on binary encoding.

Crossover rate: Crossover rate generally should be high, about 80%-95%. (However some results show that for some problems crossover rate about 60% is the best.)

Mutation rate: On the other side, mutation rate should be very low. Best rates reported are about 0.5%-1%.

Population size: It may be surprising, that very big population size usually does not improve performance of GA (in meaning of speed of finding solution). Good population size is about 20-30, however sometimes sizes 50-100 are reported as best. Some research also shows that best population size depends on encoding, on size of encoded string. It means, if you have chromosome with 32 bits, the population should be say 32, but surely two times more than the best population size for chromosome with 16 bits.

Selection: Basic roulette wheel selection can be used, but sometimes rank selection can be better. Also some more sophisticated method, which changes parameters of selection during run of GA. Basically they behave like simulated annealing. Selection again depends on the problem.

Encoding: Encoding depends on the problem and also on the size of instance of the problem. Check chapter about encoding for some suggestions or look to other resources.

Crossover and mutation type: Operators depend on encoding and on the problem.

Applications of GA

Genetic algorithm has been used for difficult problems (such as NP-hard problems), for machine learning and also for evolving simple programs. They have been also used for some art, for evolving pictures and music.

Advantage of GAs is in their parallelism. GA is travelling in a search space with more individuals (and with genotype rather than phenotype) so they are less likely to get stuck in a local extreme like some other methods.

They are also easy to implement. Once you have some GA, you just have to write new chromosome (just one object) to solve another problem. With the same encoding you just change the fitness function and it is all. On the other hand, choosing encoding and fitness function can be difficult.

Disadvantage of GAs is in their computational time. They can be slower than some other methods. But with today's computers it is not so big problem.

To get an idea about problems solved by GA, here is a short list of some applications:

- Nonlinear dynamical systems—predicting, data analysis.
- Designing neural networks, both architecture and weights.
- Robot trajectory.
- Evolving LISP programs (genetic programming).
- Strategy planning.
- Finding shape of protein molecules.
- TSP and sequence scheduling.
- Functions for creating images.

❑❑❑

QUESTION BANK

1. What is a Neural Network?

Ans. An Artificial Neural Network (ANN) is an information processing paradigm that is inspired by the way biological nervous systems, such as the brain, process information. The key element of this paradigm is the novel structure of the information processing system. It is composed of a large number of highly interconnected processing elements (neurons) working in unison to solve specific problems. ANNs, like people, learn by example. An ANN is configured for a specific application, such as pattern recognition or data classification, through a learning process. Learning in biological systems involves adjustments to the synaptic connections that exist between the neurons. This is true of ANNs as well.

2. Why use Neural Networks?

Ans. Neural networks, with their remarkable ability to derive meaning from complicated or imprecise data, can be used to extract patterns and detect trends that are too complex to be noticed by either humans or other computer techniques. A trained neural network can be thought of as an "expert" in the category of information it has been given to analyze. This expert can then be used to provide projections given new situations of interest and answer "what if" questions.

Other advantages include:

1. **Adaptive learning:** An ability to learn how to do tasks based on the data given for training or initial experience.
2. **Self-Organisation:** An ANN can create its own organisation or representation of the information it receives during learning time.
3. **Real Time Operation:** ANN computations may be carried out in parallel, and special hardware devices are being designed and manufactured which take advantage of this capability.
4. **Fault Tolerance via Redundant Information Coding:** Partial destruction of a network leads to the corresponding degradation of performance. However, some network capabilities may be retained even with major network damage.

Advantages and disadvantages :

Advantages:

- A neural network can perform tasks that a linear program can not.
- When an element of the neural network fails, it can continue without any problem by their parallel nature.
- A neural network learns and does not need to be reprogrammed.
- It can be implemented in any application.
- It can be implemented without any problem.

Disadvantages :

- The neural network needs training to operate.
- The architecture of a neural network is different from the architecture of microprocessors therefore needs to be emulated.
- Requires high processing time for large neural networks.

Another aspect of the artificial neural networks is that there are different architectures, which consequently requires different types of algorithms, but despite to be an apparently complex system, a neural network is relatively simple.

3. What is a simple neuron ?

Ans. An artificial neuron is a device with many inputs and one output. The neuron has two modes of operation; the training mode and the using mode.

1. In the training mode, the neuron can be trained to fire (or not), for particular input patterns.
2. In the using mode, when a taught input pattern is detected at the input, its associated output becomes the current output. If the input pattern does not belong in the taught list of input patterns, the firing rule is used to determine whether to fire or not.

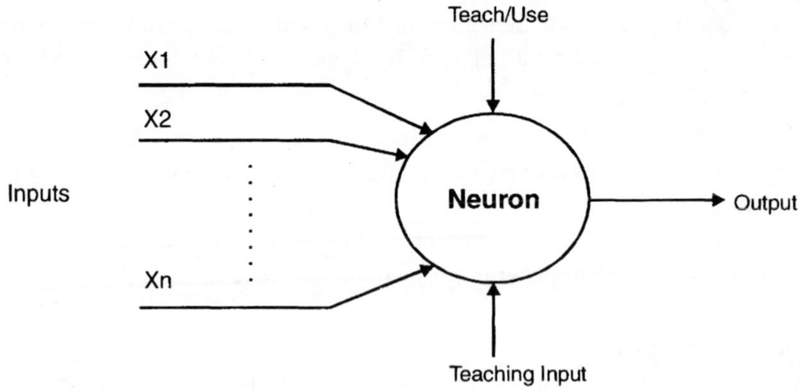

Fig. 1

4. Explain Firing rules.

Ans. The firing rule is an important concept in neural networks and accounts for their high flexibility. A firing rule determines how one calculates whether a neuron should fire for any input pattern. It relates to all the input patterns, not only the ones on which the node was trained.

A simple firing rule can be implemented by using Hamming distance technique. The rule goes as follows :

Take a collection of training patterns for a node, some of which cause it to fire (the 1-taught set of patterns) and others which prevent it from doing so (the 0-taught set). Then the patterns not in the collection cause the node to fire if, on comparison , they have more input elements in common with the 'nearest' pattern in the 1-taught set than with the 'nearest' pattern in the 0-taught set. If there is a tie, then the pattern remains in the undefined state.

For example, a 3-input neuron is taught to output 1 when the input (X1, X2 and X3) is 111 or 101 and to output 0 when the input is 000 or 001. Then, before applying the firing rule, the truth table is;

X1:	0	0	0	0	1	1	1	1
X2:	0	0	1	1	0	0	1	1
X3:	0	1	0	1	0	1	0	1
OUT:	0	0	0/1	0/1	0/1	1	0/1	1

As an example of the way the firing rule is applied, take the pattern 010. It differs from 000 in 1 element, from 001 in 2 elements, from 101 in 3 elements and from 111 in 2 elements. Therefore, the 'nearest' pattern is 000 which belongs in the 0-taught set. Thus the firing rule requires that the neuron should not fire when the input is 001. On the other hand, 011 is equally distant from two taught patterns that have different outputs and thus the output stays undefined (0/1).

By applying the firing in every column the following truth table is obtained;

X1:	0	0	0	0	1	1	1	1
X2:	0	0	1	1	0	0	1	1
X3:	0	1	0	1	0	1	0	1
OUT:	0	0	0	0/1	0/1	1	1	1

The difference between the two truth tables is called the *generalisation of the neuron.* Therefore the firing rule gives the neuron a sense of similarity and enables it to respond 'sensibly' to patterns not seen during training.

A more complicated neuron

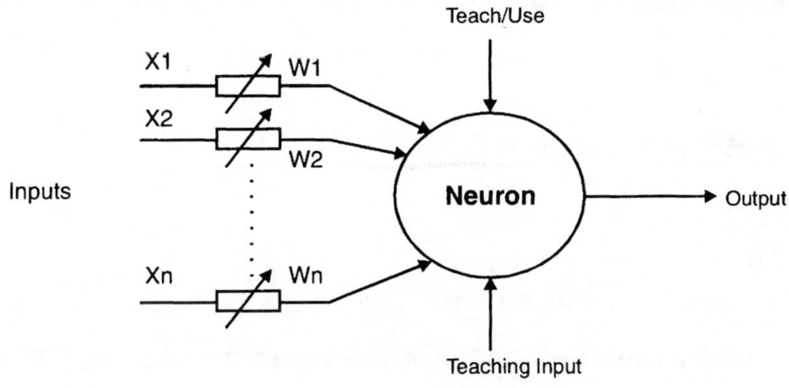

Fig. 2. An MCP Neuron

A more sophisticated neuron (Fig. 2) is the McCulloch and Pitts model (MCP). The difference from the previous model is that the inputs are 'weighted', the effect that each input has at decision making is dependent on the weight of the particular input. The weight of an input is a number which when multiplied with the input gives the weighted input. These weighted inputs are then added together and if they exceed a pre-set threshold value, the neuron fires. In any other case the neuron does not fire.

In mathematical terms, the neuron fires if and only if;

$$X1W1 + X2W2 + X3W3 + \ldots > T$$

The addition of input weights and of the threshold makes this neuron a very flexible and powerful one. The MCP neuron has the ability to adapt to a particular situation by changing its weights and/or threshold. Various algorithms exist that cause the neuron to 'adapt'; the most used ones are the Delta rule and the back error propagation. The former is used in feed-forward networks and the latter in feedback networks.

5. List at least 3 areas of application where ANN is used.

Ans. Most applications of neural networks fall into the following categories:

- ***Prediction:*** Use input values to predict some output.
  Weather prediction, word/letter prediction, behaviour prediction.
- ***Classification:*** Use input values to determine the classification.
  Fish on a conveyor belt, medical (cancer/non cancer)
- ***Data Association:*** Like classification, but it also recognizes data that contains errors (classification using association rules (*e.g.*, extract from a database the names of those most likely to buy a particular product)).

6. Define the taxonomy of artificial neural network architecture.

Ans. Various ANN systems have been designed since its evolution. Some of ANN systems are:

1. Perceptron
2. ADALINE
3. MADALINE
4. Neocognitron
5. Hopfield network
6. Hamming Network
7. Cauchy machine
8. Radial Basis Function
9. Adaptive Resonance Theory
10. Associative Memory
11. Boltzmann Machine
12. Self Organizing feature Map
13. Bidirectional Associative Memory
14. Recurrent Neural Network
15. Cascade correlation
16. Learning Machine
17. Multilayer feedforward network

6. Explain the logic function (truth table) performed by following MP neurons.

Ans. Logic function: $x + \overline{x}$

Input	Output
0	1
1	1

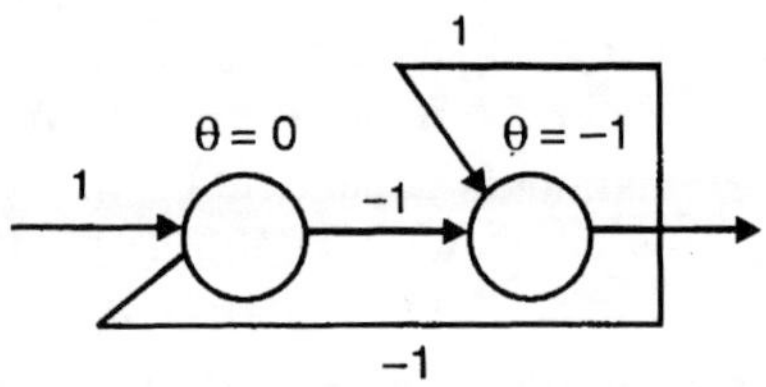

7. List out various ANN architectures. What is the significance of having recurrent type architecture.

Ans. Various ANN architectures are:

- Feed-forward Networks (Single layer and multilayer feed forward network)
- Feedback Networks (Single layer and multilayer feedback network)

Recurrent networks contain feedback connections. They are very powerful and can get extremely complicated. Feedback networks are dynamic; their 'state' continues to change until they reach an equilibrium point. They remain at the equilibrium point until the input changes and a new equilibrium needs to be found.

8. Design networks of McCulloch-Pitts neurons that implement logical AND and OR gates. Draw each network and label all the weight and threshold values.

Ans.

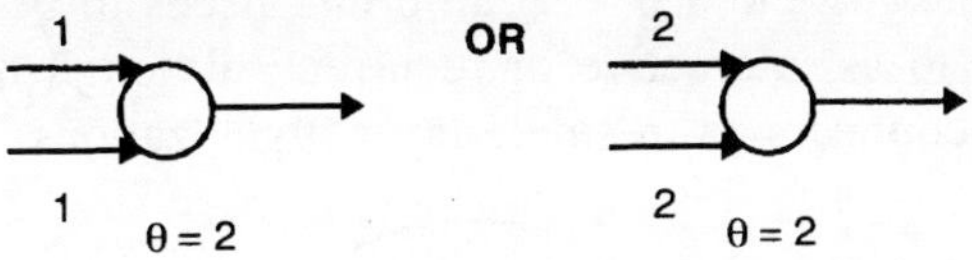

Fig. 3

9. Construct a 3-2-1 fully connected feed forward Neural network.

Ans.

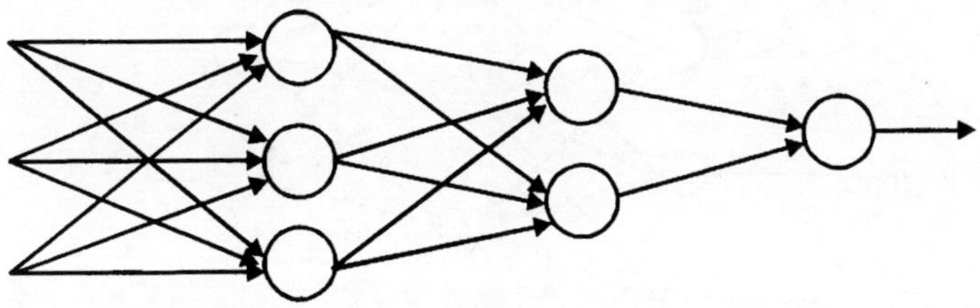

Fig. 4

10. Construct a fully connected single layer recurrent network with 3 neurons.

Ans.

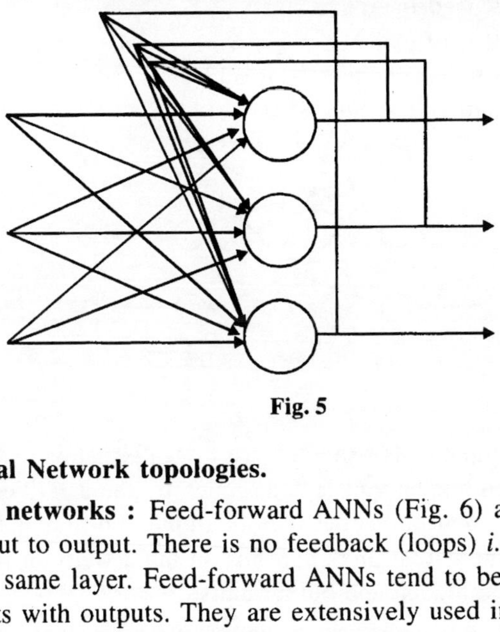

Fig. 5

11. Explain Neural Network topologies.

Ans. Feed-forward networks : Feed-forward ANNs (Fig. 6) allow signals to travel one way only; from input to output. There is no feedback (loops) *i.e.*, the output of any layer does not affect that same layer. Feed-forward ANNs tend to be straight forward networks that associate inputs with outputs. They are extensively used in pattern recognition. This type of organization is also referred to as bottom-up or top-down.

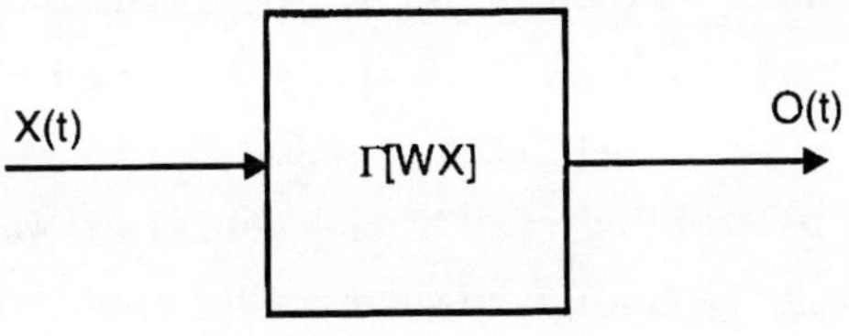

Fig. 6. Block diagram.

Feedback networks : Feedback networks (Fig. 7) can have signals traveling in both directions by introducing loops in the network. Feedback networks are very powerful and can get extremely complicated. Feedback networks are dynamic; their 'state' is changing continuously until they reach an equilibrium point. They remain at the equilibrium point until the input changes and a new equilibrium needs to be found. Feedback architectures are also referred to as interactive or recurrent, although the latter term is often used to denote feedback connections in single-layer organizations.

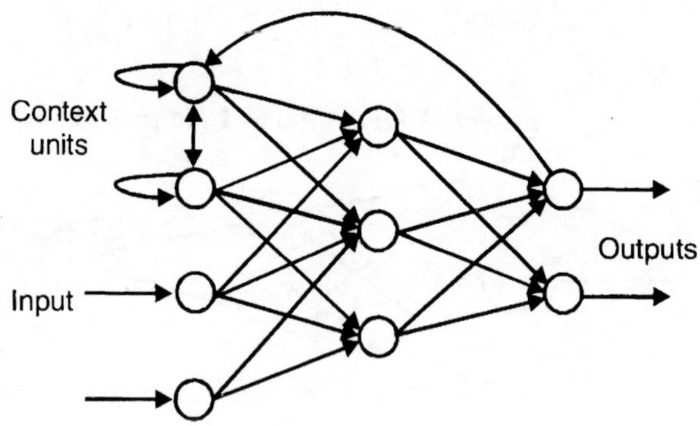

Fig. 7. Feedback neural networks.

12. Explain various layers used in ANN.

Ans. The commonest type of artificial neural network consists of three groups, or layers, of units: a layer of "**input**" units is connected to a layer of "**hidden**" units, which is connected to a layer of **"output"** units. (see Fig. 8)

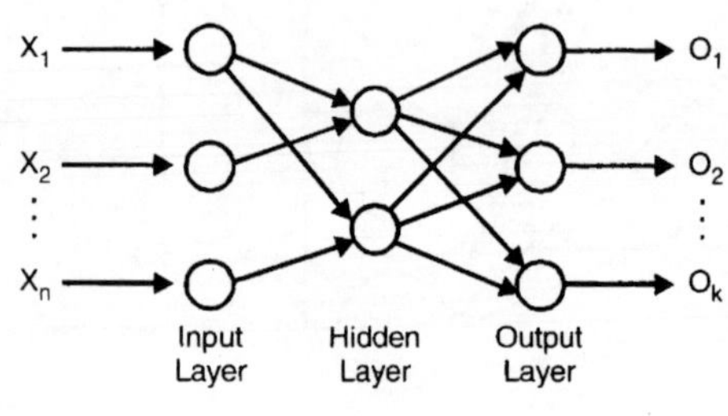

Fig. 8

- The activity of the input units represents the raw information that is fed into the network.
- The activity of each hidden unit is determined by the activities of the input units and the weights on the connections between the input and the hidden units.
- The behaviour of the output units depends on the activity of the hidden units and the weights between the hidden and output units.

13. For a perceptron with binary inputs x_1, x_2 and x_3 (whose values are either 0 or 1), whose output is x_3 if $x_1 = 1$ and $x_2 = 0$ and is 0 otherwise. With bias $b = -1.5$ find the weights.

Ans. For this problem, the output will be 1 only if $x_1 = 1, x_2 = 0$ and $x_3 = 1$. A possible solution is

$$W = [1 \ \ -1 \ \ 1],\ b = -1.5.$$

14. What do you mean by training of artificial neural networks.

Ans. A **neural network** has to be configured such that the application of a set of inputs produces (either 'direct' or via a relaxation process) the desired set of outputs. Various methods to set the strengths of the connections exist. One way is to set the weights explicitly, using a priori knowledge. Another way is to **'train' the neural network** by feeding it teaching patterns and letting it change its weights according to some learning rule.

We can categories the learning situations in two distinct sorts. These are:

- **Supervised learning** or Associative learning in which the network is trained by providing it with input and matching output patterns. These input-output pairs can be provided by an external teacher, or by the system which contains the neural network (self-supervised).
- **Unsupervised learning** or Self-organization in which an (output) unit is trained to respond to clusters of pattern within the input. In this paradigm the system is supposed to discover statistically salient features of the input population. Unlike the supervised learning paradigm, there is no a priori set of categories into which the patterns are to be classified; rather the system must develop its own representation of the input stimuli.

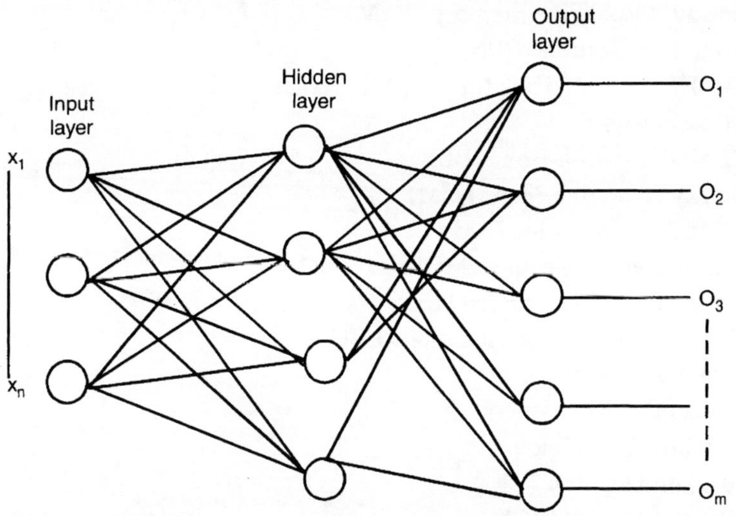

Fig. 9. Different layers used in ANN.

- **Reinforcement Learning :** This type of learning may be considered as an intermediate form of the above two types of learning. Here the learning machine does some action on the environment and gets a feedback response from the environment. The learning system grades its action good (rewarding) or bad (punishable) based on the environmental response and accordingly adjusts its parameters. Generally, parameter adjustment is continued until an equilibrium state occurs, following which there will be no more changes in its parameters. The self organizing neural learning may be categorized under this type of learning.

UNSUPERVISED LEARNING (*i.e.,* **without a "teacher") :**

- **Feedback Nets :**
  - Additive Grossberg (AG)
  - Shunting Grossberg (SG)
  - Binary Adaptive Resonance Theory (ART1)
  - Analog Adaptive Resonance Theory (ART2, ART2a)

- Discrete Hopfield (DH)
- Continuous Hopfield (CH)
- Discrete Bidirectional Associative Memory (BAM)
- Temporal Associative Memory (TAM)
- Adaptive Bidirectional Associative Memory (ABAM)
- Kohonen Self-organizing Map (SOM)
- Kohonen Topology-preserving Map (TPM)

➡ **Feedforward-only Nets :**

- Learning Matrix (LM)
- Driver-Reinforcement Learning (DR)
- Linear Associative Memory (LAM)
- Optimal Linear Associative Memory (OLAM)
- Sparse Distributed Associative Memory (SDM)
- Fuzzy Associative Memory (FAM)
- Counterprogation (CPN)

2. **SUPERVISED LEARNING (*i.e.,* with a "teacher"):**

➡ **Feedback Nets :**

- Brain-State-in-a-Box (BSB)
- Fuzzy Congitive Map (FCM)
- Boltzmann Machine (BM)
- Mean Field Annealing (MFT)
- Recurrent Cascade Correlation (RCC)
- Learning Vector Quantization (LVQ)

➡ **Feedforward-only Nets:**

- Perceptron
- Adaline, Madaline
- Backpropagation (BP)
- Cauchy Machine (CM)
- Adaptive Heuristic Critic (AHC)
- Time Delay Neural Network (TDNN)
- Associative Reward Penalty (ARP)
- Avalanche Matched Filter (AMF)
- Backpercolation (Perc)
- Artmap
- Adaptive Logic Network (ALN)
- Cascade Correlation (CasCor)

15. What is Perceptron ? Explain.

Ans. The most influential work on neural nets in the 60's went under the heading of 'perceptrons' a term coined by Frank Rosenblatt. The perceptron (Fig. 4.4) turns out to be an MCP model (neuron with weighted inputs) with some additional, fixed, pre—processing. Units labeled A_1, A_2, A_j, A_p are called association units and their task is to extract specific, localized featured from the input images They were mainly used in pattern recognition even though their capabilities extended a lot more.

16. What do you mean by Learning Process.

Ans. The memorisation of patterns and the subsequent response of the network can be categorised into two general paradigms:

- **Associative mapping** in which the network learns to produce a particular pattern on the set of input units whenever another particular pattern is applied on the set of input units. The associative mapping can generally be broken down into two mechanisms:
- **Auto-association:** An input pattern is associated with itself and the states of input and output units coincide. This is used to provide pattern completion, *i.e.*, to produce a pattern whenever a portion of it or a distorted pattern is presented. In the second case, the network actually stores pairs of patterns building an association between two sets of patterns.
- **Hetero-association:** is related to two recall mechanisms
- **Nearest-neighbour** recall, where the output pattern produced corresponds to the input pattern stored, which is closest to the pattern presented, and
- **Interpolative** recall, where the output pattern is a similarity dependent interpolation of the patterns stored corresponding to the pattern presented. Yet another paradigm, which is a variant associative mapping is classification, i.e., when there is a fixed set of categories into which the input patterns are to be classified.
- **Regularity detection** in which units learn to respond to particular properties of the input patterns. Whereas in associative mapping the network stores the relationships among patterns, in regularity detection the response of each unit has a particular 'meaning'. This type of learning mechanism is essential for feature discovery and knowledge representation.

  Every neural network possesses knowledge which is contained in the values of the connections weights. Modifying the knowledge stored in the network as a function of experience implies a learning rule for changing the values of the weights.

  Information is stored in the weight matrix W of a neural network. Learning is the determination of the weights. Following the way learning is performed, we can distinguish two major categories of neural networks:
- **Fixed networks** in which the weights cannot be changed, *i.e.*, $dW/dt = 0$. In such networks, the weights are fixed a priori according to the problem to solve.
- **Adaptive networks** which are able to change their weights, *i.e.*, dW/dt not $= 0$.

All learning methods used for adaptive neural networks can be classified into two major categories :

- **Supervised learning** which incorporates an external teacher, so that each output unit is told what its desired response to input signals ought to be. During the learning process global information may be required. Paradigms of supervised learning include error-correction learning, reinforcement learning and stochastic learning.

  Important issue concerning supervised learning is the problem of error convergence, *i.e.*, the minimization of error between the desired and computed unit values. The aim is to determine a set of weights which minimizes the error. One well-known method, which is common to many learning paradigms, is the least mean square (LMS) convergence.
- **Unsupervised learning** uses no external teacher and is based upon only local information. It is also referred to as self-organization, in the sense that it self-organizes

data presented to the network and detects their emergent collective properties. Paradigms of unsupervised learning are Hebbian learning and competitive learning. From Human Neurons to Artificial Neurons aspect of learning concerns the distinction or not of a separate phase, during which the network is trained, and a subsequent operation phase. We say that a neural network learns off-line if the learning phase and the operation phase are distinct. A neural network learns on-line if it learns and operates at the same time. Usually, supervised learning is performed off-line, whereas unsupervised learning is performed on-line.

17. List various types of Transfer Functions used in ANN.

Ans. The behavior of an ANN (Artificial Neural Network) depends on both the weights and the input-output function (transfer function) that is specified for the units. This function typically falls into one of three categories:

- linear (or ramp)
- threshold
- sigmoid

For **linear units**, the output activity is proportional to the total weighted output.

For **threshold units**, the output is set at one of two levels, depending on whether the total input is greater than or less than some threshold value.

For **sigmoid units**, the output varies continuously but not linearly as the input changes. Sigmoid units bear a greater resemblance to real neurones than do linear or threshold units, but all three must be considered rough approximations.

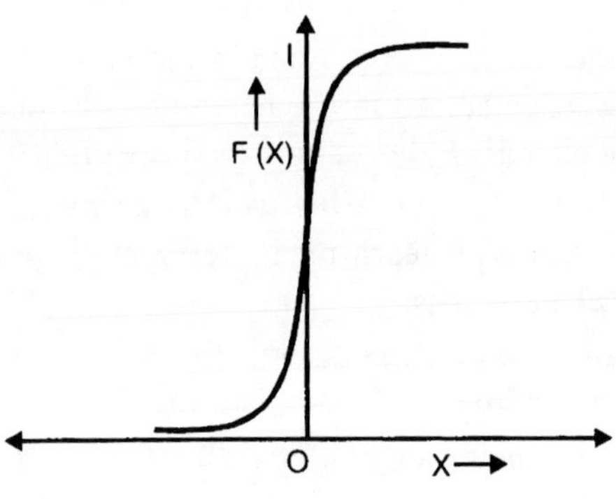

Fig. 10. The Sigmoid Function

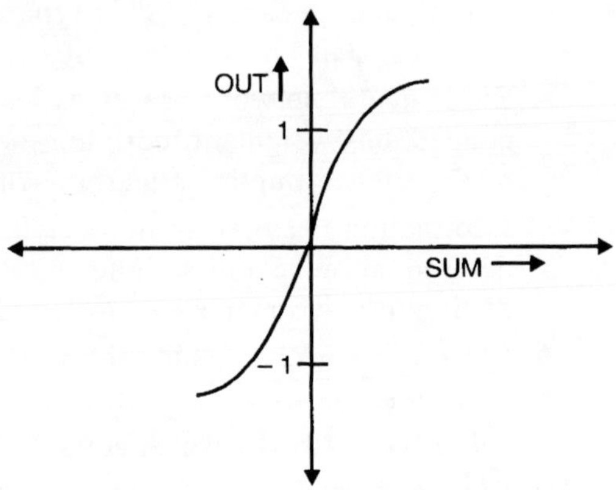

Fig. 11. Hyperbolic Tangential Function

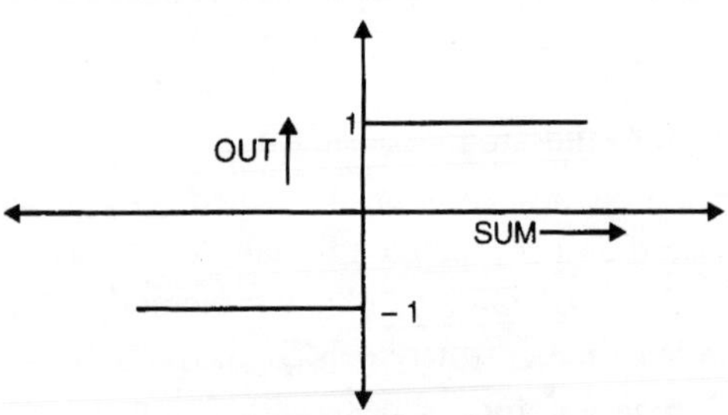

Fig. 12. Signum Function.

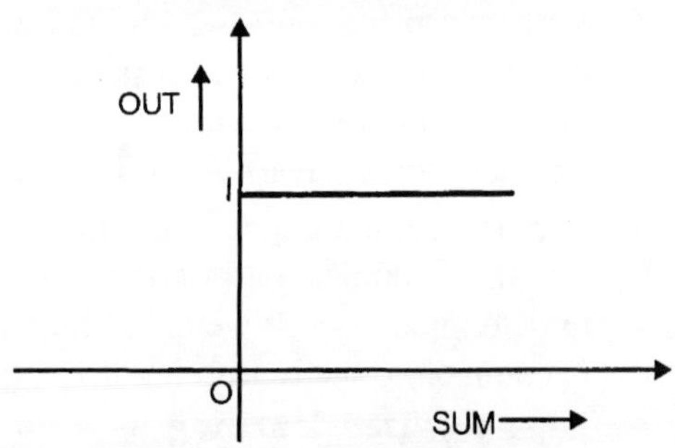

Fig. 13. Threshold Function.

To make a neural network that performs some specific task, we must choose how the units are connected to one another , and we must set the weights on the connections appropriately. The connections determine whether it is possible for one unit to influence another. The weights specify the strength of the influence.

We can teach a three-layer network to perform a particular task by using the following procedure:

1. We present the network with training examples, which consist of a pattern of activities for the input units together with the desired pattern of activities for the output units.
2. We determine how closely the actual output of the network matches the desired output.
3. We change the weight of each connection so that the network produces a better approximation of the desired output.

An Example to illustrate the above teaching procedure:

Assume that we want a network to recognise hand-written digits. We might use an array of, say, 256 sensors, each recording the presence or absence of ink in a small area of a single digit. The network would therefore need 256 input units (one for each sensor), 10 output units (one for each kind of digit) and a number of hidden units.

For each kind of digit recorded by the sensors, the network should produce high activity in the appropriate output unit and low activity in the other output units.

To train the network, we present an image of a digit and compare the actual activity of the 10 output units with the desired activity. We then calculate the error, which is defined as the square of the difference between the actual and the desired activities. Next we change the weight of each connection so as to reduce the error. We repeat this training process for many different images of each different images of each kind of digit until the network classifies every image correctly.

To implement this procedure we need to calculate the error derivative for the weight (EW) in order to change the weight by an amount that is proportional to the rate at which the error changes as the weight is changed. One way to calculate the EW is to perturb a weight slightly and observe how the error changes. But that method is inefficient because it requires a separate perturbation for each of the many weights.

Another way to calculate the EW is to use the Back-propagation algorithm which is described below, and has become nowadays one of the most important tools for training neural networks. It was developed independently by two teams, one (Fogelman-Soulie, Gallinari and Le Cun) in France, the other (Rumelhart, Hinton and Williams) in U.S.

18. What is Fuzzy Logic ?

Ans. Fuzzy Logic is an area of research based on the work of L.A. Zadeh. It is a departure from classical two-valued sets and logic, that uses "soft" linguistic (*e.g.*, large, hot, tall) system variables and a continuous range of truth values in the interval [0,1], rather than strict binary (True or False) decisions and assignments.

Fuzzy logic is used where a system is difficult to model exactly (but an inexact model is available), is controlled by a human operator or expert, or where ambiguity or vagueness is common. A typical fuzzy system consists of a rule base, membership functions, and an inference procedure.

19. What is meant by stability and plasticity in ANN?

Ans. The learning system to remain plastic or adaptable enough to learn new inputs whenever they might appear and yet remain stable enough to preserve the prestored information is called stability and plasticity in ANN.

20. Which is the fundamental model in ANN ?

Ans. Mc-Cullouch Pitts Model.

21. How learning rate is chosen.

Ans. Choice of learning rate:

- If η is too large: existing weights are overtaken by Δw
- If η is too small (≈ 0): very slow to converge
- Common choice : $\eta = 1$.

22. Define characteristics of Neural Networks

- number of layers
- number of nodes per layer
- activation function (linear, binary, softwmax)
- error function (mean squared error (MSE), cross entropy)
- type of learning algorithms (gradient descent, perceptron, delta rule)

23. Discuss the limitation of Hebbian learning rule?

Ans. This rule does not learn if the input and output patterns are in binary form. This is the limitation of this rule.

24. What are features of perceptron learning rule?

Ans.
- The perceptron learning rule is a method for finding the weights in a network.
- We consider the problem of **supervised learning** for **classification** although other types of problems can also be solved.
- A nice feature of the perceptron learning rule is that *if there exist* a set of weights that solve the problem, then the perceptron will find these weights. This is true for either binary or bipolar representations.

25. What are the limitations of perceptron learning rule?

Ans. In the general for the 2- and 3- layers cases, there is no simple way to determine the weights.

26. What are the applications of perceptron network?

Ans. 1. Pattern classification

2. Pattern recognition

3. Image processing

27. What are other names used for delta rule?

Ans.
- Adaline Rule
- Widrow-Hoff Rule
- Least Mean Squares (LMS) Rule

28. How Delta rule is different from Perceptron learning rule.

Ans.
- Replace the step function in the with a continuous (differentiable) activation function, *e.g.*, linear
- For classification problems, use the step function only to determine the class and not to update the weights.

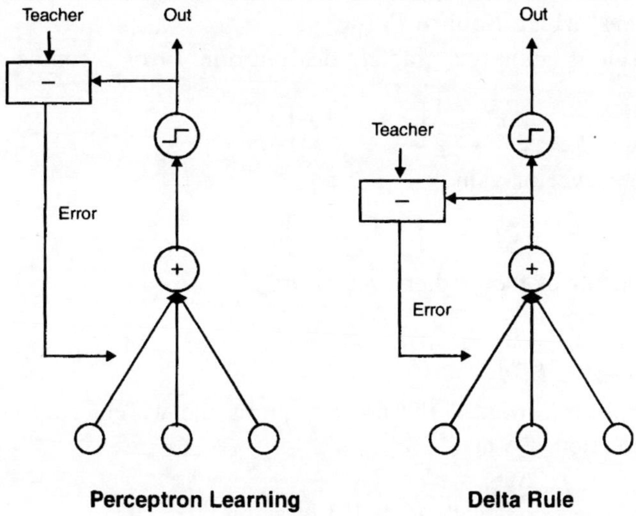

Fig. 14

29. What is adaline ?

Ans. The adaline (adaptive linear element) developed by Widrow is similar to the perceptron. The adaline has a single output (a single processing element) which receives multiple inputs.

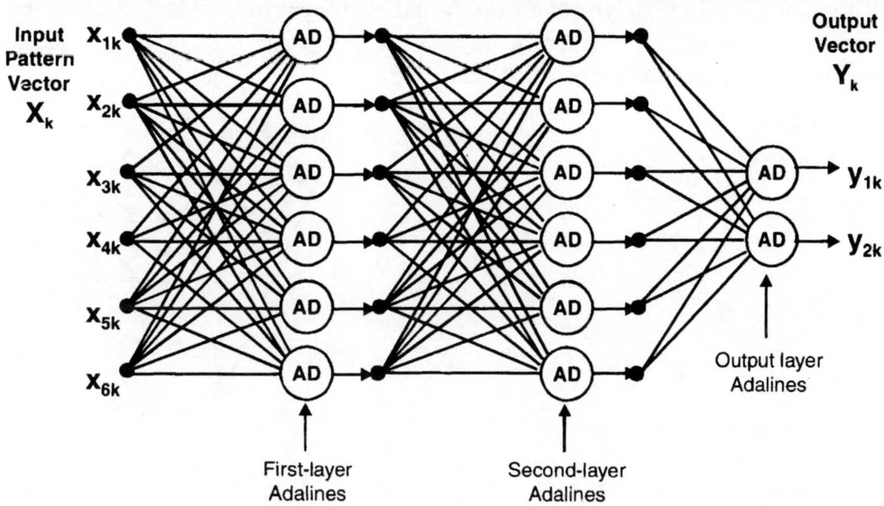

Fig. 15. Adaline

The processing element takes the weighted linear sum of the inputs and passes it to a bipolar function, which produces either +1 or –1, depending upon the polarity of the sum. A system with multiple outputs is implemented by using multiple adalines in parallel.

30. What is madaline ?

Ans. The madaline refers to the multiple adalines which can be organized into multiple layers. The adaline and the madaline have found many useful applications in adaptive signal processing.

31. Discuss the problem with adaline.

Ans. The problem dealt with by adaline is to determine the weights in such a way that the output response is correct or the error is minimized. The training rule of the adaline is the LMS (least mean square) or Widrow-Hoff rule.

32. Explain Least Mean Square Error.

Ans. The LMS rule is actually a gradient descent rule. Error E can be represented as

$$E \cong \frac{1}{2}\left[d_i - (w_i^t x)\right]^2$$

Error gradient vector value

$$\nabla E = -\left[d_i - (w_i^t x)\right]X$$

The components of the gradient vector are

$$\frac{\partial E}{\partial w_{ij}} = -[d_i - (w_i^t x)]\,x_j \qquad \text{for } j = 1, 2, \ldots n$$

Since the minimization of the error requires the weight changes to be in the negative gradient direction, we take

$$\Delta w_i = \eta \nabla E$$

where η is a +ve constant and called learning rate

$$\Delta w_i = \eta[d_i - (w_i^t x)]\,X$$

which is LMS rule.

33. Discuss Kohonen's Self-Organizing Map (SOM).

Ans. Kohonon's SOMs are a type of unsupervised learning. The goal is to discover some underlying structure of the data. Consider the physical arrangement of the nodes. Nodes that are "close" together are going to interact differently than nodes that are "far" apart.

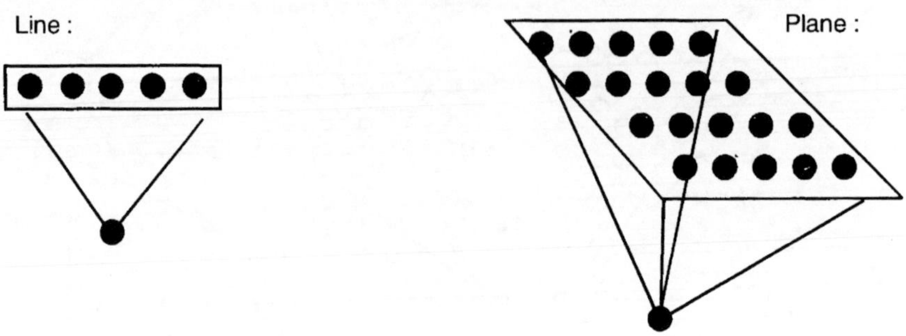

Fig. 16.

We can think of organizing the output nodes in a line or in a planar configuration.

The goal is to train the net so that nearby outputs correspond to nearby inputs.

For example, if x_1 and x_2 are two input vectors and t_1 and t_2 are the locations of the corresponding winning output nodes, then t_1 and t2 should be close if x_1 and x_2 are similar. A network that performs this kind of mapping is called a feature map.

In the brain, neurons tend to cluster in groups. The connections within the group are much greater than the connections with the neurons outside of the group. Kohonen's network tries to mimick this in a simple way.

34. Explain WINNER-TAKE-ALL-LEARNING (Kohonen Network)

Ans. (1) It consists of single layer of nodes plus an input layer.

(2) Each layer receives inputs from environment and from other nodes within the layer.

(3) Important point about this network is that both weight vector and input vectors are normalized to a constant value before learning

$$\hat{W}_i = \frac{(W_i)}{|W_i|} \qquad \text{for } i = 1, 2, \ldots P$$

(4) Each node computes by taking the dot product of its weight vector and input vector. $O_j = X.W_j$ where O_j is output or activation level of unit j.

(5) Node with the largest activation level is declared the winner in the competition. This **winning** node is the only node that will generate an output signal, the activation level of other neurons is suppressed to zero. The Winning node with largest net_i is allowed to change the weight that is weight adjustment is for winning neuron only while weights of other neurons remain unaffected. The weight adjustment criteria for this mode of training is such that

$$|X - W_m| = \min\{|X - W_i|\} \qquad \ldots(iii)$$

where $i = 1, 2, \ldots P$ and m is winning neuron no.

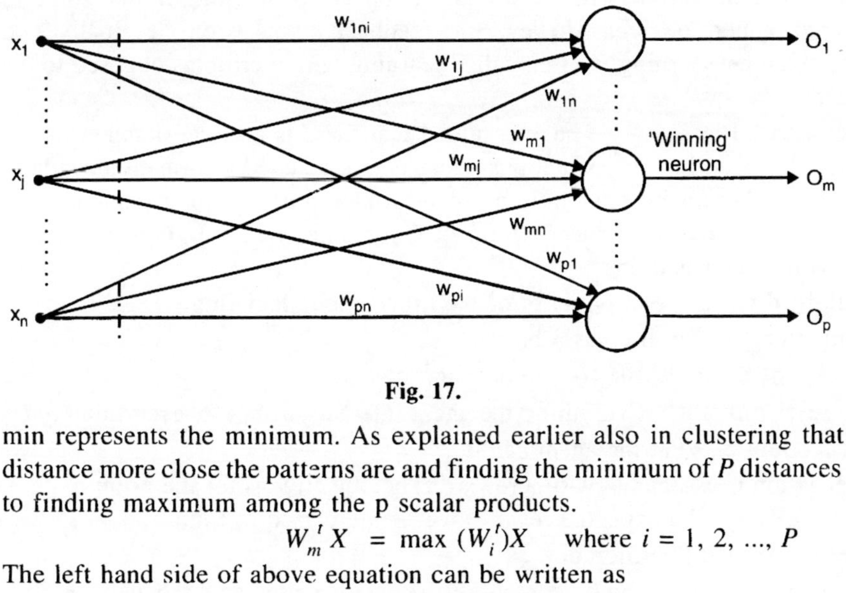

Fig. 17.

min represents the minimum. As explained earlier also in clustering that smaller the distance more close the patterns are and finding the minimum of P distances corresponds to finding maximum among the p scalar products.

$$W_m^t X = \max (W_i^t)X \quad \text{where } i = 1, 2, \ldots, P$$

The left hand side of above equation can be written as

$$|X - W_m| = (X^t X - 2W_m^t X - 1)^{1/2} \qquad \ldots(iv)$$

It is clear that searching for the minimum of p distances as on right hand side of (iii) Corresponds to finding maximum among p scalar products

$$W_m^t X = \max(W_i^t X) \qquad \ldots(v)$$

For $i = 1, 2, \ldots, p$

After the winning neuron has been identified and declared a winner, its weights must be adjusted so that the distance is reduced in the current training step. Thus $|X - W_m|$ must be reduced, preferably along the gradient direction in the weight space $w_{m1}, w_{m2}, \ldots, w_{mn}$

$$\nabla w_m \|x - w_m\|^2 = -2(x - w_m) \qquad \ldots(vi)$$

It seems reasonable to reward the weights of the $X - W_m$ winning neuron with an increment of weight in the negative gradient direction, thus in the direction $x - w_m$. We thus have

$$\Delta\hat{w}'_m = \alpha(x - \Delta\hat{w}_m) \qquad ...(vii)$$

where α is a small learning constant selected heuristically, usually between 0.1 and 0.7.

Learning according to equation (*vi*) and (*vii*) is called Winner –take –All learning and it is a common competitive and unsupervised learning technique. The winning neuron with the largest net is rewarded with a weight adjustment, while the weights of others remain unaffected.

Another modification of the winner-take-all learning rule is that both the winners' and losers' weights are adjusted in proportion to their level of responses. This may be called leaky competitive learning and should provide more subtle learning in the cases for which clusters may be hard to distinguish.

35. Explain Reinforcement Learning.

Ans. Reinforcement learning (RL) is a kind of supervised learning in that some feedback from the environment is given. However the feedback signal is only evaluative, not instructive. Reinforcement learning is often called learning with a critic as opposed to *learning with a teacher*.

In supervised learning we have assumed that there is a target output value for each input value. However, in many situations, there is less detailed information available. In extreme situations, there is only a single bit of information after a long sequence of inputs telling whether the output is right or wrong. Reinforcement learning is one method developed to deal with such situations.

36. Explain different components of Reinforcement learning.

Ans. Reinforcement learning has 3 basic

1. **Agent :** The learner or the decision maker.
2. **Environment :** Everything the agent interacts with, *i.e.,* everything *outside* the agent.
3. **Actions :** What the agent can do.

Each action is associated with a reward. The objective is for the agent to choose actions so as to maximize the expected reward over some period of time.

37. Explain Associative memory ?

Ans. A memory unit accessed by content of the data is called an Associative Memory or content addressable memory (CAM). This type of memory is accessed simultaneously and in parallel on the basis of data content rather than by specific address or location. Associative memory usually enables a parallel search within a stored data file. The purpose of the search is to output one or all stored data items that match the given search argument and to retrieve it entirely or partially. It is also believed that biological memory operates according to associative memory principles. No memory locations have addresses; storage is distributed over a large, densely interconnected assembly of neurons.

Retrieving the information is content addressable, which for a computer system can be sequential or indexed.

ANN plays the role of an associative memory in which data is collectively stored in the form of a weight matrix or memory and the process of developing the weight matrix is referred to as **learning or storing the desired patterns**, while the **retrieval or recall refers** to the generation of output pattern when an input is presented to the network.

There are two types of associative memory: **Autoassociative** and **Heteroassociative**. An **autoassociative** memory retrieves a previously stored pattern that most closely resembles the current pattern. In a **heteroassociative** memory, the retrieved pattern is, in general, different from the input pattern not only in content but possibly also in type and format.

A neural network that supports autoassociation will pass a pattern directly from its input neurons to the output neurons. No change occurs; to the causal observer it appears as if no work has taken place.

38. Define Storage capacity.

Ans. It refers to the quantity of information that can be stored in the network in such a way that it can be retrieved without error. The storage capacity can be measured as:

C = (Number of stored pattern)/ (Number of neurons in the network).

If the network is not fully connected then

C = (Number of stored pattern)/ (Number of connection weight in the network).

39. Define Pattern.

Ans. A pattern is a quantitative description of any object, event or phenomenon. A pattern may be Spatial or temporal. Examples of Spatial pattern are: images, fingerprints etc.

And examples of temporal pattern are: signals, signals vs. time etc.

40. Define Euclidean Distance.

Ans. The absolute difference between input and each of the weights is called Euclidean distance. The weights are updated for the neurons whose Euclidean distance is minimum.

41. Define Clustering and Calibration.

Ans. **Clustering**: Clustering is defined as the grouping of similar objects and separating of dissimilar ones. The objective of clustering neural networks discussed in this section is to categorize or cluster data. The classes must first be found from the correlations of an input data stream. Since the network actually deals with unlabeled data, the clustering should be followed by labeling clusters with appropriate category names or numbers. This process of providing the category of objects with a label is usually termed as ***calibration***

42. Define lateral inhibition.

Ans. In neurobiology, **lateral inhibition** is the capacity of an excited neuron to reduce the activity of its neighbors. The kohonen network uses intra-layer connections to moderate the competition. The output of each node acts as an inhibitory input to the other nodes but is actually excitatory in its neighborhood. Thus , even there is only one winning neuron , more than one neurons are allowed to change their weights. This complex scheme for adjusting the weights within the layer is known as lateral inhibition.

43. Define leaky competitive learning.

Ans. If in winner-take-all learning rule both the winners' and losers' weights are adjusted in proportion to their level of responses. This may be called leaky competitive learning and should provide more subtle learning in the cases for which clusters may be hard to distinguish.

44. What is Cognitron ?

Ans. Cognitron is a self-organizing multilayered neural network and the neocognitron has been extended from the cognitron,

45. What is Biocybernetics?

Ans. Biocybernetics is a cojoined word from bio and cybernetics. Cybernetics is the interdisciplinary study of the structure of regulatory systems. Cybernetics is closely related to control theory and systems theory.

Biocybernetics is the application of cybernetics to biological science, composed of biological disciplines that benefit from the application of cybernetics: neurology, multicellular systems and others. Biocybernetics plays a major role in systems biology, seeking to integrate different levels of information to understand how biological systems function.

46. Explain Radial Basis Function.

Ans. Radial basis function is a 2 layer network whose output units form a linear combination of the basis functions computed by the hidden units. They are used in function approximation, time series prediction, and control. The most common basis function chosen is a Gaussian function, in which the activation level O_j of hidden unit j is calculated by

$$O_j = \exp\left[\frac{-(X - W_j).(X - W_j)}{2\sigma j^2}\right]$$

47. Explain the Perceptron Algorithm.

Ans. 1. Weight Initialization :

Set all the weights and node thresholds to small random numbers.

2. Calculation of Activation :

(*a*) The activation level of an input unit is determined by the instance presented to the network.

(*b*) The activation level O_j of an output unit is determined by

$$O_j = F_h(W_{ji}X_i - \theta_j) \quad ...(1)$$

Where F_h is hard-limiting function:

$$F_h(a) = \begin{cases} 1 & \text{if } a > 0 \\ 0 & \text{Else} \end{cases}$$

3. Weight Training :

(*a*) Adjust weights by

$$W_{ji}(t+1) = W_{ji}(t) + \Delta W_{ji} \quad ...(2)$$

where $W_{ji}(t)$ is the weight from unit i to unit j at time t and ΔW_{ji} is the weight adjustment.

(*b*) The weight change may be computed by Delta Learning Rule:

$$\Delta W_{ji} = \eta\delta_j X_i \quad ...(3)$$

where η is trial–independent learning rate ($0 < \eta < 1$)

and δ_j is the error at unit j and $\delta_j = T_j - O_j$

where T_j is the desired output activation and

O_j is the actual output activation at output unit j

(*c*) Repeat the iterations till convergence.

48. What are the limitations of Perceptron networks.

Ans. 1. First, the output values of a perceptron can take on only one of two values (True or False).

2. Second, perceptrons can only classify linearly separable sets of vectors. If a straight line or plane can be drawn to separate the input vectors into their correct categories, the input vectors are linearly separable and the perceptron will find the solution. If the vectors are not linearly separable learning will never reach a point where all vectors are classified properly.

49. Give the most famous example of the perceptrons inability to solve problems with linearly nonseparable vectors.

Ans. Boolean EX-OR problem.

50. What is Multilayer Perceptron.

Ans. An MLP is a network of simple neurons called perceptrons. The basic concept of a single perceptron was introduced by Rosenblatt in 1958. A Multilayer Perceptron is a feedforward neural network with at least one hidden layer. It can deal with nonlinear classification problems.

51. What is Backpropagation algorithm?

Ans.

- ➠ Backpropagation network is **Multilayer feedforward network** with different transfer function in ANN and a more powerful learning rule. The learning rule is called **"Backpropagation"**, which is a kind of **gradient descent technique** with backward error propagation.
- ➠ It is a **Supervised Learning** method, and is an implementation of the **Delta rule**. It requires a teacher that knows, or can calculate, the desired output for any given input.

52. Is there any difference between RBF and Backpropagation.

Ans. The main difference is in their Basis Function. The radial basis function is used in RBF and it can cover only small regions, whereas the sigmoid function assumes large number of nonzero values over an infinitely large region of the input space. Depending upon the problem to be solved either can be used.

53. What are various learning factors. Explain.

Ans. 1. Initialization of weights

2. Learning constant
3. Momentum
4. Steepness of activation function
5. Network Size
6. Necessary number of Hidden neurons
7. Stopping Criteria

54. Differentiate between Artificial Intelligence and Artificial Neural Networks.

Ans. Traditional artificial intelligence and neural networks are generally considered appropriate for solving different types of problems. On thesurface these two approaches appear to be very different, but agrowing body of current research is focused on how the strengths of each can be incorporated into the other and built into systems that include the best features of both.

Artificial Neural Network	Artificial Intelligence
In ANN the knowledge is represented by weights on the connections between the nodes.	In AI knowledge is represented at higher level *i.e.*, explicit knowledge.
Intelligence is obtained by training.	Intelligence is obtained by designing.
Processing in ANN is faster because of parallel processing of the nodes. Therefore ANN systems are good for real time systems.	Comparatively not so good for real time systems.
Numeric representation is used. ANN good for fault tolerant systems.	Symbolic representation is used. AI not good for fault tolerant systems.
There is no explanation for any output or response because knowledge is encoded in the form of nodes	There is proper explanation for any response or output, that it is derived from these facts and inputs.

55. Explain Relationship between Statistics and ANN.

Ans. ➠ Major points of *differences* are:

- While statistics is concerned with data analysis, supervised ANN emphasizes statistical inference.
- Some neural networks are not concerned with data analysis (*e.g.*, those intended to model biological systems)
- Some neural networks do not learn (*e.g.*, Hopfield nets) and therefore have little to do with statistics.
- Some neural networks can learn successfully only from noise-free data (*e.g.*, ART or the perceptron rule) and therefore would not be considered statistical methods

➠ Most neural networks that can learn to generalize effectively from noisy data are similar or identical to statistical methods.

➠ Major points of *similarities* are:

- Feedforward nets with no hidden layer (including functional-link neural nets and higher-order neural nets) are basically generalized linear models.
- Probabilistic neural nets are identical to kernel discriminant analysis.
- Kohonen nets for adaptive vector quantization are very similar to k-means cluster analysis.
- Hebbian learning is closely related to principal component analysis.

➠ Some neural network areas that appear to have no close relatives in the existing statistical literature are:

- Kohonen's self-organizing maps.
- Reinforcement learning (although this is treated in the operations research literature on Markov decision processes

56. Name the networks based on Feedback and Feedforward connections

Ans. The following shows some types in each category

➠ ***Unsupervised Learning***

Feedback Networks:

- Binary Adaptive Resonance Theory (ART1)
- Analog Adaptive Resonance Theory (ART2, ART2a)

- Discrete Hopfield (DH)
- Continuous Hopfield (CH)
- Discrete Bidirectional Associative Memory (BAM)
- Kohonen Self-organizing Map/Topology-preserving map (SOM/TPM)

➠ *Feedforward-only Networks:*

- Learning Matrix (LM)
- Sparse Distributed Associative Memory (SDM)
- Fuzzy Assocriative Memory (FAM)
- Counterprogation (CPN)

➠ *Supervised Learning*

Feedback Networks:

- Brain-State-in-a-Box (BSB)
- Fuzzy Congitive Map (FCM)
- Boltzmann Machine (BM)
- Backpropagation through time (BPTT)

Feedforward-only Networks:

- Perceptron
- Adaline, Madaline
- Backpropagation (BP)
- Artmap
- Learning Vector Quantization (LVQ)
- Probabilistic Neural Network (PNN)
- General Regression Neural Network (GRNN)

57. Explain the meaning of Training, Testing and Validation Datasets.

Ans. ➠ ***Training set:*** A set of examples used for learning that is to fit the parameters [weights] of the classifier.

➠ ***Validation set:*** A set of examples used to tune the parameters of a classifier, for example to choose the number of hidden units in a neural network.

➠ ***Test set:*** A set of examples used only to assess the performance [generalization] of a fully-specified classifier.

58. Write Short note on Fuzzy ARTMAP.

Ans. ➠ This is a supervised neural network architecture that is based on" Adaptive Resonance Theory", proposed by Stephen Grossberg in 1976.

➠ ART encompasses a wide variety of neural networks based explicitly on human information processing and neurophysiology.

➠ ART networks are defined algorithmically in terms of detailed differential equations intended as plausible models of biological neurons.

➠ In practice, ART networks are implemented using analytical solutions or approximations to these differential equations.

➠ ART is capable of developing stable clusterings of arbitrary sequences of input patterns by self-organisation.

➠ Fuzzy ARTMAP is based on ART.

- Fuzzy ARTMAP's internal control mechanisms create stable recognition categories of optimal size by maximizing code compression while minimizing predictive error during on-line learning.

- Fuzzy ARTMAP incorporates fuzzy logic in its ART modules
- Fuzzy ARTMAP has fuzzy set-theoretic operations instead of binary set-theoretic operations.
- It learns to classify inputs by a fuzzy set of features (or a pattern of fuzzy membership values between 0 and 1)

59. What are the main applications of Hopfield Network ?

Ans. 1. Analog Converter

2. Travelling Salesman problem.

60. Explain optical neural networks.

Ans. An **optical neural network** is a physical implementation of an artificial neural network with optical components.

Some artificial neural networks that have been implemented as optical neural networks include the Hopfield neural network and the Kohonen self-organizing map with liquid crystals.

Biological neural networks function on an electrochemical basis, while optical neural networks use electromagnetic waves. Optical interfaces to biological neural networks can be created with optogenetics, but is not the same as an optical neural networks. In biological neural networks there exist a lot of different mechanisms for dynamically changing the state of the neurons, these include short-term and long-term synaptic plasticity. Synaptic plasticity is among the electrophysiological phenomena used to control the efficiency of synaptic transmission, long-term for learning and memory, and short-term for short transient changes in synaptic transmission efficiency. Implementing this with optical components is difficult, and ideally requires advanced photonic materials. Properties that might be desirable in photonic materials for optical neural networks include the ability to change their efficiency of transmitting light, based on the intensity of incoming light.

61. List various types of activation functions used.

Some of the activation functions are discussed here:

1. Sigmoid Function : F is called the "Squashing function" which is often chosen to be the logistic function or "sigmoid function' as shown in Fig. 18. The function F is expresses mathematically as

$$F(x) = \frac{1}{(1 + e^{-x})}$$

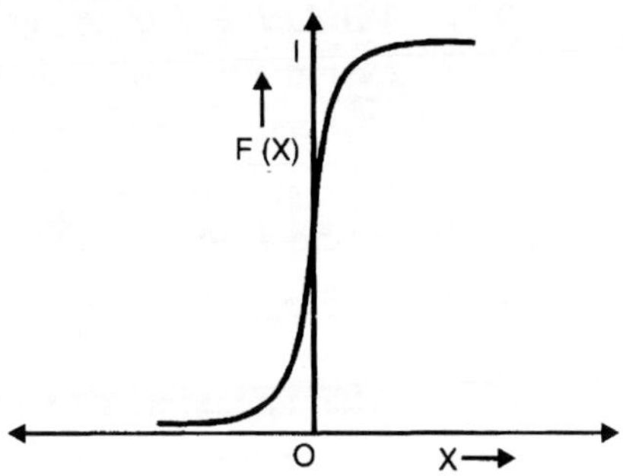

Fig. 18. The Sigmoid Function

By analogy to analog electronic system, Activation function is defined as a non-linear gain for the artificial neuron. This gain is calculated by finding the ratio of the change in $F(X)$ to a small change in X. Thus gain is the slope of the wave at a specific excitations level.

Figure 19 describes the block labeled F that accepts the SUM output and produces the signal OUT. The SUM signal is produced by an activation function F to produce the neuron output signal, OUT. This may be a simple linear function.

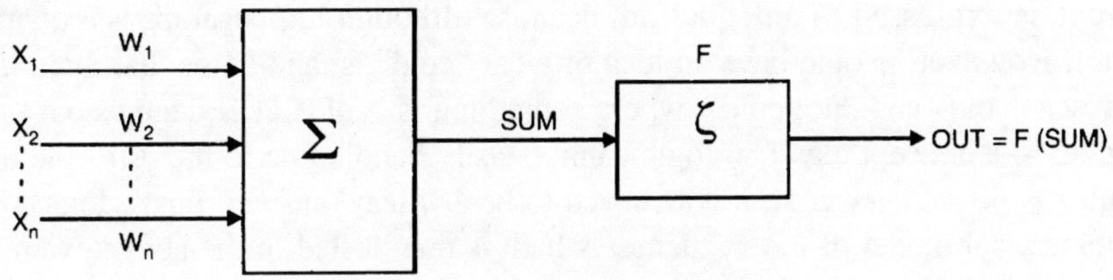

Fig. 19. ANN's Working Model.

```
OUT=F (SUM)
OUT=1 if SUM >T
OUT=0 if otherwise.
```

where K is a constant and T is a threshold constant value.

2. Hyperbolic Tangential Functions

This function can be expressed as

```
OUT= tanh (SUM)
```

This is also S-shaped function, as shown in Fig. 20, but it is symmetrical about origin. This function confines the output between 1 and –1.

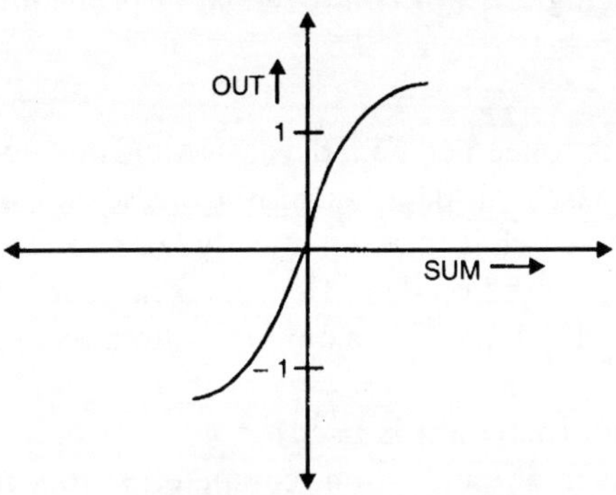

Fig. 20. Hyperbolic Tangential Function.

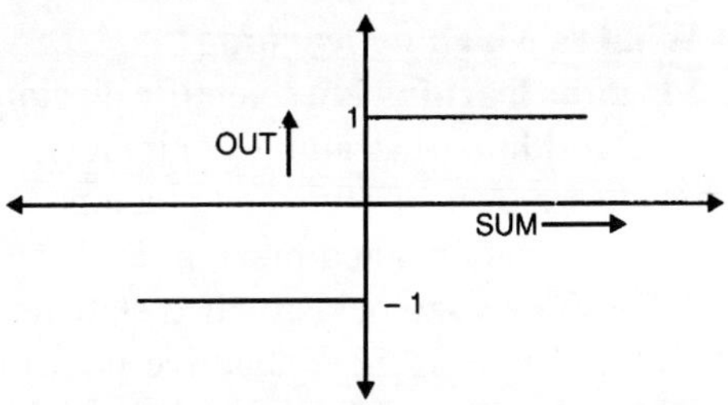

Fig. 21. Signum Function.

3. Signum Function : Signum function can be expressed as

```
OUT= Sgn (SUM)
```

This function takes only two values for any range of inputs.

Mathematically it is expressed as

```
Sgn (SUM) =1 for SUM >0
Sgn (SUM) =-1 for SUM <0
```

It is graphically represented as shown in Fig. 21.

4. Threshold Function : This function output is either 0 or 1 and given by the function as

```
OUT    =1 if SUM > 0
       =0 if SUM < 0
```

And its graphical representation is shown in Fig. 22.

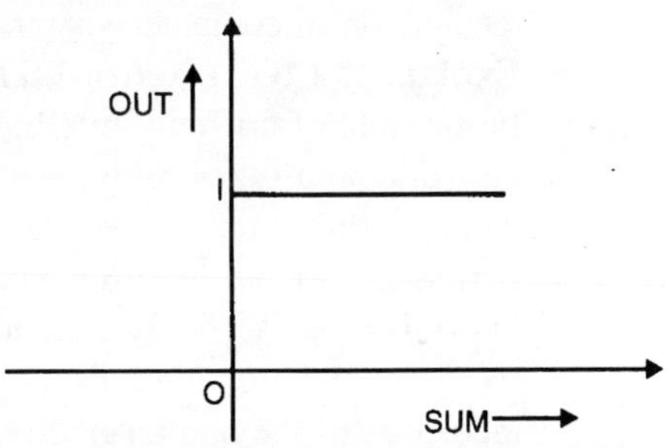

Fig. 22. Threshold Function.

62. Is Fuzzy logic is the same as "imprecise logic"?

Ans. Fuzzy logic is not any less precise than any other form of logic: it is an organized and mathematical method of handling *inherently* imprecise concepts. The concept of "coldness"

cannot be expressed in an equation, because although temperature is a quantity, "coldness" is not. However, people have an idea of what "cold" is, and agree that there is no sharp cutoff between "cold" and "not cold", where something is "cold" at N degrees but "not cold" at N+1 degrees —a concept classical logic cannot easily handle due to the principle of bivalence. The result has no set answer so it is believed to be a 'fuzzy' answer. Fuzzy logic simply provides a mathematical model of the vagueness which is manifested in the above example.

63. Discuss that Fuzzy logic is a new way of expressing probability.

Ans. Fuzzy logic and probability are different ways of expressing uncertainty. While both fuzzy logic and probability theory can be used to represent subjective belief, fuzzy set theory uses the concept of fuzzy set membership (*i.e., how much* a variable is in a set), probability theory uses the concept of subjective probability (*i.e., how probable* do I think that a variable is in a set). While this distinction is mostly philosophical, the fuzzy-logic-derived possibility measure is inherently different from the probability measure, hence they are not *directly* equivalent.

But Lotfi Zadeh argues that fuzzy logic is different in character from probability, and is not a replacement for it. He fuzzified probability to fuzzy probability and also generalized it to what is called possibility theory. Other approaches to uncertainty include Dempster-Shafer theory and rough sets. However, fuzzy logic is not controversial to probability but rather complementary.

64. What is Machine learning?

Ans. Machine learning is a scientific discipline that is concerned with the design and development of algorithms that allow computers to learn based on data, such as from sensor data or databases. A major focus of machine learning research is to automatically learn to recognize complex patterns and make intelligent decisions based on data. Hence, machine learning is closely related to fields such as statistics, probability theory, data mining, pattern recognition, artificial intelligence, adaptive control etc.

65. What are the applications in which Machine Learning is used?

Ans. Applications for machine learning include machine perception, computer vision, natural language processing, syntactic pattern recognition, search engines, medical diagnosis, bioinformatics, brain-machine interfaces and cheminformatics, detecting credit card fraud, stock market analysis, classifying DNA sequences, speech and handwriting recognition, object recognition in computer vision, game playing, software engineering and robot locomotion.

66. Explain the term Neuro-fuzzy.

Ans. In the field of artificial intelligence, **neuro-fuzzy** refers to combinations of artificial neural networks and fuzzy logic. Neuro-fuzzy hybridization results in a hybrid intelligent system that synergizes these two techniques by combining the human-like reasoning style of fuzzy systems with the learning and connectionist structure of neural networks. Neuro-fuzzy hybridization is widely termed as Fuzzy Neural Network (FNN) or Neuro-Fuzzy System (NFS) in the literature. Neuro-fuzzy system (the more popular term is used henceforth) incorporates the human-like reasoning style of fuzzy systems through the use of fuzzy sets and a linguistic model consisting of a set of IF-THEN fuzzy rules. The main strength of neuro-fuzzy systems is that they are universal approximators with the ability to solicit interpretable IF-THEN rules. Although generally assumed to be the realization of a fuzzy system through connectionist networks, this term is also used to describe some other configurations including:

- Deriving fuzzy rules from trained RBF networks.

- Fuzzy logic based tuning of neural network training parameters.
- Fuzzy logic criteria for increasing a network size.
- Realising fuzzy membership function through clustering algorithms in unsupervised learning in SOMs and neural networks.
- Representing fuzzification, fuzzy inference and defuzzification through multi-layers feed-forward connectionist networks.

67. What do you mean by Rough Set?

Ans. A **rough set** is a formal approximation of a crisp set (*i.e.*, conventional set) in terms of a pair of sets which give the *lower* and the *upper* approximation of the original set. In the standard version of rough set theory (Pawlak 1991), the lower- and upper-approximation sets are crisp sets, but in other variations, the approximating sets may be fuzzy sets.

68. Explain the term Fuzzification.

Ans. The fuzzification comprises the process of transforming crisp values into grades of membership for linguistic terms of fuzzy sets. The membership function is used to associate a grade to each linguistic term.

69. Explain Defuzzyfication.

Ans. This transformation from a fuzzy set to a crisp number is called a *defuzzification*. It is not a unique operation as different approaches are possible.

70. What are various components of a fuzzy system ?

Ans. A fuzzy system is drawn as a black box with some inputs and an output. Figure shows the contents of a fuzzy system.

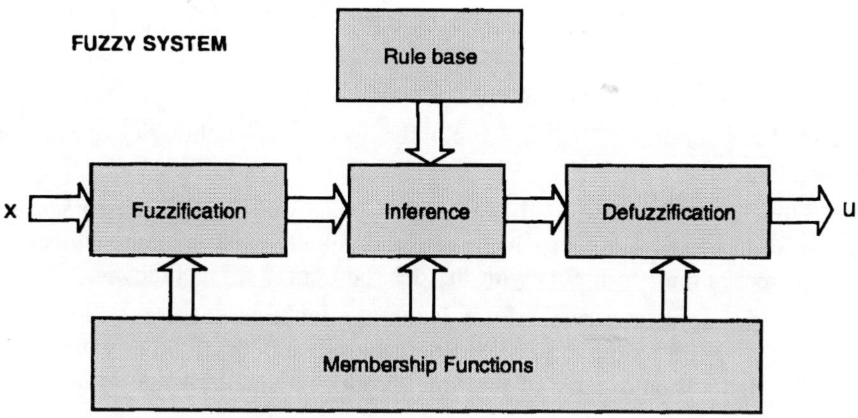

Fig. 23. Components of a fuzzy system.

Now it is clear, what is in this black box. The input signals combined to the vector $x = [x_1, x_2, ..., x_n]^T$ comes out directly from the defuzzification block, which transforms an output fuzzy set back to a crisp value using different methods. The set of membership functions responsible for the transforming part and the rule base as the relational part contain as a whole the modeling information about the system, which is processed by the inference machine. This rule-based fuzzy system is the basis of a fuzzy controller.

❑❑❑

MODEL TEST PAPERS

MODEL TEST PAPER–1

B.E. 7th Semester Examination
NEURAL NETWORKS (CSE-407-F)

Note: Answer all the Sections.

SECTION A

1. Attempt all the parts: **(10*2=20)**

(*a*) Why NNs are called as Parallel Distributed Processing?

(*b*) Define Gradient descent Processing.

(*c*) Name all types of error based learning algorithms.

(*d*) Justify- " The use of adaptive coefficient where the value of the learning coefficient is the function of error derivative on successive updates."

(*e*) List all the tuning parameters of Back propagation Networks.

(*f*) Differentiate between auto associative and hetero associative memory.

(*g*) What is ART?

(*h*) How is generalization possible?

Hint: three conditions that are typically necessary—although not sufficient—for good generalization:

1. The first necessary condition is that the inputs to the network contain sufficient information pertaining to the target, so that there exists a mathematical function relating correct outputs to inputs with the desired degree of accuracy.
2. The second necessary condition is that the function you are trying to learn (that relates inputs to correct outputs) be, in some sense, smooth. In other words, a small change in the inputs should, most of the time, produce a small change in the outputs.
3. The third necessary condition for good generalization is that the training cases be a sufficiently large and representative subset of the set of all cases that we want to generalize.

(*i*) Define kohonen network.

(*j*) What is overfitting and how it can be avoided.

SECTION B

2. Attempt any two: **(10*2=20)**

(*a*) Explain briefly the operation of a biological NN and give its features.

(*b*) What are the basic learning laws? Identify supervised and unsupervised basic learning laws?

(*c*) Give the output of the network for the input [111]:

SECTION C

3. Attempt any two: **(10*2=20)**

(*a*) Discuss backpropagation algorithm for a Multilayer network.

(*b*) Explain the learning factors of Backpropagation network.

(*c*) What do u mean by Hopfield Networks. Explain in detail.

SECTION D

4. Attempt any two: **(10*2=20)**

(*a*) Explain the perceptron Networks along with Perceptron learning rule.

(*b*) How single layer continuous perceptrons are used for classifying the linearly separable problems.

(*c*) What is Kohonen SOM? How is it useful? Illustrate through the examples.

SECTION E

5. Attempt any two: **(10*2=20)**

(*a*) Explain how an auto-correlator results in the refinement of the pattern or removal of the noise to retrieve the closest matching stored pattern.

(*b*) Explain BAM architecture employ bipolar /binary encoding of patterns.

(*c*) Explain:

(*i*) Initialization of weights

(*ii*) Recall mode

(*iii*) Separability Limitations.

MODEL TEST PAPER–2

B.E. 7th Semester Examination
NEURAL NETWORKS (CSE-407-F)

Note: Answer all the Sections.

SECTION A

1. Attempt all the parts: **(10*2=20)**

(*a*) Make a 3-2-1 feedforward network.

(*b*) A 3-input neuron is trained to output a zero when the input is 110 and a one when the input is 111. After generalisation, the output will be zero when and only when the input is: find the answer.

Hint: The truth table before generalisation is:

Inputs	*Output*
000	$
001	$
010	$
011	$
100	$
101	$
110	0
111	1

where $ represents don't know cases and the output is random.

After generalisation, the truth table becomes:

Inputs	*Output*
000	0
001	1
010	0
011	1
100	0
101	1
110	0
111	1

Therefore, the output will be zero when the input is 000 or 010 or 110 or 100

(*c*) What do u mean by perceptrons.

(*d*) What are the advantages of neural networks over conventional computers?

(*e*) Explain ADALINE and MADALINE.

(*f*) How learning rate is chosen in BAckpropagation Network.

(*g*) What do u mean by clustering.

(*h*) Explain gradient descend rule.

(*i*) What is the difference between the competitive learning and the conventional learning.

(*j*) Write the definition of Linear Associator.

SECTION B

2. Attempt any two: **(10*2=20)**

(*a*) Explain Augmented BP networks with its architecture and transfer function.

(*b*) What are the characteristics of NN? Explain three fundamentally classes of networks.

(*c*) Implement a MADALINE network to solve the XOR Problem.

Hint: Madaline (Multiple Adaline) is a two layer neural network with a set of ADALINEs in parallel as its input layer and a single PE (processing element) in its output layer. For problems with multiple input variables and one output, each input is applied to one Adaline. For similar problems with multiple outputs, madalines in parallel can be used. The madaline network is useful for problems which involve prediction based on multiple inputs, such as weather forecasting (Input variables: barometric pressure, difference in pressure. Output variables: rain, cloudy, sunny).

A MADALINE (many Adaline) network is created by combining a number of Adalines. The network of ADALINES can span many layers. The use of Multiple Adalines help counter the problem of non linear separability. For example, the MADALINE network with two units exhibits the capability to solve the XOR problem.

SECTION C

3. Attempt any two: **(10*2=20)**

(*a*) Explain the following:

(*i*) Hebbian learning rule

(*ii*) Widrow-Hoff learning rule

(*iii*) Delta learning Rule

(*b*) What do you mean by linearly separable and Non linearly separable problems. What is the role of Single layer and Multilayer perceptrons.

(*c*) Derive Backpropagation learning algorithm and the factors affecting it.

SECTION D

4. Attempt any two: **(10*2=20)**

(*a*) Explain the basic structures and dynamic behaviors of HOPFIELD Networks.

(*b*) Differentiate between the support vector machine and RBF.

(*c*) How decision regions are made.

SECTION E

3. Attempt any two: **(10*2=20)**

(*a*) How patterns are stored and retrieved from the autoassociative memory.

(*b*) Write Note on:

(*i*) unsupervised learning of clusters

(*ii*) Separability limitations.

(*iii*) Initialization of weights.

(*c*) Explain winner-take-all learning in details and also explain how the weights are modified.

MODEL TEST PAPER–3

1. Explain the structure of biological Neuron and ANN. In which manner biological neurons are relevant to ANN? Explain.
2. (*a*) What do you mean by activation functions. Explain different types of activation functions used.
   (*b*) Explain McCulloch Pitts Model.
3. (*a*) What do you mean by Pattern. How patterns are classified. Discuss one Technique.
   (*b*) Is there any difference between classification and recognition?
4. (*a*) Explain Backpropagation algorithm with the help of flowchart.
   (*b*) What are various learning factors used while ANN starts learning.
5. (*a*) Describe the training of Hopfield Network.
   (*b*) How Hopfield Networks are used to optimize the traveling salesman problem.
6. Discuss storage and retrieval algorithm for associative memory .
7. Explain Unsupervised Learning of Clusters. What is Kohonen Self Organizing network? How is it useful? Illustrate through examples.
8. (*a*) Using Hebbian Learning with binary and continuous activation functions of a very simple network. Find out the updated weights $w^1 = \begin{bmatrix} 1 \\ -1 \\ 0 \\ 0.5 \end{bmatrix}$
   (*b*) Needs to be trained using set of three input vectors as below

$$x_1 = \begin{bmatrix} 1 \\ -1 \\ 1.5 \\ 0 \end{bmatrix},\ x_2 = \begin{bmatrix} 1 \\ -0.5 \\ -1 \\ 0 \end{bmatrix},\ x_3 = \begin{bmatrix} 0 \\ 1 \\ -1 \\ 1.5 \end{bmatrix}$$

   For an arbitrary choice of learning constant $c = 1$.

MODEL TEST PAPER–4

1. (*a*) Why Use Neural Network ? Write its applications
   (*b*) How the information flow takes place in brain?
2. Write short notes on
   (*a*) Feed forward networks (*b*) Hebbian learning
   (*c*) Delta learning rule (*d*) Widrow-Hoff learning rule
3. Explain Single Discrete Perceptron Training Algorithm
4. (*a*) Derive $\Delta W_{ji} = \eta \delta_j O_i$.
   (*b*) What are Linearly separable problems? What kind of perception is required to handle the problems?
5. Discuss mathematical foundation of Discrete–time Hopfield networks.
6. (*a*) Explain the architecture of associative memory
   (*b*) Discuss the types of BAM.
7. (*a*) What is competitive learning? How it is different from conventional learning? Discuss its significance and usefulness in NN.
   (*b*) Discuss Self organizing mapping.
8. Write short note on :
   (*a*) Supervised and unsupervised learning (*b*) Recall mode
   (*c*) Initialization of weights (*d*) Perceptron

❑❑❑

PREVIOUS YEARS SOLVED EXAMINATION PAPERS

B.E.

7TH SEMESTER EXAMINATION, MAY 2006

NEURAL NETWORKS

Q. 1. (*a*) How many neurons are there in an average human being's brain and how many synaptic junctions in its brain and information flows in human brain.

Ans. Refer to Chapter 1, Art. 1.5.1, page 7.

Q. 1. (*b*) Draw a simple model of artificial neuron. Describe its working and explain the various terms used in it.

Ans. Refer to Chapter 2, Art. 2.2, pages 17-19.

Q. 2. (*a*) Describe Widrow-Hoff learning rule.

Ans. Refer to Chapter 2, Art. 2.5.4.4, pages 45-46.

Q. 2. (*b*) What do you mean by learning in context of ANN.

Ans. Refer to Chapter 2, Art. 2.5, pages 33-35.

Q. 3. What are Linearly separable problems? What kind of perception is required to handle the problems?

Ans. Refer to Chapter 4, Art. 4.1.1, pages 76-78.

Q. 4. (*a*) What is the basic principles of feedback networks?

Ans. Refer to Chapter 2, Art. 2.4.3, pages 32-33.

Q. 4. (*b*) Describe the training of Hopfield Network.

Ans. Hopfield Network : Unlike the linear associator model which consists of two layers of processing units, one serving as the input layer while the other as the output layer, **the Hopfield model consists of a single layer of processing elements where each unit is connected to every other unit in the network other than itself.** The connection weight matrix W of this type of network is square and symmetric, *i.e.*, $w_{ij} = w_{ji}$ for $i, j = 1, 2, ..., m$. Each unit has an extra external input I_i. This extra input leads to a modification in the computation of the net input to the units:

$$input_j = \sum_{i=1}^{m} x_i w_{ij} + I_j$$

for $j = 1, 2, ..., m$.

Unlike the linear associator, the units in the Hopfield model act as both input and output units. But just like the linear associator, a single associated pattern pair is stored by computing the weight matrix as follows:

$$W_k = X_k^T Y_k$$

where $$Y_k = X_k$$

to store p different associated pattern pairs. Since the Hopfield model is an autoassociative memory model, patterns, rather than associated pattern pairs, are stored in memory.

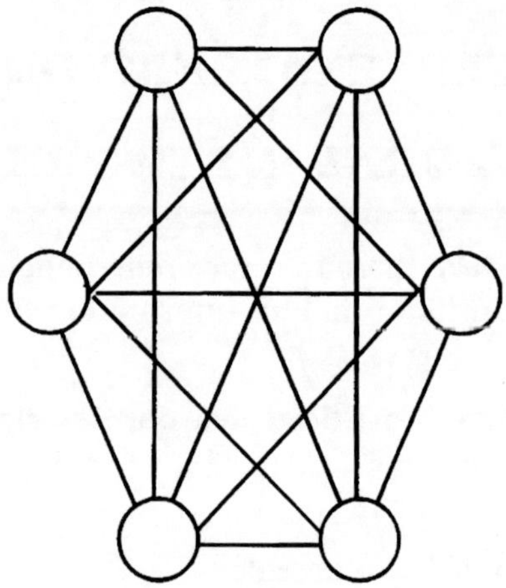

Fig. 1. Hopfield Network.

Hopfield (1982) demonstrated that the maximum number of patterns that can be stored in the Hopfield model of m nodes before the error in the retrieved pattern becomes severe is around $0.15m$.

Training of Hopfield Network : Training a Hopfield net involves lowering the energy of states that the net should "remember". This allows the net to serve as a content addressable memory system, that is to say, the network will converge to a "remembered" state if it is given only part of the state. The net can be used to recover from a distorted input the trained state that is most similar to that input. This is called associative memory because it recovers memories on the basis of similarity. For example, if we train a Hopfield net with five units so that the state (1, 0, 1, 0, 1) is an energy minimum, and we give the network the state (1, 0, 0, 0, 1) it will converge to (1, 0, 1, 0, 1). Thus, the network is properly trained when the energy of states which the network should remember are local minima.

Q. 5. Explain Unsupervised Learning of Clusters.

Ans. Refer to Chapter 7, Art. 7.4, pages 148-150.

Q. 6. What is the basic principle of associative memories. Discuss the storage and retrieval algorithm.

Ans. Refer to Chapter 6, Art. 6.1, pages 121-123.

Storage and Retrieval Algorithm : As explained earlier also storage algorithm is called encoding and the retrieval algorithm is called decoding process.

Given : P bipolar binary vector $\left\{S^{(1)}, S^{(2)}, \ldots, S^{(P)}\right\}$

where S^m is $n \times 1$ for $m = 1, 2, \ldots, P$

Initializing vector V_0 is $n \times 1$.

Storage Algorithm :

Step 1 : Weight matrix W is $(n \times n)$: $W \leftarrow 0,\ m \leftarrow 1$

Step 2 : Vector S^m is stored. Therefore the storage algorithm for calculating the weight matrix is

$$W \leftarrow S^{(m)} S^{(m)t} - I$$

or $$w_{ij} = (1 - S_{ij}) \sum_{m=1}^{P} S_i^{(m)} S_j^{(m)} \quad [S_{ij} \to \text{Kronecker function}]$$

Step 3 : If $m < P$ then $m \in m + 1$ and go to step 2, otherwise go to step 4.

Step 4 : Storage is complete.

Note. If instead of **bipolar** binary vectors, unipolar binary vectors $S(m)$ for $m = 1, 2, \ldots, P$ are used then entries of original unipolar vector $S(m)$ are replaced by $2S_i^{(m)} - 1$ where $i = 1, 2, \ldots, n$

$$\therefore \quad w_{ij} = (1 - S_{ij}) \sum_{m=1}^{P} (2S_i^{(m)} - 1)(2S_j^{(m)} - 1)$$

Retrival Algorithm

1. Cycle counter K is initialized, $K \leftarrow 1$. Within the cycle counter i is initialized, $i \leftarrow 1$ and network is initialized as $V \leftarrow V(0)$.
2. Integers 1, 2, ..., n are arranged in an ordered random sequences $\alpha_1, \alpha_2, \ldots, \alpha_{\underline{n}}$. If randomized sequence of updation is not used then $\alpha_1 = 1, \alpha_2 = 2, \ldots \alpha_m = n$ or $\alpha_i = i$ at each update cycle.
3. Neuron i is updated by computing $V_{\text{new}}\ \alpha_i$

$$\text{net}_{\alpha_i} = \sum_{j=1}^{n} W_{\alpha_i j} V_j$$

$$V_{\text{new}\,\alpha_i} = \text{sgn}\,(\text{net}_{\alpha_i}) \qquad [\text{where net} = w_{ij}^k V_j^k]$$

4. If $i < n$ then $i \leftarrow i + 1$ and goto step 3, otherwise go to step 5.
5. If $V_{\text{new}\,\alpha_i} = V_{\alpha_i}$ for $i = 1, 2, \ldots, n$

$\Rightarrow$ no updation and retrieval is complete output is K and $V_{\text{new}_1}, V_{\text{new}_2}, V_{\text{new}_n}$ otherwise $K \leftarrow K + 1$, go to step 2.

Q. 7. (*a*) What are the guidelines with regard to choice of neurons in different layers and choice of no. of hidden layers?

Ans. Refer to Chapter 4, Art. 4.6, pages 94-96.

Q. 7. (*b*) Discuss the architecture of associative memory.

Ans. Refer to Chapter 6, Art. 6.1, pages 121-124.

Q. 8. Write short note on :

(*a*) Hebbian learning rule.

(*b*) Decisions region for single layer perceptron.

Ans. (*a*) Refer to Chapter 2, Art. 2.5.4.1, pages 36-37.

(*b*) **Decision Functions :**

1. The classification at the system's output is obtained by the classifier implementing the decision function $i_0(x)$.
2. **Responses** represent the categories into which the patterns should be placed.

3. Classification (decision) function is provided by mapping of n-component vector X into one of the category numbers.

$$i_0 = i_0(x)$$

where $$X = \begin{bmatrix} X_1 \\ X_2 \\ \vdots \\ X_n \end{bmatrix}$$

4. In spatial object pattern vector X generated is Fig. 15.

$$X = [X_1, X_2, \ldots.., X_n]^t$$

In this the cell is assigned a value +1 if cell contains a portion of a spatial object, otherwise a value of 0 or –1 is assigned \ here

$$X = [\,0\ \ 1\ \ 1\ \ 0\ \ 0\ \ 1\ \ 1\ \ 0\ \ 1\ \ 0\ \ 1\ \ 0\ \ 1\ \ 0\ \ 0\ \ 1]^t$$

and in case of temporal object (Fig. 16), being a continuous function of time t, the pattern vector may be formed discrete time instants t_i

$$\therefore \qquad x_i = f(t_i) \text{ for } i = 1, 2, \ldots., n$$

5. Any pattern can be represented by a point in n-dimensional space. This n-dimension space generally called "Euclidean space E^n or pattern space."
6. Regions that separates the different classes called decision regions.

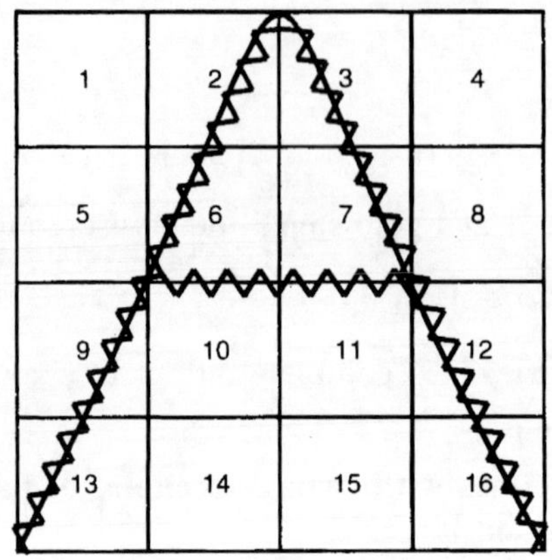

Fig. 2. Spatial object.

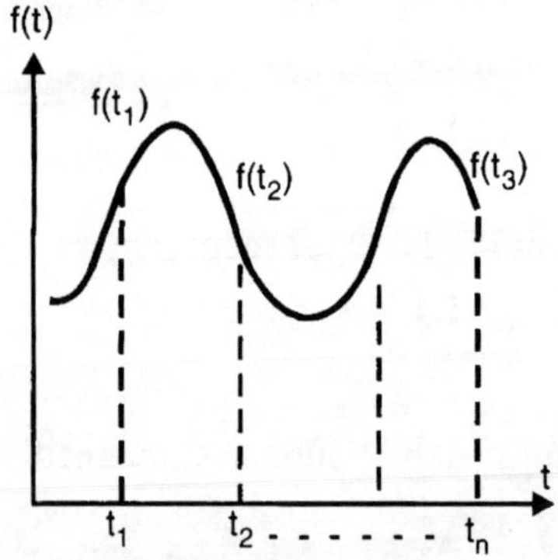

Fig. 3. Temporal object.

B.E.
7TH SEMESTER EXAMINATION, DEC. 2006
NEURAL NETWORKS

Q. 1. What is competitive learning? How it is different from conventional learning? Discuss its significance and usefulness in NN.

Ans. Conventional learning: The process of adjusting the weights to make the network learn the relationship between the inputs and targets is called learning or training. A prescribe set of well defined rules for the solution of a learning problem is called learning algorithm. Learning algorithms differ from each other in the way in which the adjustment to a synaptic weight of a neuron is formulated. In principle, almost all learning algorithms for neural networks are iterative optimization algorithms. There are two main categories for learning, namely supervised learning and unsupervised learning.

Competitive learning. Competitive learning occurs when the processing elements compete; only the processing element yielding the strongest response to a given input can modify itself, becoming more like the input. In all cases, the final values of the weighting functions constitute the "memory" of the ANN.

Learning Materials For ANNs

Learning in neural networks obviously implies that some learning material should be provided to it to learn from. In case of ANN any non-numerical data must be processed in such a way to obtain a numeric representation. A "**data set**" is a matrix containing several samples. In Ann methodology, the data set is often sub divided into "training" and "test" sets. A **training set** is a set of examples used for learning, which is to fit the parameters [*i.e.*, weights] of the classifier. A **validation set** is a set of examples used to tune the parameters [*i.e.*, architecture, not weights] of a classifier, for example, number of hidden units.

A **test set** is a set of examples used only to assess the performance [generalization] of a fully specified classifier. **Training sets** needs to be fairly large to contain all the needed information if the network is to learn the features and relationship that are important. Not only do the sets have to be large but also the training session must include a wide variety of data.

Network Validation

The primary risk in developing a model is that of over training, a situation in which the neural network starts to reproduce the noise specific to a particular sample in the training data, which may cause it to lose its ability to predict accurately. There are certain techniques that are used to avoid this; the most popular of this is ***network validation***. The testing of the network predictivity is done by reserving some of the data, which is excluded from the training data sets. The network is used to predict the outputs for these reserve data records, and the calculated outputs are compared with the observed values. If they are found to be sufficiently close, the network is considered to be sufficiently predictive and the network is said to be validated.

Generalization. Generalization is the ability of a neural network to produce reasonable responses to input patterns that are similar, but not identical, to training patterns. Once it is claimed that neural network has been trained properly or it is said that the neural network has learned the concept. Then it is expected that neural network produces the correct output for previously unknown inputs, but belonging to the same concept class. Then, the trained neural network has the capability to generalize. For example, if we have trained a neural network using several images (poses) of the leopard. If the neural network has been well trained it is expected that when one presents another image (pose) of leopard, which was not a part of training set, the neural network should be able to recognize the presented image as a image of leopard. This is what is meant by generalization.

Q. 2. Explain :

(*a*) Backpropagation

(*b*) Perceptron training rule.

Ans. (*a*) Refer to Chapter 4, Art. 4.3, pages 80-86.

(*b*) Refer to Chapter 2, Art. 2.5.4.2, pages 40-41.

Q. 3. What are biological neurons? How they help in creating ANN models? Discuss the significance of such models.

Ans. (*a*) Refer to Chapter 1, Arts. 1.5.2 and 1.3.2, pages 5, 8.

Q. 4. What do you mean by Hopfield networks? Discuss their significance and usefulness in NN.

Ans. Hopfield Network : Unlike the linear associator model which consists of two layers of processing units, one serving as the input layer while the other as the output layer, **the Hopfield model consists of a single layer of processing elements where each unit is connected to every other unit in the network other than itself.** The connection weight matrix W of this type of network is square and symmetric, *i.e.*, $w_{ij} = w_{ji}$ for $i, j = 1, 2, ..., m$. Each unit has an extra external input I_i. This extra input leads to a modification in the computation of the net input to the units:

$$input_j = \sum_{i=1}^{m} x_i w_{ij} + I_j$$

for $j = 1, 2, ..., m$.

Unlike the linear associator, the units in the Hopfield model act as both input and output units. But just like the linear associator, a single associated pattern pair is stored by computing the weight matrix as follows:

$$W_k = X_k^T Y_k$$

where $$Y_k = X_k$$

to store p different associated pattern pairs. Since the Hopfield model is an autoassociative memory model, patterns, rather than associated pattern pairs, are stored in memory.

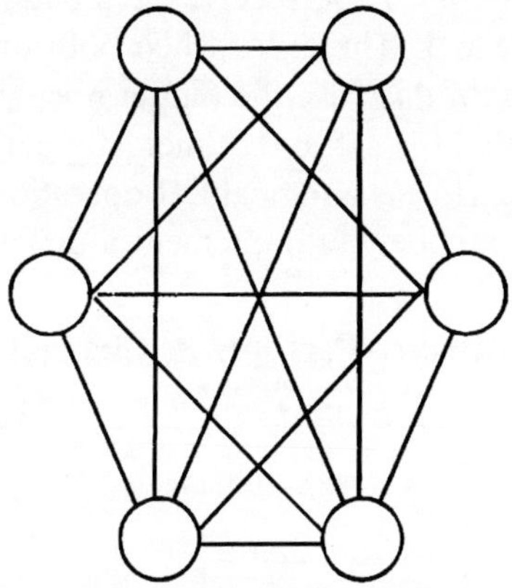

Fig. 1. Hopfield Network.

Hopfield (1982) demonstrated that the maximum number of patterns that can be stored in the Hopfield model of m nodes before the error in the retrieved pattern becomes severe is around $0.15m$.

Significance and usefulness of Hopfield network

Hopfield showed that models of physical systems could be used to solve computational problems. Such systems could be implemented in hardware by combining standard components such as capacitors and resistors.

The importance of the different Hopfield networks in practical application is limited due to theoretical limitations of the network structure but, in certain situations, they may form interesting models. Hopfield networks are typically used for classification problems with binary pattern vectors.

The Hopfield network is created by supplying input data vectors, or pattern vectors, corresponding to the different classes. These patterns are called *class patterns*. In an n-dimensional data space the class patterns should have n binary components $\{1,-1\}$; that is, each class pattern corresponds to a corner of a cube in an n-dimensional space. The network is then used to classify distorted patterns into these classes. When a distorted pattern is presented to the network, then it is associated with another pattern. If the network works properly, this associated pattern is one of the class patterns. In some cases (when the different class patterns are correlated), spurious minima can also appear. This means that some patterns are associated with patterns that are not among the pattern vectors.

Hopfield networks are sometimes called *associative networks* since they associate a class pattern to each input pattern.

Q. 5. Explain the following :

(*a*) Multilayer Networks

Ans. An MLP is a network of simple neurons called perceptrons. The basic concept of a single perceptron was introduced by Rosenblatt in 1958. A Multilayer Perceptron is a feedforward neural network with at least one hidden layer. It can deal with nonlinear classification problems.

EXAMPLE: Suppose an output node receives 2 inputs, its threshold is set to 1.5 and its input weights are both set to 1. Therefore when both the input units are active (i.e.1) the output node will be active. In this case the output node performs a logical AND operation. On the other hand if threshold is set to 0.5, then any active input can activate the node. In this case the output node performs a logical OR operation. Thus by choosing a different set of weights and threshold, a node can implement a different logical operation.

Multilayer-Perceptron Characteristics

Sample Structure	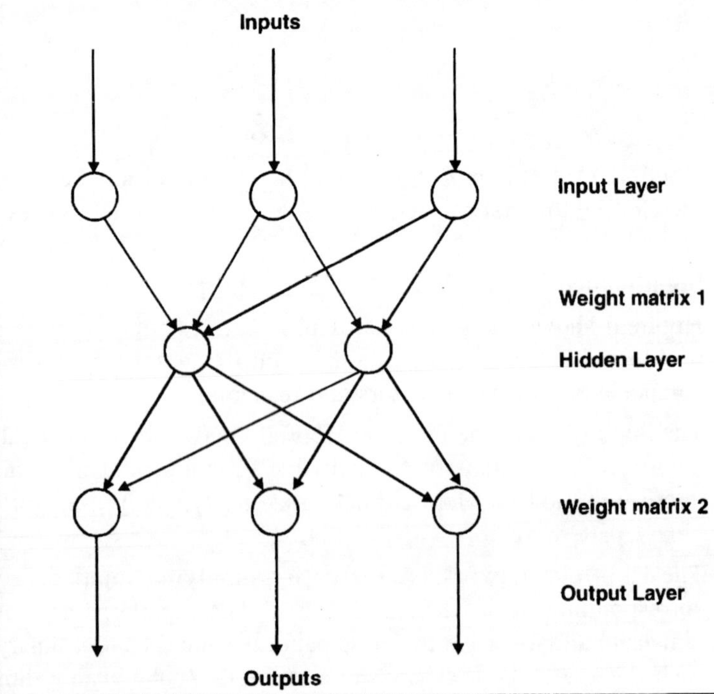
Type	Feed-forward
Neuron Layers	1 input layer 1 or more hidden layers 1 output layer
Input value types	Binary
Activation Functions	Hard limiter/sigmoid
Learning Method	Supervised
Learning algorithm	Delta learning rule Back-propagation (mostly used)

Q. 6. What are Neural networks? Discuss the applications.

Ans. (*a*) Refer to Chapter 1, Art. 1.3.1, pages 4-5 and Chapter 1, Art. 1.6, pages 10-13.

Q. 7. Explain the following :

(*a*) Linear Separability

Ans. A Single Layer Perceptron consists of an input layer and an output layer. The activation function applied is hard-limiting function as discussed in chapter 2. An output unit will assume the value 1 if the sum of weighted inputs is greater than its threshold. *i.e.*

$W_{ji}X_i > \theta_j$ where W_{ji} is weight from unit i to unit j.

X_i is the input from unit i and Δ_j is the threshold on unit j

Let there are 2 classes A and B. If $W_{ji}X_i > \theta_j$ then object will be classified as Class A otherwise Class B.

Suppose there are n inputs then the equation $W_{ji}X_i = \theta_j$ where $i = 1, 2 \ldots n$ forms a hyperplane, dividing the space in 2 halves.

Linear Separability : When a linear hyperplane exists to place the instances of one class on one side and those of other class on the other side of plane. Then this is called Linear Separability. For example in case of OR gate we can draw a linear hyperplane to separate the inputs whose corresponding outputs are 1 on one side and the inputs whose corresponding outputs are 0 on other side of the plane.

TABLE 1. OR Gate.

Input *A*	Input *B*	Output
0	0	0
0	1	1
1	0	1
1	1	1

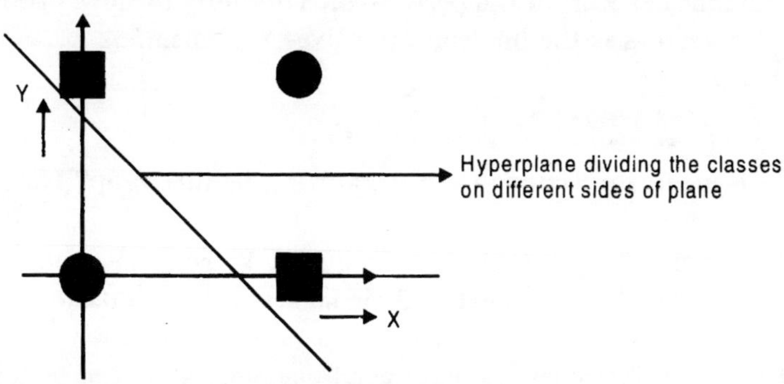

Fig. 2. The OR Gate.

But all the classification problems are not linear separable. For example the EX-OR problem is not linear separable because EX-OR is non equality gate.

TABLE 2. The EX-OR Gate.

Input *A*	Input *B*	Output
0	0	0
0	1	1
1	0	1
1	1	0

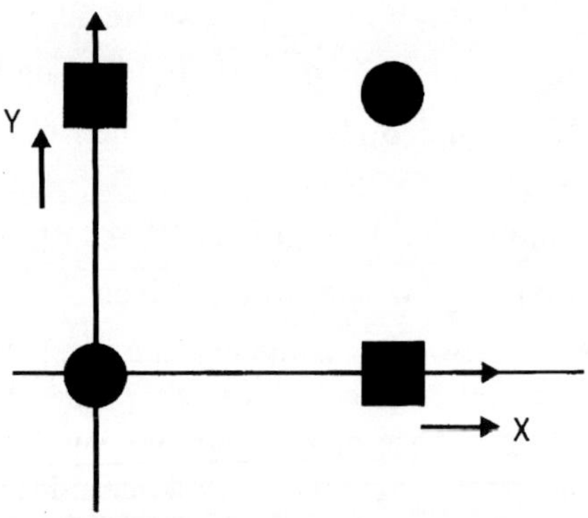

Fig. 3. The EX-OR Function.

Limitations : Perceptron networks have several limitations.

First, the output values of a perceptron can take on only one of two values (True or False). Second, perceptrons can only classify linearly separable sets of vectors. If a straight line or plane can be drawn to separate the input vectors into their correct categories, the input vectors are linearly separable and the perceptron will find the solution. If the vectors are not linearly separable learning will never reach a point where all vectors are classified properly.

The most famous example of the perceptron's inability to solve problems with linearly nonseparable vectors is the boolean exclusive-or problem.

Q. 7. (*b*) Delta Rule.

Ans. Refer to Chapter 2, Art. 2.5.4.3, pages 42-43.

Q. 8. What do you mean by Kohonen self-organizing feature map? How are these useful?

Ans. Kohonen Network :

The **Self-Organizing Map (SOM)**, commonly also known as **Kohonen** network (Kohonen 1982, Kohonen 2001) is a computational method for the visualization and analysis of high-dimensional data, especially experimentally acquired information.

The Self-Organizing Map defines an ordered mapping, a kind of projection from a set of given data items onto a regular, usually two-dimensional grid. A *model* m_i is associated with each grid node. These models are computed by the SOM algorithm. A data item will be mapped into the node whose model which is most similar to the data item, e.g., has the smallest distance from the data item in some metric.

(1) It consists of single layer of nodes plus an input layer.

(2) Each layer receives inputs from environment and from other nodes within the layer.

(3) Important point about this network is that both weight vector and input vectors are normalized to a constant value before learning

$$\hat{W}_i = \frac{(W_i)}{|W_i|} \qquad \text{for } i = 1, 2, \ldots P$$

(4) Each node computes by taking the dot product of its weight vector and input vector. $O_j = X.W_j$ where O_j is output or activation level of unit j.

(5) Node with the largest activation level is declared the winner in the competition. This **winning** node is the only node that will generate an output signal, the activation level of other neurons is suppressed to zero. The Winning node with largest net_i is allowed to change the weight that is weight adjustment is for winning neuron only while weights of other neurons remain unaffected. The weight adjustment criteria for this mode of training is such that

$$|X - W_m| = \min\{|X - W_i|\} \qquad \ldots(iii)$$

where $i = 1, 2, \ldots P$ and m is winning neuron number.

min represents the minimum. As explained earlier also in clustering that smaller the distance more close the patterns are and finding the minimum of P distances corresponds to finding maximum among the p scalar products.

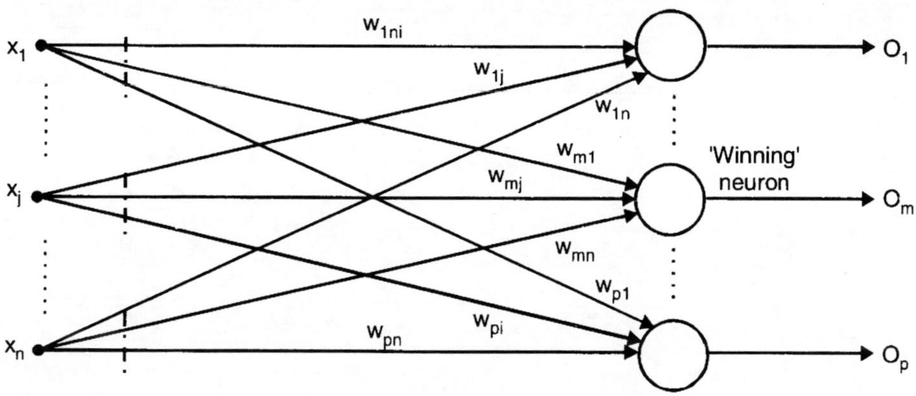

Fig. 4.

$$W_m^t X = \max\,(W_i^t)X \qquad \text{where } i = 1, 2, \ldots, P$$

The left hand side of above equation can be written as

$$|X - W_m| = (X^t X - 2W_m^t X - 1)^{1/2} \qquad \ldots(iv)$$

It is clear that searching for the minimum of p distances as on right hand side of (iii) Corresponds to finding maximum among p scalar products

$$W_m^t X = \max(W_i^t X) \qquad \ldots(v)$$

For $i = 1, 2, \ldots, p$

After the winning neuron has been identified and declared a winner, its weights must be adjusted so that the distance is reduced in the current training step. Thus $|X - W_m|$ must be reduced, preferably along the gradient direction in the weight space $w_{m1}, w_{m2}, \ldots, w_{mn}$

$$\nabla w_m \, \|x - w_m\|^2 \;=\; -2(x - w_m) \qquad \text{...}(vi)$$

It seems reasonable to reward the weights of the $X - W_m$ winning neuron with an increment of weight in the negative gradient direction, thus in the direction x - w_m. We thus have

$$\Delta\hat{w}'_m \;=\; \alpha(x - \Delta\hat{w}_m) \qquad \text{...}(vii)$$

where α is a small learning constant selected heuristically, usually between 0.1 and 0.7.

Learning according to equation (*vi*) and (*vii*) is called Winner –take –All learning and it is a common competitive and unsupervised learning technique. The winning neuron with the largest net is rewarded with a weight adjustment, while the weights of others remain unaffected.

Another modification of the winner-take-all learning rule is that both the winners' and losers' weights are adjusted in proportion to their level of responses. This may be called leaky competitive learning and should provide more subtle learning in the cases for which clusters may be hard to distinguish.

B.E.
7TH SEMESTER EXAMINATION, DEC. 2007

NEURAL NETWORKS

Q. 1. With the help of suitable diagram explain the structure of biological neurons? Also show how information flows in neural systems.

Ans. (*a*) Refer to Chapter 1, Art. 1.5.2, pages 8-9; Chapter 1, Art. 1.5.1, pages 7-8.

Q. 2. Discuss the following learning rules:

(*a*) Widrow-Hoff learning rule

(*b*) Correction learning rule

Ans. (*a*) Refer to Chapter 2, Art. 2.5.4.4, pages 45-46.

(*b*) Refer to Chapter 2, Art. 2.5.4.5, page 46.

Q. 3. Write and explain the "Winner-Take-All" algorithm for cluster classifications? What are its separability limitations ?

Ans. Clustering : Clustering is understood to be the grouping of similar objects and separating of dissimilar ones. The objective of clustering neural networks discussed in this section is to categorize or cluster data. The classes must first be found from the correlations of an input data stream. Since the network actually deals with unlabeled data, the clustering should be followed by labeling clusters with appropriate category names or numbers. This process of providing the category of objects with a label is usually termed as ***calibration.***

WINNER-TAKE-ALL-LEARNING (Kohonen Network)

(1) It consists of single layer of nodes plus an input layer.

(2) Each layer receives inputs from environment and from other nodes within the layer.

(3) Important point about this network is that both weight vector and input vectors are normalized to a constant value before learning

$$\hat{W}_i = \frac{(W_i)}{|W_i|} \qquad \text{for } i = 1, 2, \ldots P$$

(4) Each node computes by taking the dot product of its weight vector and input vector. $O_j = X.W_j$ where O_j is output or activation level of unit j.

(5) Node with the largest activation level is declared the winner in the competition. This **winning** node is the only node that will generate an output signal, the activation level of other neurons is suppressed to zero. The Winning node with largest net_i is allowed to change the weight that is weight adjustment is for winning neuron only while weights of other neurons remain unaffected. The weight adjustment criteria for this mode of training is such that

$$|X - W_m| = \min\{|X - W_i|\} \qquad \ldots(iii)$$

where $i = 1, 2, \ldots P$ and m is winning neuron no.

min represents the minimum. As explained earlier also in clustering that smaller the distance more close the patterns are and finding the minimum of P distances corresponds to finding maximum among the p scalar products.

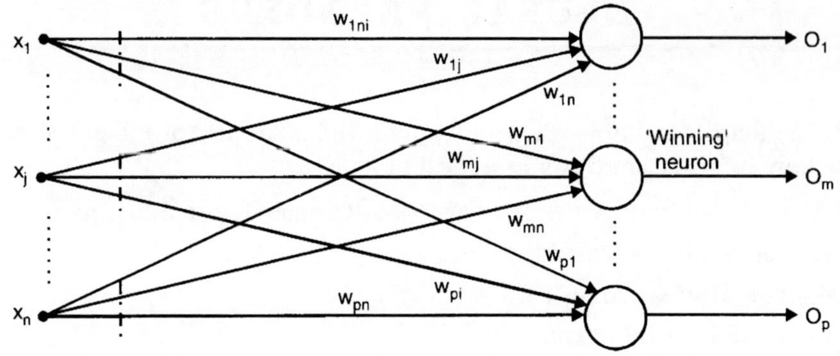

Fig. 1.

$$W_m^t X = \max (W_i^t)X \quad \text{where } i = 1, 2, ..., P$$

The left hand side of above equation can be written as

$$|X - W_m| = (X^t X - 2W_m^t X - 1)^{1/2} \qquad ...(iv)$$

It is clear that searching for the minimum of p distances as on right hand side of (*iii*) Corresponds to finding maximum among p scalar products

$$W_m^t X = \max(W_i^t X) \qquad ...(v)$$

For $i = 1, 2, ..., p$

After the winning neuron has been identified and declared a winner, its weights must be adjusted so that the distance is reduced in the current training step. Thus $|X - W_m|$ must be reduced, preferably along the gradient direction in the weight space w_{m1}, $w_{m2}, ..., w_{mn}$

$$\nabla w_m \|x - w_m\|^2 = -2(x - w_m) \qquad ...(vi)$$

It seems reasonable to reward the weights of the $X - W_m$ winning neuron with an increment of weight in the negative gradient direction, thus in the direction $x - w_m$. We thus have

$$\Delta\hat{w}_m' = \alpha(x - \Delta\hat{w}_m) \qquad ...(vii)$$

where α is a small learning constant selected heuristically, usually between 0.1 and 0.7.

Learning according to equation (*vi*) and (*vii*) is called Winner –take –All learning and it is a common competitive and unsupervised learning technique. The winning neuron with the largest net is rewarded with a weight adjustment, while the weights of others remain unaffected.

Another modification of the winner-take-all learning rule is that both the winners' and losers' weights are adjusted in proportion to their level of responses. This may be called leaky competitive learning and should provide more subtle learning in the cases for which clusters may be hard to distinguish.

Separability Limitations : A Single Layer Perceptron consists of an input layer and an output layer. The activation function applied is hard-limiting function as discussed in chapter 2. An output unit will assume the value 1 if the sum of weighted inputs is greater than its threshold. *i.e.*

$$\Sigma W_{ji} X_i > \theta_j$$

where W_{ji} is weight from unit i to unit j.

X_i is the input from unit i and θ_j is the threshold on unit j

Let there are 2 classes A and B. If $\theta W_{ji} X_i > \theta_j$ then object will be classified as Class A otherwise Class B.

Suppose there are n inputs then the equation $\Sigma W_{ji} X_i = \theta_j$ where $i = 1, 2 \ldots n$ forms a hyperplane, dividing the space in 2 halves.

Linear Separability: When a linear hyperplane exists to place the instances of one class on one side and those of other class on the other side of plane. Then this is called Linear Separability. For example in case of OR gate we can draw a linear hyperplane to separate the inputs whose corresponding outputs are 1 on one side and the inputs whose corresponding outputs are 0 on other side of the plane.

TABLE 1. OR Gate.

Input *A*	Input *B*	Output
0	0	0
0	1	1
1	0	1
1	1	1

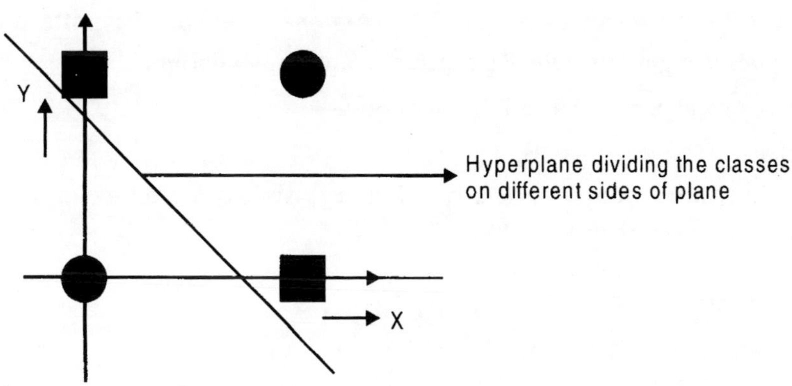

Fig. 2. The OR Gate.

But all the classification problems are not linear separable. For example the EX-OR problem is not linear separable because EX-OR is non equality gate,

TABLE 2. The EX-OR Gate.

Input *A*	Input *B*	Output
0	0	0
0	1	1
1	0	1
1	1	0

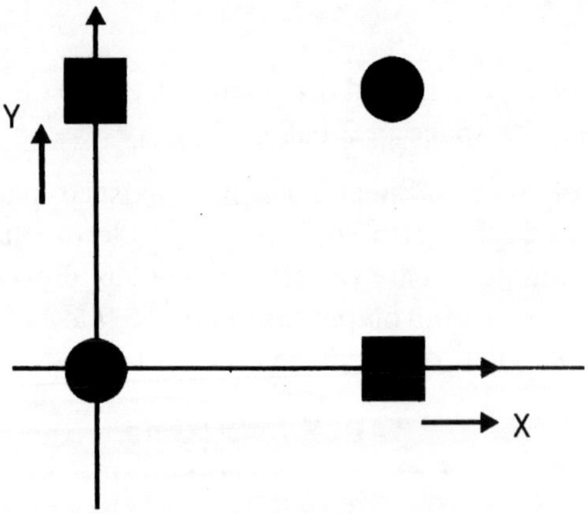

Fig. 3. The EX-OR Function.

In this case we can't draw any hyperplane that divides the space so that different classes are on different sides. To cope up with a problem which is not linear separable, a Multilayer Perceptron is required.

Q. 4. (*a*) What do you mean by pattern? Discuss one technique for pattern classification.

(*b*) Explain Single Discrete Perceptron training algorithm.

Ans. (*a*) Refer to Chapter 3, Art. 3.1.1, pages 56-60.

(*b*) Refer to Chapter 3, Art. 3.4, pages 61-63.

Q. 5. For BAM (Bidirectional Associative Memory) discuss the following :

(*a*) Memory Architecture

(*b*) Association Encoding and Decoding.

Ans. Bidirectional Associative Memory (BAM)

The network structure of BAM is similar to that of linear associator but connections are bidirectional that is

$$w_{ij} = w_{ji} \qquad \text{for} \begin{cases} i = 1, 2, ..., m \\ j = 1, 2, ..., n \end{cases}$$

i.e., BAM allows **forward and backward** flow of information between the layers. In BAM the units in both layers serve as both input and output units depending on the direction of propagation.

Propagating the signals from X layer to Y layer makes the units in X layer to act as input units while units in Y layer act as output units. Same is true for other direction because memory is bidirectional.

BAM model can perform both auto and hetero associative recall of stored information.

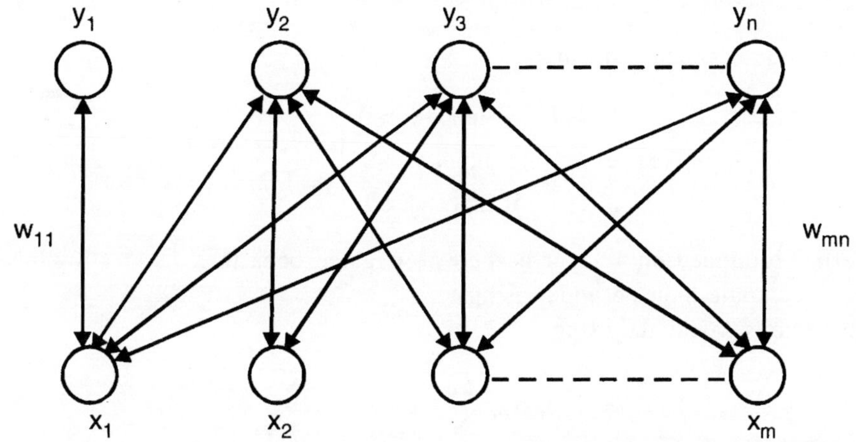

Fig. 4. BAM model (arrows are bidirectional).

Like is linear associator and hepfield network encoding in BAM can be carried out by using

$$w_k = x_K^T y_K \quad \text{(to store a single associated pattern pair)}$$

and

$$w = \alpha \sum_{K=1}^{P} w_K \quad \text{(to store simultaneously several associated pattern pairs)}$$

This process is generally called storage or encoding.

After encoding, network can be used for decoding. The decoding process is as follows :

1. An input pattern can be applied either on X layer or on the Y layer.
2. When input pattern is applied, network will propagate the input pattern to other layer allowing the units in other layer to compute their output values.
3. Pattern that was produced by other layer is then propagated back to original layer and units in the original layer compute the output values. New pattern that was produced of the original layer is again propagated to other layer.
4. The process is repeated until further propagations and computations do not result in a change in the associates of units in both layers where final pattern pair is one of the stored associated pattern pairs.
5. Final pattern pair that will be produced by the network depends on initial pattern pair and connection weight matrix.

BAM also uses Hebb's learning rule to build the connection weight matrix to store the associated pattern pairs.

Discrete Bidirectional Associative Memory

Network propagates an input pattern X to Y layer and Y layer compute the net input using :

Input value is calculated using

$$y_j = \sum_{i=1}^{m} x_i \, w_{ij}$$

Output value is calculated using

$$y_i(t+1) = \begin{cases} +1 & \text{if input } y_i > 0 \\ y_i(t) & \text{if input } y_i = 0 \\ -1 & \text{if input } y_i < 0 \end{cases} \quad \begin{cases} i = 1, 2, \ldots., m \\ j = 1, 2, \ldots., n \end{cases}$$

pattern Y produced on Y layer is then propagated back to X layer allowing units in X-layer to compute their net input using :

Input value is calculated using

$$x_j = \sum_{i=1}^{n} y_i \, w_{ij}$$

Output value is calculated using

$$x_i(t+1) = \begin{cases} +1 & \text{if input } x_i > 0 \\ x_i(t) & \text{if input } x_i = 0 \\ -1 & \text{if input } x_i < 0 \end{cases}$$

Energy function characterize the state of discrete Bidirectional Associative Memory and given by :

$$E = -\sum_{i=1}^{m} \sum_{j=1}^{n} x_i \, w_{ij} \, y_j$$

Energy of discrete BAM is bounded and given below

$$E = -\sum_{i=1}^{m} \sum_{j=1}^{n} \left| w_{ij} \right|$$

which guarantees that the network will converge to a local minimum corresponding to a stored associated pattern pair.

Continuous BAM : In this, the units use signoid or hyperbolic tangent output function. The units in X-layer have an extra external input I_i while units in Y-layer have an extra external input J_j for i = 1, 2, ..., m and j = 1, 2, ..., n

These extra input lead to a modification in the computation of net input to units :

$$\text{Input } x_i = \sum_{j=1}^{n} y_i \, w_{ij} + I_i$$

$$\text{Input } y_i = \sum_{j=1}^{m} x_i w_{ij} + j_i$$

Energy function for continous BAM is given by

$$E = \sum_{i=1}^{m} \sum_{j=1}^{n} f(x_i) f(y_j) w_{ij} - \sum_{i=1}^{m} f(x_i) I_i - \sum_{j=1}^{n} f(y_j) J_j$$

$$+ \sum_{i=1}^{m} \int_0^{x_i} f(x_i) x_i \, dx_i + \sum_{j=1}^{n} \int_0^{y_i} f(y_j) y_j \, dy_j$$

Recurrent Autoassociative Memory : In order to produce an improved association there is a need for suppressing the output noise at memory output in linear associative memory. This can be done by thresholding the output and by recycling of the output to input.

The repetitive process of recycling of the output to the input through the network can be performed by a **Recurrent Neural Network (RNN).**

RNN is similar to linear associative memory. But it has feedback is dynamical.

There are two modes of operation of updation.

1. Under the asynchronous update mode, only one neuron is allowed to compute or change state at a time and then all outputs are delayed by a time ú that is produced by unity delay element in the feedback loop.
2. In synchronous mode

$$X = \begin{bmatrix} x_1 \\ x_2 \\ \vdots \\ x_n \end{bmatrix}$$

Explanation of figure → dots represents the neurons TLU → threshold logic unit and these are used because associative memories are updated in discrete time.

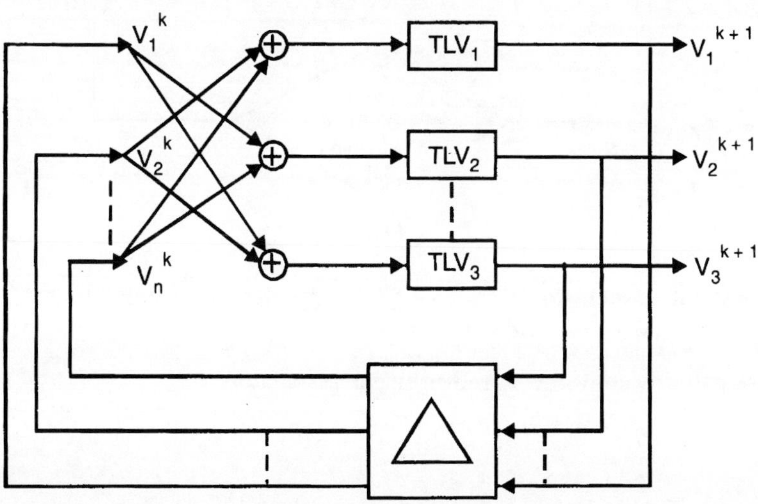

Fig. 5. Hepfield model autoassociative memory (Recurrent autoassociative memory).

Q. 6. What do you mean by attractor? Explain.

Ans. (*a*) Refer to Chapter 5, Art. 5.1, page 106.

Q. 7. Write the algorithm and draw the flowchart for "error backpropagation training algorithm" for 2 layer network.

Ans.

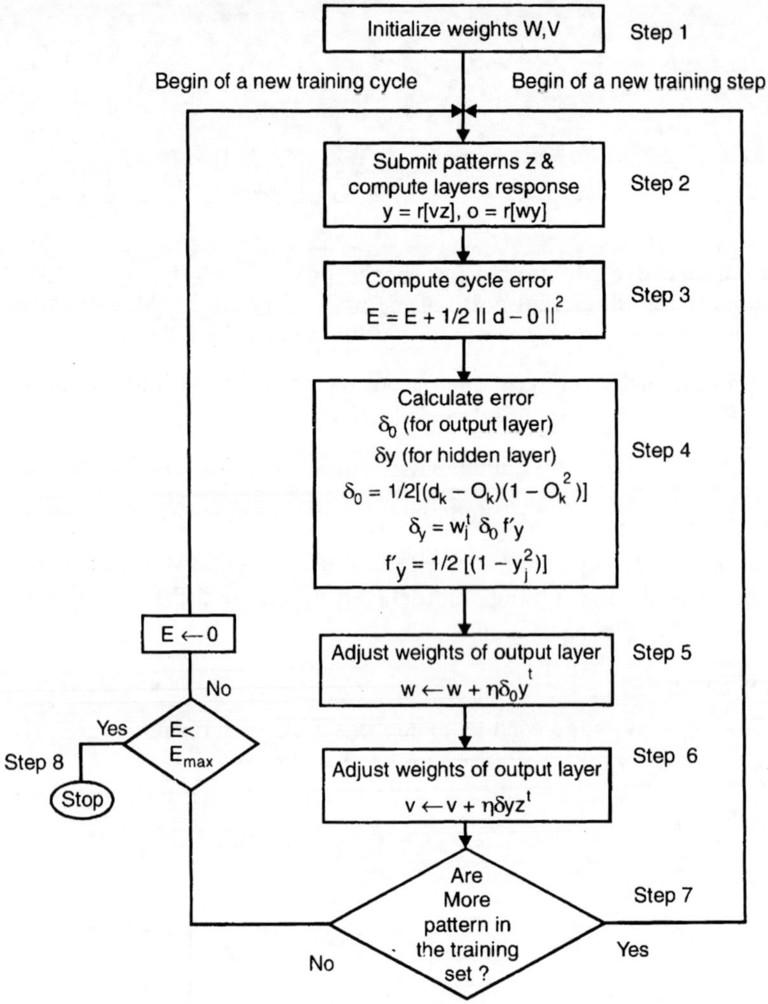

Fig. 6.

Q. 8. Prove that the association $S^{(p+1)} \rightarrow f^{(p+1)}$ results in no noise term.

Ans. The patterns are $\{S^{(P+1)}, f^{(P+1)}\}$

Let these patterns are to be stored in linear associator.

$$\therefore \quad S^{(P+1)} = \{S_1^{(P+1)} S_2^{(P+1)} - S_n^{(P+1)}\}^t$$

$$f^{(P+1)} = \left\{f_1^{(P+1)} f_2^{(P+1)} \dots f_m^{(P+1)}\right\}^t$$

For linear associative mapping, an input pattern X is presented and mapped to the output by simple rule

$$V = WX$$

or $$f^{(P+1)} + \eta^{(P+1)} = WS^{(P+1)}$$

or using the mapping symbol

$$S^{(P+1)} \rightarrow f^{(P+1)} + \underset{\substack{\uparrow \\ \text{noise term and length of} \\ \text{noise term vector is minimized}}}{\eta^{(P+1)}}$$

The weight update rule for ith output node and jth input node can be expressed as

$$w'_{ij} = w_{ij} + f_i S_j \quad \text{for } i = 1, 2, ..., m, \ j = 1, 2, ... n$$

or $$w'_{(P+1)} = w_{(P+1)_j} + f_{(P+1)} S_j$$

or $$W' = W + fS^t$$

Let initializing weights in unbiased position $W_0 = 0$

∴ By outer product learning rule

$$W' = f^{(P+1)} S^{(P+1)t}$$

or $$W' = FS^t$$

Now recall of vector $f^{(P+1)}$ when S(P + 1) is applied as stimuli.

$$V = \left(\sum_{i=1}^{P+1} f^{(P+1)} S^{(P+1)t} \right) S^{(j)}$$

Expanding the formula

$$V = f^{(1)} S^{(1)t} S^{(j)} + ... + f^{(P+1)} S^{(P+1)t} S^{(j)}$$

According to ideal mapping $S^{(j)} \rightarrow f^{(i)}$, no noise term is present

∴ $$V = f^{(j)}$$

or $$S^{(P+1)t} S^{(j)} = 0 \ \text{ for } P + 1 \neq j$$

$$S^{(j)t} S^{(j)} = 1$$

Thus the orthogonal set of P input stimuli vectors $\{S^{(P+1)}, S^{(P+2)}, ...\}$ ensures **perfect mapping.**

B.E.
7TH SEMESTER EXAMINATION, MAY 2008
NEURAL NETWORKS

Q. 1. What is Kohonen Self Organizing network? How is it useful? Illustrate through examples.

Ans. Kohonen Network :

The **Self-Organizing Map (SOM)**, commonly also known as **Kohonen** network (Kohonen 1982, Kohonen 2001) is a computational method for the visualization and analysis of high-dimensional data, especially experimentally acquired information.

The Self-Organizing Map defines an ordered mapping, a kind of projection from a set of given data items onto a regular, usually two-dimensional grid. A *model* m_i is associated with each grid node. These models are computed by the SOM algorithm. A data item will be mapped into the node whose model which is most similar to the data item, e.g., has the smallest distance from the data item in some metric.

(1) It consists of single layer of nodes plus an input layer.

(2) Each layer receives inputs from environment and from other nodes within the layer.

(3) Important point about this network is that both weight vector and input vectors are normalized to a constant value before learning

$$\hat{W}_i = \frac{(W_i)}{|W_i|} \qquad \text{for } i = 1, 2, \ldots P$$

(4) Each node computes by taking the dot product of its weight vector and input vector. $O_j = X.W_j$ where O_j is output or activation level of unit j.

(5) Node with the largest activation level is declared the winner in the competition. This **winning** node is the only node that will generate an output signal, the activation level of other neurons is suppressed to zero. The Winning node with largest net_i is allowed to change the weight that is weight adjustment is for winning neuron only while weights of other neurons remain unaffected. The weight adjustment criteria for this mode of training is such that

$$|X - W_m| = \min\{|X - W_i|\} \qquad \ldots(iii)$$

where $i = 1, 2, \ldots P$ and m is winning neuron no.

min represents the minimum. As explained earlier also in clustering that smaller the distance more close the patterns are and finding the minimum of P distances corresponds to finding maximum among the p scalar products.

The left hand side of above equation can be written as

$$|X - W_m| = (X^t X - 2W_m^t X - 1)^{1/2} \qquad \ldots(iv)$$

It is clear that searching for the minimum of p distances as on right hand side of (*iii*) Corresponds to finding maximum among p scalar products

$$W_m{}^t X = \max(W_i{}^t X) \qquad \ldots(v)$$

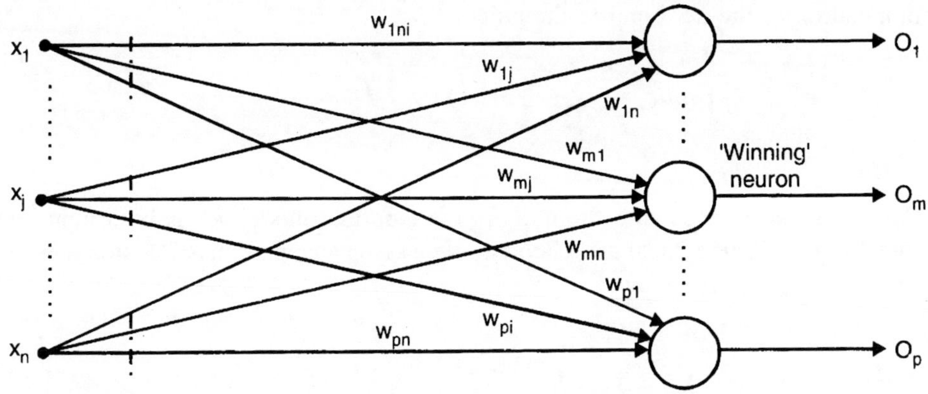

Fig. 1.

$$W_m^t X = \max (W_i^t)X \quad \text{where } i = 1, 2, ..., P$$

For $i = 1, 2, ..., p$

After the winning neuron has been identified and declared a winner, its weights must be adjusted so that the distance is reduced in the current training step. Thus $|X - W_m|$ must be reduced, preferably along the gradient direction in the weight space $w_{m1}, w_{m2}, ..., w_{mn}$

$$\nabla w_m \|x - w_m\|^2 = -2(x - w_m) \quad ...(vi)$$

It seems reasonable to reward the weights of the $X - W_m$ winning neuron with an increment of weight in the negative gradient direction, thus in the direction $x - w_m$. We thus have

$$\Delta\hat{w}'_m = \alpha(x - \Delta\hat{w}_m) \quad ...(vii)$$

where α is a small learning constant selected heuristically, usually between 0.1 and 0.7.

Learning according to equation (*vi*) and (*vii*) is called Winner –take –All learning and it is a common competitive and unsupervised learning technique. The winning neuron with the largest net is rewarded with a weight adjustment, while the weights of others remain unaffected.

Another modification of the winner-take-all learning rule is that both the winners' and losers' weights are adjusted in proportion to their level of responses. This may be called leaky competitive learning and should provide more subtle learning in the cases for which clusters may be hard to distinguish.

Q. 2. Explain the following:

(*a*) Hopfield Network

Ans. Hopfield Network : Unlike the linear associator model which consists of two layers of processing units, one serving as the input layer while the other as the output layer, **the Hopfield model consists of a single layer of processing elements where each unit is connected to every other unit in the network other than itself.** The connection weight matrix W of this type of network is square and symmetric, *i.e.*, $w_{ij} = w_{ji}$ for $i, j = 1, 2, ...,$

m. Each unit has an extra external input I_i. This extra input leads to a modification in the computation of the net input to the units:

$$input_j = \sum_{i=1}^{m} x_i w_{ij} + I_j$$

for $j = 1, 2, ..., m$.

Unlike the linear associator, the units in the Hopfield model act as both input and output units. But just like the linear associator, a single associated pattern pair is stored by computing the weight matrix as follows:

$$W_k = X_k^T Y_k$$

where $$Y_k = X_k$$

to store p different associated pattern pairs. Since the Hopfield model is an autoassociative memory model, patterns, rather than associated pattern pairs, are stored in memory.

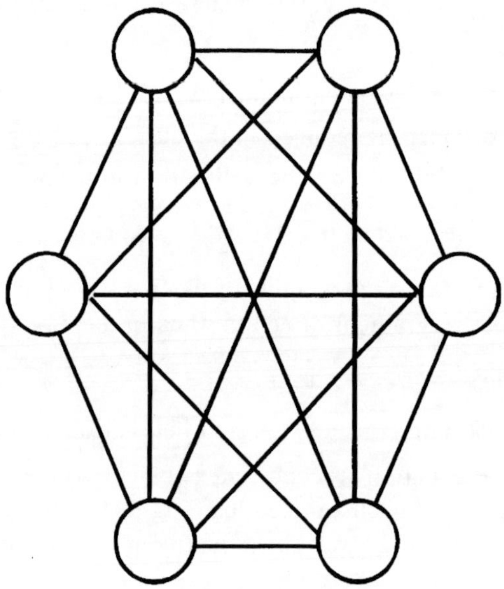

Fig. 2. Hopfield Network.

Hopfield (1982) demonstrated that the maximum number of patterns that can be stored in the Hopfield model of m nodes before the error in the retrieved pattern becomes severe is around $0.15m$.

Training of Hopfield Network

Training a Hopfield net involves lowering the energy of states that the net should "remember". This allows the net to serve as a content addressable memory system, that is to say, the network will converge to a "remembered" state if it is given only part of the state. The net can be used to recover from a distorted input the trained state that is most similar to that input. This is called associative memory because it recovers memories on the basis of similarity. For example, if we train a Hopfield net with five units so that the state (1, 0,

1, 0, 1) is an energy minimum, and we give the network the state (1, 0, 0, 0, 1) it will converge to (1, 0, 1, 0, 1). Thus, the network is properly trained when the energy of states which the network should remember are local minima.

(*b*) Delta rule

Ans. Refer to Chapter 2, Art. 2.5.4.3, pages 42-43.

Q. 3. What do you understand by biological Neuron. How they help in creating ANN. Illustrate the significance of such models.

Ans. Refer to Chapter 1, Art. 1.5.2, pages 8-9; Chapter 1, Art. 1.3.2, pages 5-6.

Q. 4. What do you mean by linear separability? Discuss its relevance in ANN.

Ans. Refer to Chapter 4, Art. 4.1, pages 75-78.

Q. 5. Explain the following :

(*a*) Backpropagation algorithm

(*b*) Perceptron training rule

Ans. (*a*) Refer to Chapter 4, Art. 4.3, pages 80-86.

(*b*) Refer to Chapter 2, Art. 2.5.4.2, pages 40-41.

Q. 6. What are ANN? Discuss the applications.

Ans. Refer to Chapter 1, Art. 1.3.1, pages 4-5; Chapter 1, Art. 1.6, pages 10-13.

Q. 7. What is competitive learning? How it is different from conventional learning? Illustrate the significance in ANN.

Ans. Conventional learning: The process of adjusting the weights to make the network learn the relationship between the inputs and targets is called learning or training. A prescribe set of well defined rules for the solution of a learning problem is called learning algorithm. Learning algorithms differ from each other in the way in which the adjustment to a synaptic weight of a neuron is formulated. In principle, almost all learning algorithms for neural networks are iterative optimization algorithms. There are two main categories for learning, namely supervised learning and unsupervised learning.

Competitive learning. Competitive learning occurs when the processing elements compete; only the processing element yielding the strongest response to a given input can modify itself, becoming more like the input. In all cases, the final values of the weighting functions constitute the "memory" of the ANN.

Learning Materials For Anns

Learning in neural networks obviously implies that some learning material should be provided to it to learn from. In case of ANN any non-numerical data must be processed in such a way to obtain a numeric representation. A "**data set**" is a matrix containing several samples. In Ann methodology, the data set is often sub divided into "training" and "test" sets. A **training set** is a set of examples used for learning, which is to fit the parameters [*i.e.*, weights] of the classifier. A **validation set** is a set of examples used to tune the parameters [*i.e.*, architecture, not weights] of a classifier, for example, number of hidden units.

A **test set** is a set of examples used only to assess the performance [generalization] of a fully specified classifier. **Training sets** needs to be fairly large to contain all the needed information if the network is to learn the features and relationship that are important. Not only do the sets have to be large but also the training session must include a wide variety of data.

Network Validation : The primary risk in developing a model is that of over training, a situation in which the neural network starts to reproduce the noise specific to a particular sample in the training data, which may cause it to lose its ability to predict accurately. There are certain techniques that are used to avoid this; the most popular of this is ***network validation***. The testing of the network predictivity is done by reserving some of the data, which is excluded from the training data sets. The network is used to predict the outputs for these reserve data records, and the calculated outputs are compared with the observed values. If they are found to be sufficiently close, the network is considered to be sufficiently predictive and the network is said to be validated.

Generalization : Generalization is the ability of a neural network to produce reasonable responses to input patterns that are similar, but not identical, to training patterns. Once it is claimed that neural network has been trained properly or it is said that the neural network has learned the concept. Then it is expected that neural network produces the correct output for previously unknown inputs, but belonging to the same concept class. Then, the trained neural network has the capability to generalize. For example, if we have trained a neural network using several images (poses) of the leopard. If the neural network has been well trained it is expected that when one presents another image (pose) of leopard, which was not a part of training set, the neural network should be able to recognize the presented image as a image of leopard. This is what is meant by generalization.

Q. 8. Explain the following :

(*a*) Multilayer Nets

Ans. An MLP is a network of simple neurons called perceptrons. The basic concept of a single perceptron was introduced by Rosenblatt in 1958. A Multilayer Perceptron is a feedforward neural network with at least one hidden layer. It can deal with nonlinear classification problems.

EXAMPLE: Suppose an output node receives 2 inputs, its threshold is set to 1.5 and its input weights are both set to 1. Therefore when both the input units are active (*i.e.* 1) the output node will be active. In this case the output node performs a logical AND operation. On the other hand if threshold is set to 0.5, then any active input can activate the node. In this case the output node performs a logical OR operation. Thus by choosing a different set of weights and threshold, a node can implement a different logical operation.

Multilayer-Perceptron Characteristics

Sample Structure	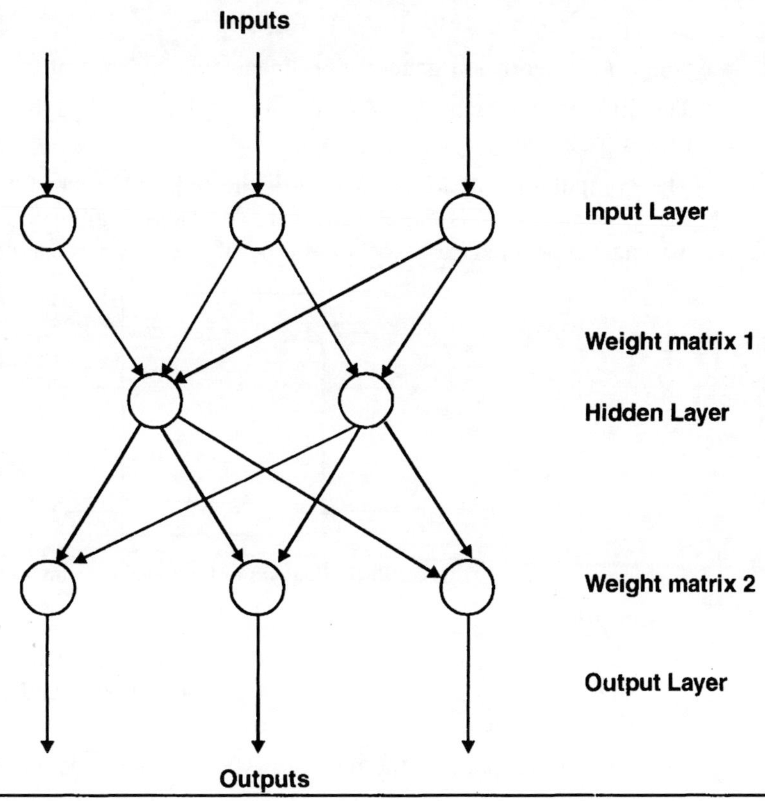
Type	Feed-forward
Neuron Layers	1 input layer 1 or more hidden layers 1 output layer
Input value types	Binary
Activation Functions	Hard limiter/sigmoid
Learning Method	Supervised
Learning algorithm	Delta learning rule Back-propagation (mostly used)

Q. 8. (*b*) Associative Memory

Ans. The block diagram of associative memory performing an associative mapping of an input vector X into an output vector V.

$$\therefore \quad V = M[X] \quad \text{...(1)}$$

where M = denotes a general non-linear diagonal operator.

The different for each type of memory model used. For dynamic memories, M also involves time variable.

The computation of M is done with the help of "recording or storage algorithm". The mapping of V to X is called retrievel. The storage algorithm depends on whether an auto associative or a hetero-associative type of memory is designed.

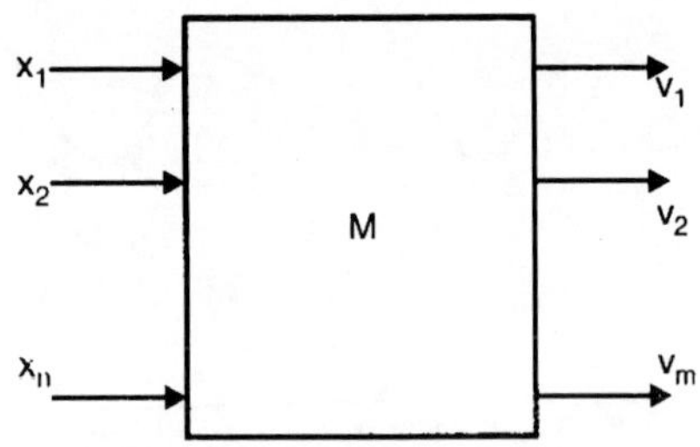

Fig. 3. Block diagram of an associative memory.

Assumption :

1. Certain prototype vector are stored in memory in such a way that once a key input has been applied, an output produced by the memory and associated with the key is the memory response.
2. Total stored pairs in the memory is P and the associations are defined as

$$X^{(i)} \rightarrow V^{(i)} \quad \text{for } i = 1, 2, ..., P \quad \text{...(2)}$$

and $$V^{(i)} \neq X^{(i)} \quad \text{for } i = 1, 2, ..., P,$$

The network can be termed as hetero associative memory.

$\because$ in hetero associative memory, there is association between pairs of 2 ordered sets of vectors $\{X^{(1)}, X^{(2)}, ..., X^{(P)}\}$ and $\{V^{(1)}, V^{(2)}, ..., V^{(P)}\}$ as explained earlier.

If $$X^{(i)} \rightarrow V^{(i)}$$
$$V^{(i)} = X^{(i)} \quad \text{for } i = 1, 2, ..., P \quad \text{...(3)}$$

Then memory is called auto associative memory.

In auto associative memory, there is association within only one set which is $\{X^{(1)}, X^{(2)}, ..., X^{(P)}\}$.

The realistic application of an auto associative mapping would be recovery of an undistorted prototype vector in response to the distorted prototype key vector.

Different Addressing Modes Which are Used for Memory Data Retrieval

1. Address - addressble memory
2. Content - addressable memory

These 2 addressing modes are commonly used for memory data retrieval. In digital computers, data can be accessed when their correct addresses in the memory are given as shown in Fig. 11.

From the figure it is clear that data have input and output lines, a word line occesses and activates the entire row of binary cells containing word data bits. This activation takes place whenever binary address is decoded by address decoder. The addressed word can be either "read" or replaced during "write" operation.

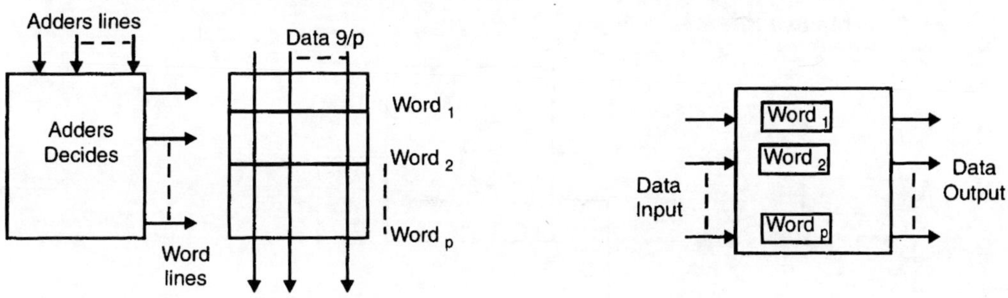

(*a*) Address-addressable memory (*b*) Content addressable memory

Fig. 4.

2nd type is CAM (content addressable memory) : In this words can be accessed based on the content of key vector. When the network excited with a portion of stored data $X(i)$, $i = 1, 2, ..., P$ the efficient response of auto associative network is the complete $X(i)$ vector. Another models of associative memory networks are :

1. Static
2. Dynamic memories

As discussed in chapter-2 the feedflow and feedback neutral networks.

Static memory menas there is one feed forward pass only and there is no delay theoretically. Static memories are also termed as non-recurrent memories. Block diagram shown in Fig. 12 and it is clear that

$$V^{(1)} = M_1[X^1] \quad \text{or} \quad V^k = M[X^k]$$

Because there is no delay.

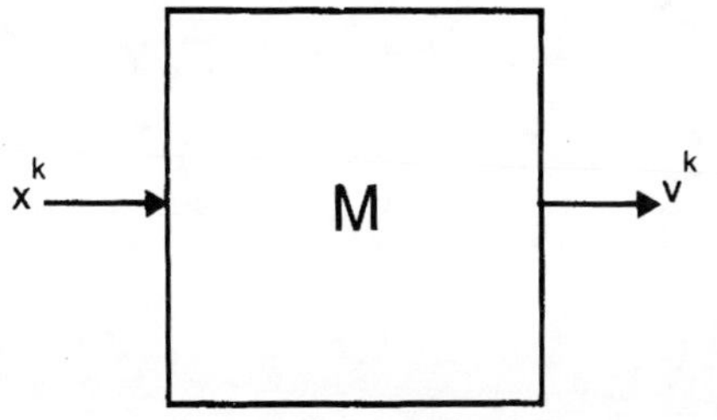

Fig. 5. Static/feedback memory.

Dynamic memory means there exists feedback property and hence time delay. Dynamic memory networks exhibit dynamic evolution means they converge to an equilibrium state according to recursive formula.

$$V^{k+1} = M_2\underbrace{[X^k, V^k]}_{\text{pair}}$$

$k + 1 \Rightarrow$Time delay of unity.

An the block diagram (0.3*b*) the delay element in the feedback loop inserts a unity delay Δ. Auto associative memory based on Hopfield model is on example of recurrent network (organic memory) for which the input X^0 is used to mitialize V^0 *i.e.* $X^0 = V^0$ and input is then removed.

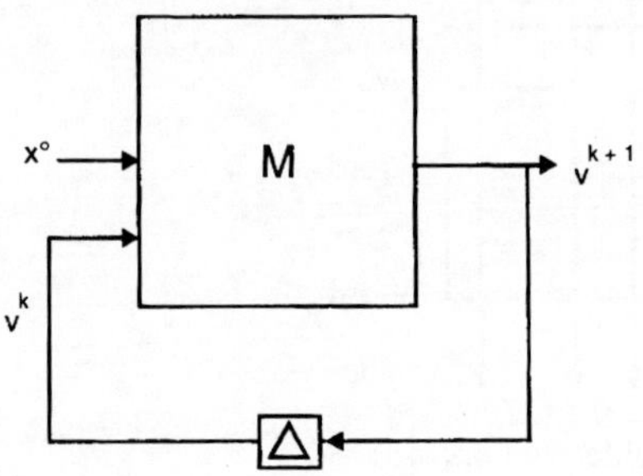

Fig. 6. Auto associative network.

For recurrent hetero-associative memory, total delay is 2Δ because the memory associates pairs of vectors $[X^{(i)}, V^{(i)}]$, $i = 1, 2, ..., P$

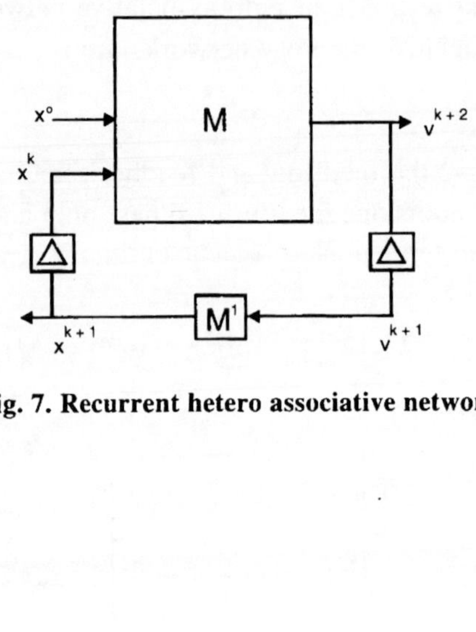

Fig. 7. Recurrent hetero associative network.

B.E.
7TH SEMESTER EXAMINATION, DEC. 2010
NEURAL NETWORKS

Q. 1. **(*a*) Describe the basic structure of biological neuron.** **(10)**

(*b*) Differentiate between supervised and unsupervised learning. **(5)**

(*c*) Describe any 2 activation functions used in ANN. **(5)**

Ans. (*a*) Refer to Chapter 1, Art. 1.5.2, page 8.

(*b*) Refer to Chapter 2, Art. 2.5, pages 33-35.

(*c*) Refer to Chapter 2, Art. 2.2.2, pages 19-22.

Q. 2. **Describe how a single perception can be used as a classifier. Draw AND and OR classification lines on a two dimensional plane and explain how a single perceptron can be used and AND and OR classification ? Also explain why a single perceptron can not be used for XOR classification.**

Ans. A Neural Network classifies a given object presented to it according to the output activation. For binary outputs, 1 corresponds to one class and 0 corresponds to the other. For continuous outputs between 0 and 1, 0.5 can be used as threshold to make a decision. In the case of multiple outputs, the object is assigned to the class corresponding to the output node with maximum activation. 2 types of classification models are discussed below:

1. Single Layer Perceptron
2. Multilayer Layer Perceptron

The Perceptron : The perceptron is a program that learns concepts, i.e. it can learn to respond with True (1) or False (0) for inputs we present to it, by repeatedly "studying" examples presented to it.

Example : For example, how do you teach a child to recognize what a chair is? You show him examples telling him "This is a chair; That one is not a chair" until the child learns the concept of what a chair is. In this stage, the child can look at the examples we have shown him and answer correctly to the question "Is this object a chair?" Furthermore, if we show to the child new objects, that he didn't see before, we could expect him to recognize correctly whether the new object is a chair or not, providing that we've given him enough positive and negative examples. This is exactly the idea behind the perceptron.

Single Layer Perceptron : A Single Layer Perceptron consists of an input layer and an output layer. The activation function applied is hard-limiting function as discussed in chapter 2. An output unit will assume the value 1 if the sum of weighted inputs is greater than its threshold. i.e.

$\Sigma W_{ji}X_i >$ θ_j where W_{ji} is weight from unit i to unit j.

X_i is the input from unit i and θ_j is the threshold on unit j

Let there are 2 classes A and B. If $\Sigma W_{ji}X_i > \theta_j$ then object will be classified as Class A otherwise Class B.

Suppose there are n inputs then the equation $\sum W_{ji} X_i = \theta_j$ where $i = 1, 2\ldots n$ forms a hyperplane, dividing the space in 2 halves.

Linear Separability : When a linear hyperplane exists to place the instances of one class on one side and those of other class on the other side of plane. Then this is called Linear Separability. For example in case of OR gate we can draw a linear hyperplane to separate the inputs whose corresponding outputs are 1 on one side and the inputs whose corresponding outputs are 0 on other side of the plane.

TABLE 1. OR Gate.

Input *A*	Input *B*	Output
0	0	0
0	1	1
1	0	1
1	1	1

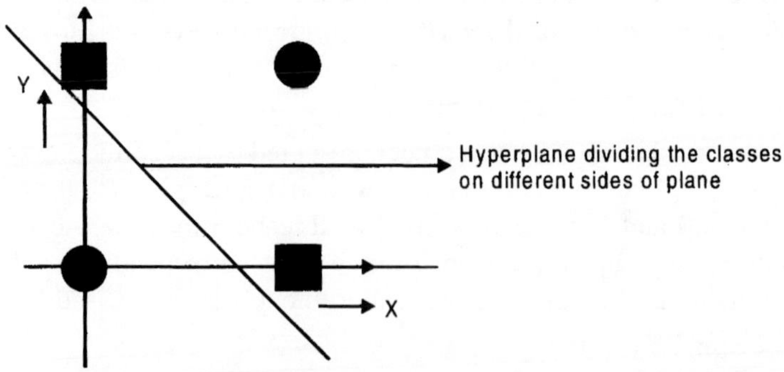

Fig. 1. The OR Gate.

But all the classification problems are not linear separable. For example the EX-OR problem is not linear separable because EX-OR is non equality gate *i.e.*

TABLE 2. The EX-OR Gate.

Input *A*	Input *B*	Output
0	0	0
0	1	1
1	0	1
1	1	0

In this case we can't draw any hyperplane that divides the space so that different classes are on different sides. To cope up with a problem which is not linear separable, a Multilayer Perceptron is required.

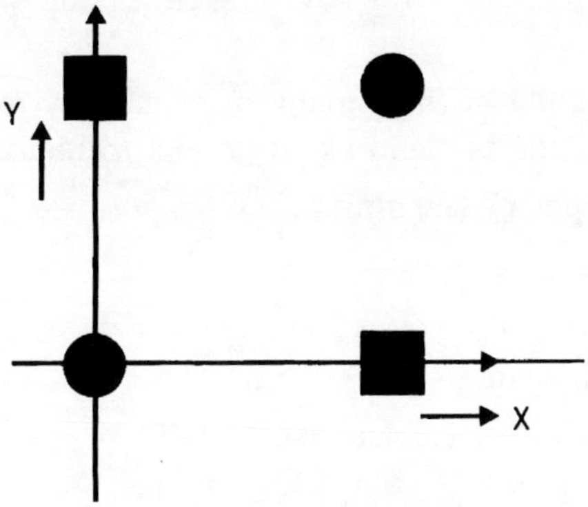

Fig. 2. The EX-OR Function.

The Perceptron Algorithm

1.Weight Initialization :

Set all the weights and node thresholds to small random numbers.

2.Calculation of Activation :

(*a*) The activation level of an input unit is determined by the instance presented to the network.

(*b*) The activation level O_j of an output unit is determined by

$$O_j = F_h(W_{ji}X_i - \theta_j) \quad (1)$$

where F_h is hard-limiting function:

$$F_h(a) = \begin{cases} 1 & \text{if } a > 0 \\ 0 & \text{Else} \end{cases}$$

3. Weight Training :

(*a*) Adjust weights by

$$W_{ji}(t + 1) = W_{ji}(t) + \Delta W_{ji} \quad ...(2)$$

where $W_{ji}(t)$ is the weight from unit i to unit j at time t and ΔW_{ji} is the weight adjustment.

(*b*) The weight change may be computed by Delta Learning Rule:

$$\Delta W_{ji} = \eta\delta_j X_i$$

where Δ is trial –independent learning rate $(0 < \eta < 1)$

and δ_j is the error at unit j and $\delta_j = T_j\text{-}O_j$

where T_j is the desired output activation and

O_j is the actual output activation at output unit j

(*c*) Repeat the iterations till convergence.

Limitations : Perceptron networks have several limitations.

First, the output values of a perceptron can take on only one of two values (True or False). Second, perceptrons can only classify linearly separable sets of vectors. If a straight line or plane can be drawn to separate the input vectors into their correct categories, the input vectors are linearly separable and the perceptron will find the solution. If the vectors are not linearly separable learning will never reach a point where all vectors are classified properly.

The most famous example of the perceptron's inability to solve problems with linearly nonseparable vectors is the **boolean exclusive-or problem.**

Q. 3. Describe following types of learning :

(*i*) Hebbian learning

(*ii*) Delta learning

(*iii*) Winner take all learning **(20)**

Ans. (*i*) Refer to Chapter 2, Art. 2.5.4.1, pages 36-37.

(*ii*) Refer to Chapter 2, Art. 2.5.4.3, pages 42-44.

(*iii*) Refer to Chapter 2, Art. 2.5.4.6, pages 46-47.

Q. 4. Explain how a neural network can be trained using back propagation. Derive equation for weight modification in a multilayer perception model. What role is played by learning rate and momentum ? **(20)**

Ans. Backpropagation network is **Multilayer feedforward network** with different transfer function in ANN and a more powerful learning rule. The learning rule is called **"Backpropagation"**, which is a kind of **gradient descent technique** with backward error propagation.

It is a **Supervised Learning** method, and is an implementation of the **Delta rule**. It requires a teacher that knows, or can calculate, the desired output for any given input.

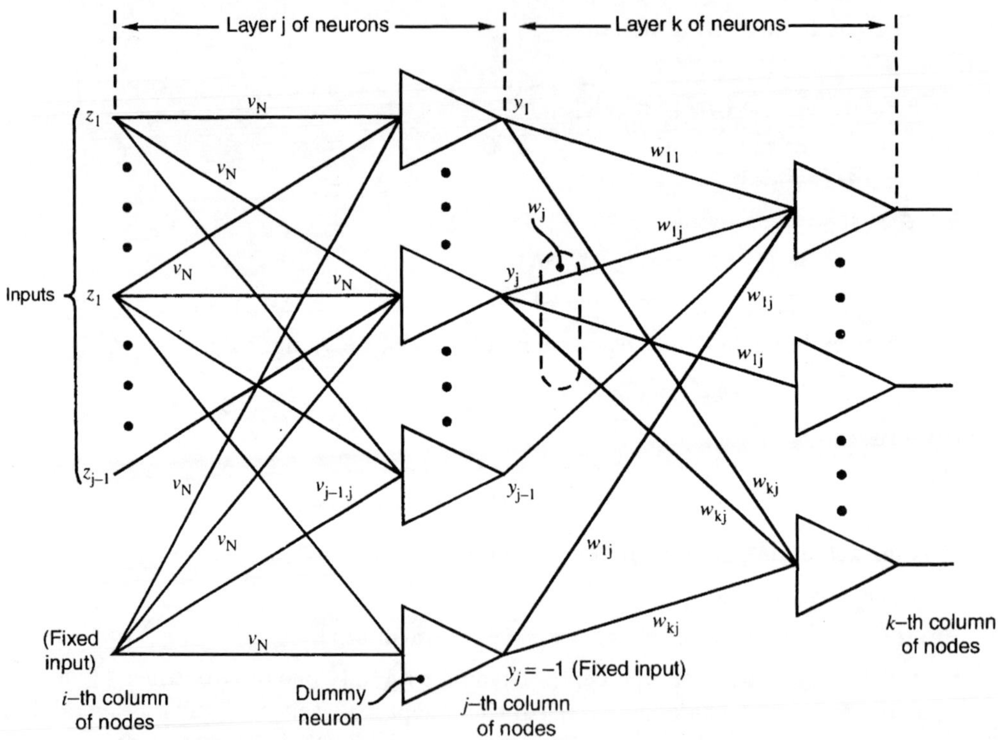

Fig. 3. Multilayer Feedforward Network.

Initialization : Set all the weights and node thresholds to small random numbers. But node threshold is the negative of the weight from the bias unit.

Calculation of Activation

1. The activation level of an input unit is determined by the instances presented to the network.
2. The activation level O_j of hidden and output unit is determined by

$$O_j = F\,(\Sigma W_{ji} O_i - \theta_j)$$

Where W_{ji} is the weight from an input O_j , θ_j is the node threshold, and F is sigmoid function.

$$F(a) = \frac{1}{(1 + e^{-a})}$$

Weight Training :

1. Start at the output units and work backward to the hidden layers recursively. Adjust weights by

$$W_{ji}(t + 1) = W_{ji}(t) + \Delta W_{ji}$$

where $W_{ji}(t)$ is the weight from unit i to unit j at time t and ΔW_{ji} is the weight adjustment.

2. The weight change is computed by $\Delta W_{ji} = \eta \delta_j O_i$

where Δ is trial –independent learning rate $(0 < \Delta < 1)$

and δ_j is the error gradient at unit j. Convergence is sometimes faster by adding a momentum term:

$$W_{ji}(t + 1) = W_{ji}(t) + \eta \delta_j O_i + \alpha[W_{ji}(t) - W_{ji}(t - 1)]$$

3. The error gradient is given by :

For the output units:

$$\delta_j = O_j(1 - O_j)(T_j - O_j)$$

Where T_j is the desired output activation and O_j isactual output activation at output unit j.

For the hidden units:

$$\delta_j = O_j(1 - O_j)\Sigma \delta_k W_{kj}$$

where δ_k is error gradient at unit k to which a connection points from hidden unit j.

4. Repeat iterations until convergence in terms of the selected error criterion. An iteration include presenting an instance, calculating activations and modifying weights.

NOTE : The name "Backpropagation" comes from the fact that the error of hidden units are derived from propagating backward the errors associated with output units since the target values for the hidden units are not given. In this activation function used is sigmoid function, which compresses the output value into the range between 0 and 1.

Why use Sigmoid Function : The sigmoid function is advantageous because it can accommodate large signals without saturation while allowing the passing of small signals without excessive attenuation. Also, it is a smooth function so that gradients can be calculated which are required for a gradient descent search.

Derivation of Backpropagation Algorithm

The negative gradient descent formula for the hidden layer is

$$\Delta v_{ji} = -\eta \frac{\partial E}{\partial v_{ji}}, \quad \text{for } j = 1, 2, ..., J \text{ and } i = 1, 2, ..., I ...(i)$$

and
$$\frac{\partial E}{\partial v_{ji}} = \frac{\partial E}{\partial (net_j)} \cdot \frac{\partial (net_j)}{\partial v_{ji}} \quad ...(ii)$$

Let us notice that the inputs to the layer are z_i, for $i = 1, 2, ..., I$. Based on relation

$$\text{net} = \text{weight} * \text{input}$$

$$\text{net}_j = v_{ji} * Z_i$$

The second term in the product (2) is equal to z_i, and we may the weight adjustment can be expressed as

$$\Delta v_{ji} = \eta \delta_{yj} z_i \quad ...(iii)$$

Where δ_{yj} is the error signal term of the hidden layer having output y. This error signal term is produced by the jth neuron of the hidden layer, where $j = 1, 2, ..., J$. The error signal term is equal to

$$\delta_{yj} \triangleq -\frac{\partial E}{\partial (net_j)}, \quad \text{for j} = 1, 2, ..., J \quad ...(iv)$$

In contrast to the output layer neurons' excitation net_k, which affected the kth neuron output only, the net, contributes now to every error component in the error sum containing ***K*** terms. The error signal terms at the node j can be computed as

$$\delta_{yj} = -\frac{\partial E}{\partial_{yj}} \cdot \frac{\partial_{yj}}{\partial (net_j)} \quad ...(v)$$

where
$$\frac{\partial E}{\partial_{yj}} = \frac{\partial}{\partial y_j}\left(\frac{1}{2}\sum_{n=-\infty}^{\infty}\{d_k - f[net_k(y)]\}^2\right) \quad ...(vi)$$

The 2nd term of above equation (v) is

$$\frac{\partial y_j}{\partial (net_j)} = f_j'(net_j) \quad ...(vii)$$

Equation (vi) will become (vii) after taking the derivative

$$\frac{\partial E}{\partial y_j} = -\sum_{n=-\infty}^{\infty}(d_k - o_k)\frac{\partial}{\partial y_j}\{f(net_k(y))\} \quad ...(viii)$$

Calculation of the derivative in braces of expression ($viii$) becomes

$$\frac{\partial E}{\partial y_j} = -\sum_{k=1}^{k}(d_k - o_k) f'(net_k)\frac{\partial (net_k)}{\partial y_j} \quad ...(ix)$$

Now after simplification of above equation, we get

$$\frac{\partial E}{\partial y_j} = -\sum_{k=1}^{K} \delta_{ok} w_{kj} \qquad \text{...}(x)$$

Combining the result from (x) and (vii)

$$\delta_{ji} = f_j'(net_j) \sum_{k=1}^{K} \delta_{ok} w_{kj}, \quad \text{for } j = 1, 2, ..., J \qquad \text{...}(xi)$$

The weight adjustment in the hidden layer now becomes

$$\Delta v_{ji} = \eta f_j'(net_j) z_i \sum_{k=1}^{K} \delta_{ok} w_{kj}, \quad \text{for } j = 1, 2, ..., J \text{ and } i = 1, 2, ..., I \qquad \text{...}(xii)$$

The equation (xii) is called generalized Delta Rule.

The adjustment of weights leading to neuron j in the hidden layer is proportional to the weighted sum of all values at the adjacent following layer of nodes connecting neuron j with the output. The weights that fan out from node j are themselves the weighting factors. The weights affecting hidden neuron have been highlighted in the output layer in Figure. The modified weights of the hidden layer can be expressed now as

$$v_{ji}' = v_{ji} + \eta f_j'(net_j) z_i \sum_{k=1}^{\infty} \delta_{ok} w_{kj}, \quad \text{for } j = 1, 2, \ldots, J \text{ and } i = 1, 2, \ldots, I \qquad \text{...}(xiii)$$

The hidden layer weight adjustment based on the generalized delta training rule. For the network can be stated in vector notation as

$$V' = V + \eta \delta_y z^t$$

where

$$z = \begin{bmatrix} z_1 \\ z_2 \\ \vdots \\ z_I \end{bmatrix}$$

$$V = \begin{bmatrix} v_{11} & v_{12} & \cdots & v_{1I} \\ v_{21} & v_{22} & \cdots & v_{2I} \\ \vdots & \vdots & \vdots & \vdots \\ v_{j1} & v_{j2} & \cdots & v_{jI} \end{bmatrix}$$

Defining now the jth column of matrix W as w_j, vector δy can be expressed compactly as follows :

$$\delta y = w_j^t \delta_0 f_y' \qquad \text{...}(xiv)$$

where f'_y is the column vector with entries f'_{yj} expressed for each hidden layer neuron 1, 2, ..., J, for unipolar and bipolar activation functions, respectively, as

$$f'_{yj} = y_j(1 - y_j)$$

$$f'_{yj} = \frac{1}{2}(1 - y_j^2)$$

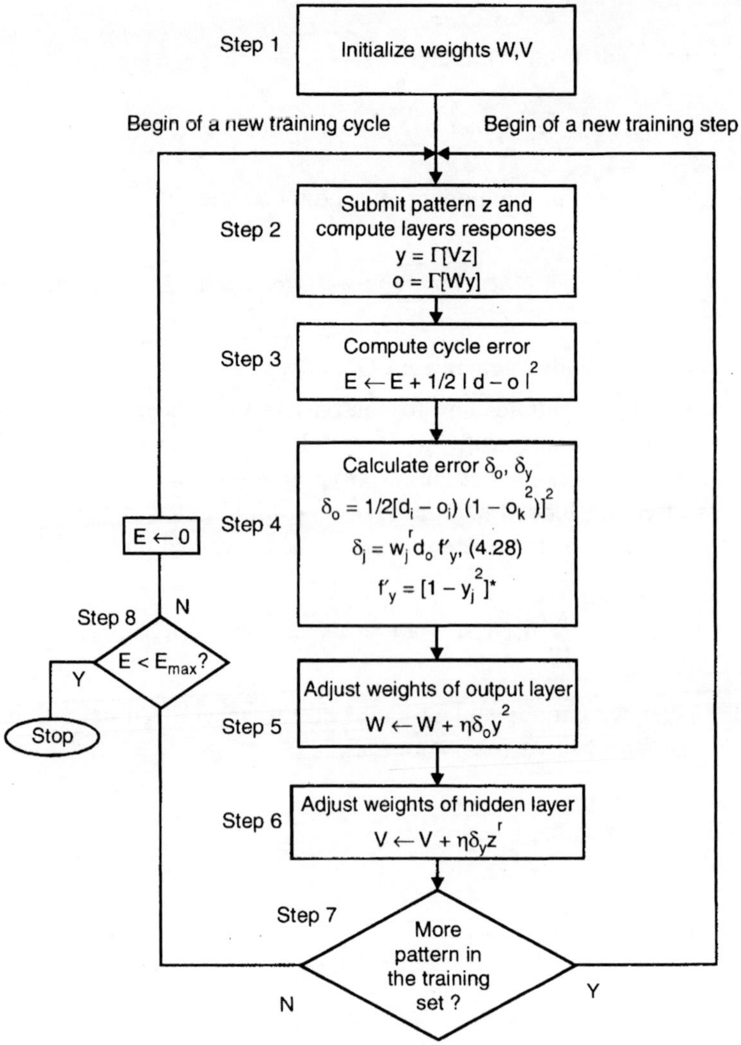

Fig. 4.

Learning Constant: Careful selection of step size (learning constant h) is often necessary to ensure the smooth convergence. However optimum value of h depends on problem being solved and there is no single learning constant value suitable for different training cases.

$$\eta = \frac{1.5}{N_1^2 + N_2^2 + \ldots + N_n^2}$$

Where N_1 is the number of patterns of type 1 and there are total n different patterns.

The learning coefficient cant be –ve because this would cause the change of weight vector to move from ideal weight vector position. If the value of η is zero then, no learning takes place and hence the value of η must be +ve. From past experiences it is clear that a large step size may cause network to become paralyzed. When network paralysis occurs, further training does little for convergence. On the other hand, if the size is too small, convergence may be too slow *i.e.* If η is too small, the algorithm will take a long time to converge (Fig. 5).

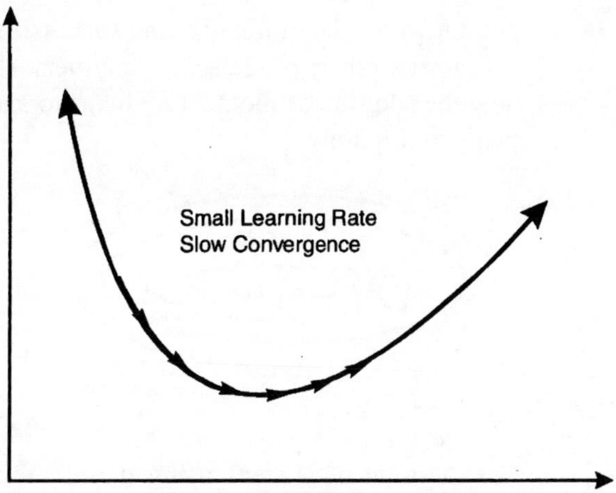

Fig. 5.

Therefore this leads to conclusion that η should be chosen experimentally for each problem.

Conversely, if η is too large, we may end up bouncing around the error surface out of control - the algorithm **diverges** (Fig. 6). This usually ends with an overflow error in the computer's floating-point arithmetic.

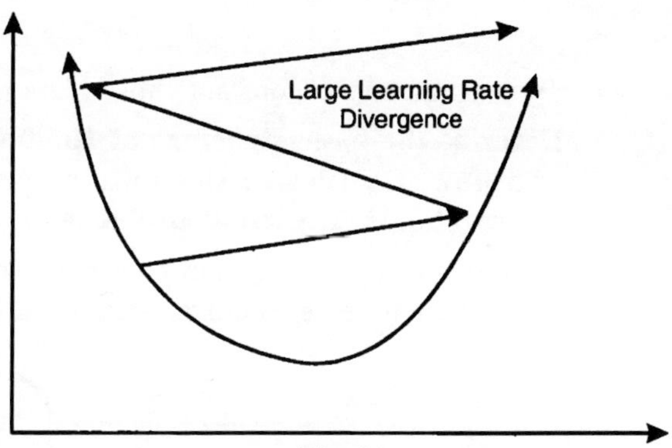

Fig. 6.

Therefore the value of learning coefficient must be between zero and one.

(3) Momentum Method : The purpose of the momentum method is to accelerate the convergence of the error back-propagation learning algorithm. The method involves supplementing the current weight adjustments with a fraction of the most recent weight adjustment. This is usually done according to the formula

$$\Sigma W(t) = (-)\delta \; \Delta E/\Delta t + \alpha \Sigma W(t-1)$$

where the arguments t and $t - 1$ are used to indicate the current and the most recent training step, respectively, and it is a user-selected positive momentum constant. The second term, indicating a scaled most recent adjustment of weights, is called the ***momentum term.*** If momentum is zero, then smoothening is minimum and the entire weight adjustment comes from newly calculated change if momentum is 1, new weight adjustment is ignored and the previous one is repeated. Between 0 and 1 is a region where the weight adjustment is smoothened by an amount proportional to the momentum factor. Typically, α is chosen between 0.1 and 0.8 to increase the speed of learning without leading to oscillations.

When the gradient keeps changing direction, momentum will smooth out the variations. This is particularly useful when the network is not well-conditioned. In such cases the error surface has substantially different curvature along different directions, leading to the formation of long narrow valleys. For most points on the surface, the gradient does not point towards the minimum, and successive steps of gradient descent can oscillate from one side to the other, progressing only very slowly to the minimum (Fig. 7). Figure 7 shows how the addition of momentum helps to speed up convergence to the minimum by damping these oscillations.

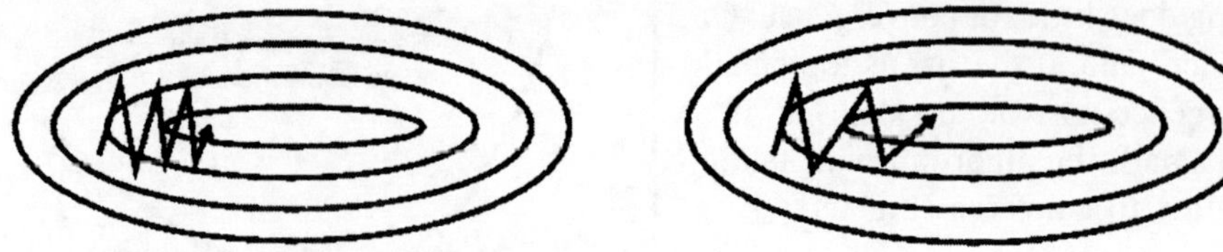

Fig. 7

For the total of N steps using the momentum method, the current weight change can be expressed as :

$$\Delta w(t) = -c\sum_{n=0}^{N} \alpha^n \nabla E(t-n)$$

Typical value of α constant chosen less than unity

Q. 5. Describe the basic structure of Hopfield network. Explain the dynamical behaviour and how it settles in the state of lowest (or nearly lowest) energy. Explain the interpretation in various problems. (20)

Ans. A **Hopfield net** is a form of recurrent artificial neural network invented by John Hopfield. Hopfield nets serve as content-addressable memory systems with binary threshold units

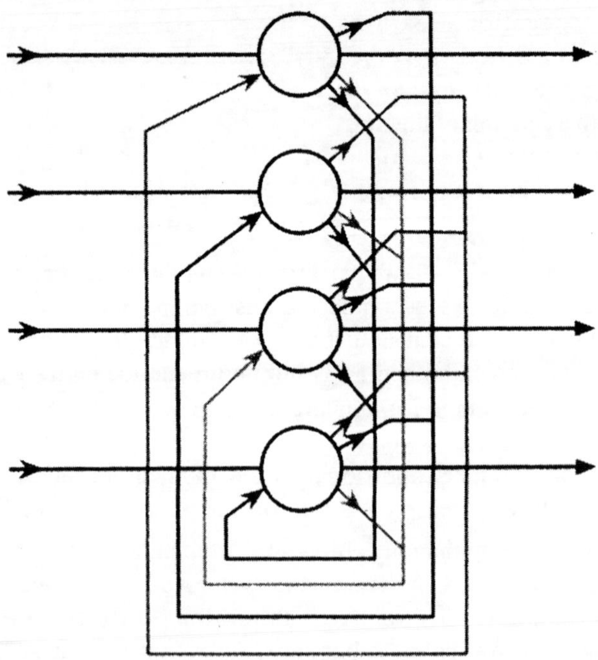

Fig. 8.

The units in Hopfield nets are binary threshold units, *i.e.,* the units only take on two different values for their states and the value is determined by whether or not the units' input exceeds their threshold. Hopfield nets can either have units that take on values of 1 or –1, or units that take on values of 1 or 0. So, the two possible definitions for unit i's activation, a_i, are:

$$a_i \leftarrow \begin{cases} 1 & \text{if } \sum_j w_{ij} s_j > \theta_i, \\ -1 & \text{otherwise} \end{cases}$$

$$a_i \leftarrow \begin{cases} 1 & \text{if } \sum_j w_{ij} s_j > \theta_i, \\ 0 & \text{otherwise} \end{cases}$$

Where:

w_{ij} is the strength of the connection weight from unit j to unit i (the weight of the connection).

s_j is the state of unit j.

α_i is the threshold of unit i.

The connections in a Hopfield net typically have the following restrictions:

$w_{ij} = 0$ (no unit has a connection with itself)

$w_{ij} = w_{ji}$ (connections are symmetric)

Symmetric weights are used because it guarantee that the energy function decreases monotonically while following the activation rules, and the network may exhibit some periodic behaviour if non-symmetric weights are used.

Hopfield nets have a scalar value associated with each state of the network referred to as the "energy", E, of the network, where

$$E = -\frac{1}{2}\sum_{ij} w_{ij} s_i s_j + \sum_{ij} \theta_i s_i$$

This value is called the "energy" because the definition ensures that if units are randomly chosen to update their activations the network will converge to states which are local minima in the energy function (which is considered to be a Lyapunov function). Thus, if a state is a local minimum in the energy function it is a stable state for the network.

Discrete Hopfield Model :

In this connection weights are not learned

Hopfield network evolves by updating the activation of each unit in turn

In final state, all units are stable according to the update rule

The units are updated at random, one unit at a time

$$\text{set } V_i \text{ to } 0 \quad \text{if } \sum_j T_{ij} V_j < \theta_i$$

$$\text{set } V_i \text{ to } 1 \quad \text{if } \sum_j T_{ij} V_j > \theta_i$$

$$\text{do not change } V_i \quad \text{if } \sum_j T_{ij} V_j < \theta_i$$

$\{Vi\}\, i = 1, \ldots, L,$

L : number of units

VI : activation level of unit I

T_{ij} : connection weight between units i and j

tetai : threshold of unit i.

Energy function :

(*i*) Units changes its activation level if and only if the energy of the network decreases by doing so:

(*ii*) Since the energy can only decrease over time and the number configuration is finite the network must converge (but not necessarily the minimum energy state)

$$E_i = -\sum_j T_{ij} V_i V_j + \theta_i V_i$$

if $V_i = 1$ then $E_i = -\sum_j T_{ij} V_j + \theta_i$

if $V_i = 0$ then $E_i = 0$

$$\Delta E_i = -\Delta V_i (\sum_j T_{ij} V_i - \theta_i).$$

Since the energy can only decrease over time and the number configuration is finite the network must converge (but not necessarily the minimum energy state)

Training a Hopfield net involves lowering the energy of states that the net should "remember". This allows the net to serve as a content addressable memory system, that is to say, the network will converge to a "remembered" state if it is given only part of the state. The net can be used to recover from a distorted input the trained state that is most similar to that input. This is called associative memory because it recovers memories on the basis of similarity. For example, if we train a Hopfield net with five units so that the state (1, 0, 1, 0, 1) is an energy minimum, and we give the network the state (1, 0, 0, 0, 1) it will converge to (1, 0, 1, 0, 1). Thus, the network is properly trained when the energy of states which the network should remember are local minima.

Q. 6. (*a*) Design an autocorrelator for encoding the patterns A1 = 1001101, A2 = 1110001 and A3 = 1000001. Check this autocorrelator for the pattern A = 00001, where it settles and why ? (10)

(*b*) Define heteroassociation and bidirectional associative memory (BAM). (5)

(*c*) What is the purpose of multiple encoding in BAM ? (5)

Ans. (*a*) $A_1 = 1001101 \quad A_2 = 1110001 \quad A_3 = 1000001$

Let bipolar patterns are stored

$$\therefore \quad A_1 = (1, -1, -1, 1, 1, -1, 1)$$
$$A_2 = (1, 1, 1, -1, -1, -1, 1)$$
$$A_3 = (1, -1, -1, -1, -1, -1, 1)$$

The connection matrix

$$T = \Sigma [A_i^T][A_i]$$

$$[A_1]^T [A_1] = \begin{bmatrix} 1 \\ -1 \\ -1 \\ 1 \\ 1 \\ -1 \\ 1 \end{bmatrix}_{7\times 1} \begin{bmatrix} 1 & -1 & -1 & 1 & 1 & -1 & 1 \end{bmatrix} \begin{bmatrix} 1 & -1 & -1 & 1 & 1 & -1 & 1 \\ -1 & +1 & +1 & -1 & -1 & +1 & -1 \\ -1 & +1 & +1 & -1 & -1 & +1 & -1 \\ 1 & -1 & -1 & 1 & 1 & -1 & 1 \\ 1 & -1 & -1 & 1 & 1 & -1 & 1 \\ 1 & +1 & +1 & -1 & -1 & +1 & -1 \\ +1 & -1 & -1 & +1 & 1 & -1 & 1 \end{bmatrix}$$

$$[A_2]^T [A_2] = \begin{bmatrix} 1 \\ 1 \\ 1 \\ -1 \\ -1 \\ -1 \\ 1 \end{bmatrix} \begin{bmatrix} 1 & 1 & 1 & -1 & -1 & -1 & 1 \end{bmatrix} = \begin{bmatrix} 1 & 1 & 1 & -1 & -1 & -1 & 1 \\ 1 & 1 & 1 & -1 & -1 & -1 & 1 \\ 1 & 1 & 1 & -1 & -1 & -1 & 1 \\ -1 & -1 & -1 & +1 & +1 & +1 & -1 \\ -1 & -1 & -1 & +1 & +1 & +1 & -1 \\ -1 & -1 & -1 & +1 & +1 & +1 & -1 \\ +1 & +1 & +1 & -1 & -1 & -1 & 1 \end{bmatrix}$$

$$[A_3]^T [A_3] = \begin{bmatrix} -1 \\ -1 \\ -1 \\ -1 \\ -1 \\ -1 \\ +1 \end{bmatrix} \begin{bmatrix} 1 & -1 & -1 & -1 & -1 & -1 & 1 \end{bmatrix} = \begin{bmatrix} 1 & -1 & -1 & -1 & -1 & -1 & +1 \\ -1 & +1 & +1 & 1 & 1 & 1 & -1 \\ -1 & 1 & 1 & 1 & 1 & 1 & -1 \\ -1 & 1 & 1 & 1 & 1 & 1 & -1 \\ -1 & 1 & 1 & 1 & 1 & 1 & -1 \\ -1 & 1 & 1 & 1 & 1 & 1 & -1 \\ 1 & -1 & -1 & -1 & -1 & -1 & +1 \end{bmatrix}$$

$$T = \sum_{i=1}^{3} [A_i]^T [A_i] = \begin{bmatrix} +1+1+1 & -1+1-1 & -1+1-1 & 1-1-1 & 1-1-1 & -1-1-1 & 1+1+1 \\ -1+1-1 & 1+1+1 & +1+1+1 & -1-1+1 & -1-1+1 & 1-1+1 & -1+1-1 \\ -1+1-1 & 1+1+1 & 1+1+1 & -1-1+1 & -1-1+1 & 1-1+1 & -1+1-1 \\ 1-1-1 & -1-1+1 & -1-1+1 & 1+1+1 & 1+1+1 & -1+1+1 & 1-1-1 \\ 1-1-1 & -1-1+1 & -1-1+1 & 1+1+1 & 1+1+1 & -1+1+1 & 1-1-1 \\ -1-1-1 & 1-1+1 & 1-1+1 & -1+1+1 & -1+1+1 & 1+1+1 & -1-1-1 \\ -1+1+1 & -1+1-1 & 1-1-1 & 1-1-1 & 1-1-1 & -1-1-1 & 1+1+1 \end{bmatrix}$$

$$T = \begin{bmatrix} 3 & -1 & -1 & -1 & -1 & -3 & 3 \\ -1 & 3 & 3 & -1 & -1 & +1 & -1 \\ +1 & 3 & 3 & -1 & -1 & 1 & -1 \\ 1 & -1 & -1 & 3 & 3 & 1 & -1 \\ 1 & -1 & -1 & 3 & 3 & 1 & -1 \\ -3 & 1 & 1 & 1 & 1 & 3 & -3 \\ 3 & -1 & -1 & -1 & -1 & -3 & 3 \end{bmatrix}$$

Recognition of Stored Patterns

$$A = 0000001$$

Recall equation is

$$a_j^{new} = f(a_i\, t_{ij}, a_j^{old})\ \forall j = 1, 2, \ldots \rho$$

where $A_i = (a_1, a_2, \ldots a_n)$

bipolar threshold for is

$$f(a,\ b) = \begin{cases} 1 \text{ if } & \alpha > 0 \\ \beta & \alpha = 0 \\ -1 & \alpha < 0 \end{cases}$$

A will become (–1, –1, –1, –1, –1, –1, 1)

$$a_1^{new} = f(-3+1+1+1+1+3+3, -1) = 1$$

$$a_2^{new} = f(+1-3-3+1+1-1+1, -1) = -1$$

$$a_3^{new} = f(-1-3-3+1+1-1+1, -1) = -1$$

$$a_4^{new} = f(-1+1+1-3-3-1+1, -1) = -1$$

$$a_5^{new} = (-1+1+1-3-3-1+1, -1) = -1$$

$$a_6^{new} = f(3-1-1-1-1-3+3, -1) = -1$$

$$a_7^{new} = f(3-1-1-1-1-3+3, 1) = 1$$

Now $\left(a_1^{new}, a_2^{new}, a_3^{new}, a_4^{new}, a_5^{new}, a_6^{new}, a_7^{new}\right) = (1, -1, -1, -1, -1, -1, 1)$ where is vector A_3.

(*b*) **(1) Auto Associative Memory :** An auto-associative memory, also know as auto-associative correlator, is used to retrieve a previously stored pattern that most closely resembles the current pattern.

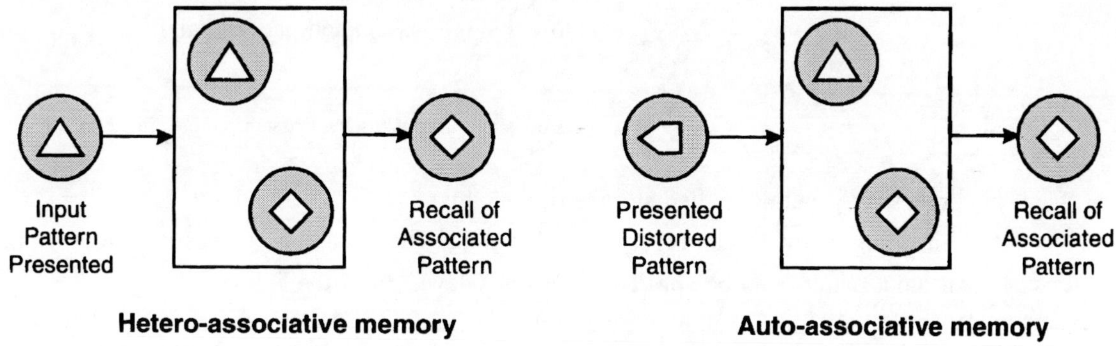

Fig. 9. Hetero and Auto Associative Memory Correlators.

(2) Hetero Associative Memory : A hetero-associative memory, also known as hetero-associative correlator, is used to rerieve pattern in general, different from the input pattern not only in content but possibly also different in type and format.

(c) **BAM** is a hetero asosciative, content-addressable memory consisting of 2 layers.

The network structure of BAM is similar to that of linear associator but connections are bidirectional that is

$$w_{ij} = w_{ji} \qquad \text{for} \begin{cases} i = 1, 2, \ldots., m \\ j = 1, 2, \ldots., n \end{cases}$$

i.e., BAM allows **forward and backward** flow of information between the layers. In BAM the units in both layers serve as both input and output units depending on the direction of propagation.

Propagating the signals from X layer to Y layer makes the units in X layer to act as input units while units in Y layer act as output units. Same is true for other direction because memory is bidirectional.

BAM model can perform both auto and hetero associative recall of stored information.

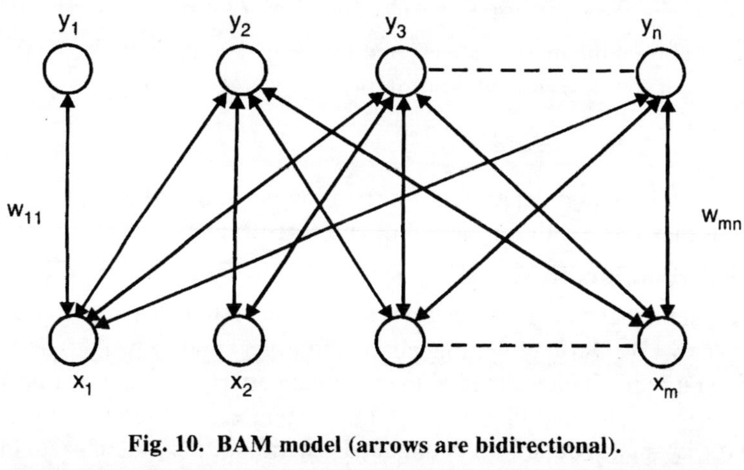

Fig. 10. BAM model (arrows are bidirectional).

Like is linear associator and hopfield network encoding in BAM can be carried out by using

$$w_k = x_K^T y_K \text{ (to store a single associated pattern pair)}$$

and
$$w = \alpha \sum_{K=1}^{P} w_K \text{ (to store simultaneously several associated pattern pairs)}$$

This process is generally called storage or encoding.

After encoding, network can be used for decoding. The decoding process is as follows :

1. An input pattern can be applied either on X layer or on the Y layer.
2. When input pattern is applied, network will propagate the input pattern to other layer allowing the units in other layer to compute their output values.
3. Pattern that was produced by other layer is then propagated back to original layer and units in the original layer compute the output values. New pattern that was produced of the original layer is again propagated to other layer.
4. The process is repeated until further propagations and computations do not result in a change in the associates of units in both layers where final pattern pair is one of the stored associated pattern pairs.
5. Final pattern pair that will be produced by the network depends on initial pattern pair and connection weight matrix.

BAM also uses Hebb's learning rule to build the connection weight matrix to store the associated pattern pairs.

Addition and Deletion of Pattern Pairs : Given a set of pattern pairs (X_i, Y_i), for $i = 1, 2, ..., n$ and a set of correlation matrix M :

- a new pair (X', Y') can be added or
- an existing pair (X_j, Y_j) can be deleted from the memory model.

Addition : Add a new pair (X', Y'), to existing correlation matrix M, them the new correlation matrix M_{new} is given by

$$M_{\text{new}} = X_1^T Y_1 + X_1^T Y_1 + ... + X_n^T Y_n + X'^T Y'$$

Deletion : Subtract the matrix corresponding to an existing pair (X_j, Y_j) from the correlation matrix M, them the new correlation matrix M_{new} is given by

$$M_{\text{new}} = M - \left(X_j^T Y_j\right)$$

Note : The addition and deletion of information is similar to the functioning of the system as a human memory exhibiting learning and forgetfulness.

Energy Function for BAM :

Note : A system that changes with tim is a dynamic system. There are two types of dynamics in a neural network. During tranining phase it itratively update weights and during production phase it asymptotically converges to the solution patterns. State is a collection of qualitative and qualitative items that characterize the system e.g. weights, data flows. The Energy function (or Lyapunov function) is bounded function of the system state that decreases with time and the system is th minimum energy.

Let a pair (A, B) defines the state of a BAM

• To store a pattern, the value of the energy function for that pattern has to occupy a minimum point in the energy landscape.

Let a pair (A, B) defines the state of a BAM

- to store a pattern, the value of the energy function for that pattern has to occupy a minimum point in the energy landscape.
- also adding a new pattern must not destroy the previously stored patterns.

The stabi!ity of a BAM can be proved by identifying the energy function E with each state (A, B).

For auto-associative memory : the energy function is

$$E(A) = -AM\ AT$$

For auto-associative memory : the energy function is

$$E(A) = -AM\ AT$$

For bidirectional hetero associative memory : the energy function is

$(A, B) = -AM\ BT$; for a particular case $A = B$, it corresponds to Hopfield auto associative function.

We wish to retrieve the nearest of $(A_i\ B_i)$ pair, when any (α, β) pair is presented as initial condition to BAM. The neurons change their states until a bidirectional stable state (A_f, B_f) is reached. Kosko has shown that such stable state is reached for any matrix M when it corresponds to local minimum of the energy function. Each cycle of decoding lowers the energy E if the energy function for any point (α, β) is given by $E = \alpha\ M\ \beta T$.

If the energy $E = A_i\ M\ B_i^T$ evaluated using coordinates of the pair (A_i, B_i) does not constitute a local minimum, then the point cannot be recalled, even though one starts with $\alpha = A_i$. Thus Kosko's encoding method does not ensure that stored pairs are at a local minimum.

Q. 7. Write short notes on following :

(*a*) Kohonen networks **(10)**

(*b*) ART Networks. **(10)**

Ans. Kohonen SOMs (Self-Organizing Maps) are unsupervised learning neural networks widely used to reduce dimensionality of input space preserving its topological structure. A typical Kohonen SOM architecture is shown below. It consists of an input layer connected to a

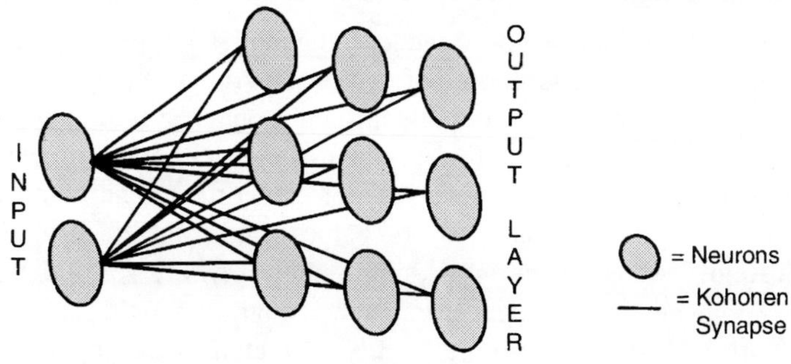

Fig. 11.

output layer (**Two-dimensional Kohonen layer**) via a **Kohonen Connector** consisting of **Kohonen Synapses**. Each neuron in a Kohonen Layer is associated with a unique set of coordinates in two-dimensional space, and hence is referred to as a **Position Neuron**. The input layer with 'n' input neurons is fed with n-dimensional input data one by one. The output layer organizes itself to represent the inputs. Hence the name 'self-organizing map'.

The objective of SOM training is to ensure that different parts of the network respond similarly to similar input vectors. So, the training mainly involves analysing the behaviour of the network for a training sample and adjusting the weights of synapses to ensure that the network exhibits a similar behaviour for a similar input. The training procedure involves following steps.

- Initialize the weights to small random values
- Choose a random training sample, assign the input vector to the input neurons and run the network.
- The output of a neuron in output layer represents the similarity between its weight vector (source synapse weights) and the input vector. The output neuron which has the highest output value is declared as the **winner neuron** for current input.
- Calculate the distance of each output neuron from the winner using a Neighborhood function
- Update the weights of synapses using following formulae. (This adjusts the weights corresponding to winner and its neighbors such that, a neuron close to the current winner wins for a similar input)

  Similarity = Difference between source neurons output and the synapse weight
  Weight Change = (Learning Rate) * (Neighborhood value of target neuron) * Similarity
- Similarly, train all samples in a random order. This completes one **training cycle** (or Training Epoch)
- Repeat the steps to complete the specified number of training epochs.

The trained SOM maps any input vector to a winner neuron, which can be interpreted as the position of vector in two-dimensional space.

(*b*) **ART Networks :** Adaptive resonance theory (ART) networks are most useful for pattern clustering, classification and recognition. They can also perform pattern association with some modifications. These networks work on binary or analog- valued input. Their ability to generalize is limited because ART networks can lack the hidden layer of neurons which perform feature recognition or pattern recognition in the backpropagation network.

The basic ART system is an unsupervised learning model. It typically consists of a comparison field and a recognition field composed of neurons, a vigilance parameter, and a reset module. The vigilance parameter has considerable influence on the system: higher vigilance produces highly detailed memories (many, fine-grained categories), while lower vigilance results in more general memories (fewer, more-general categories). The comparison field takes an input vector (a one-dimensional array of values) and transfers it to its best match in the recognition field. Its best match is the single neuron whose set of weights (weight vector) most closely matches the input vector. Each recognition field neuron outputs a negative signal (proportional to that neuron's quality of match to the input vector) to each of the other recognition field neurons and inhibits their output accordingly. In this way the recognition field exhibits lateral inhibition, allowing each neuron in it to represent a category to which input vectors are classified. After the input vector is classified, the reset module compares the strength of the recognition match to the vigilance parameter. If the vigilance threshold is met, training commences. Otherwise, if the match level does not meet the vigilance parameter, the firing recognition neuron is inhibited until a new input vector is applied; training commences only upon completion of a search procedure. In the search procedure, recognition neurons are disabled one by one by the reset function until the vigilance parameter is satisfied by a recognition match. If no committed recognition neuron's match meets the vigilance threshold, then an uncommitted neuron is committed and adjusted towards matching the input vector.

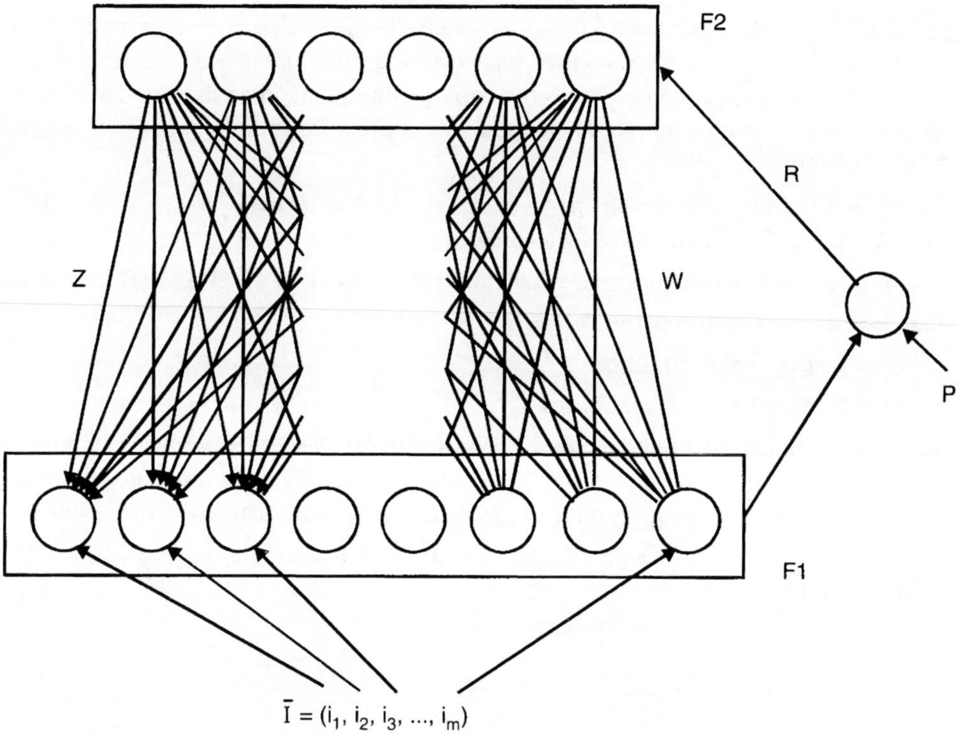

Fig. 12.

Training : There are two basic methods of training ART-based neural networks: slow and fast. In the slow learning method, the degree of training of the recognition neuron's weights towards the input vector is calculated to continuous values with differential equations and is thus dependent on the length of time the input vector is presented. With fast learning, algebraic equations are used to calculate degree of weight adjustments to be made, and binary values are used. While fast learning is effective and efficient for a variety of tasks, the slow learning method is more biologically plausible and can be used with continuous-time networks (i.e. when the input vector can vary continuously).

ART 1 : ART 1is the simplest variety of ART networks, accepting only binary inputs.

ART 2 : ART 2 extends network capabilities to support continuous inputs.

ART 2A : ART 2-A is a streamlined form of ART-2 with a drastically accelerated runtime, and with qualitative results being only rarely inferior to the full ART-2 implementation.

ART 3 : ART 3 builds on ART-2 by simulating rudimentary neurotransmitter regulation of synaptic activity by incorporating simulated sodium (Na+) and calcium (Ca2+) ion concentrations into the system's equations, which results in a more physiologically realistic means of partially inhibiting categories that trigger mismatch resets.

Fuzzy ART : Fuzzy ART implements fuzzy logic into ART's pattern recognition, thus enhancing generalizability. An optional (and very useful) feature of fuzzy ART is complement coding, a means of incorporating the absence of features into pattern classifications, which goes a long way towards preventing inefficient and unnecessary category proliferation.

ARTMAP : ARTMAP, also known as **Predictive ART**, combines two slightly modified ART-1 or ART-2 units into a supervised learning structure where the first unit takes the input data and the second unit takes the correct output data, then used to make the minimum possible adjustment of the vigilance parameter in the first unit in order to make the correct classification.

Fuzzy ARTMAP : Fuzzy ARTMAP is merely ARTMAP using fuzzy ART units, resulting in a corresponding increase in efficiency.

Q. 8. **(*a*) Explain the problem areas where neural network is a better tool than others. (10)**

(*b*) Write short note on the following :

(*i*) Radial basis function **(5)**

(*ii*) Support vector machines. **(5)**

Ans. (*b*) (*i*) A Radial Basis Function (RBF) neural network has an input layer, a hidden layer and an output layer. The neurons in the hidden layer contain Gaussian transfer functions whose outputs are inversely proportional to the distance from the center of the neuron.

RBF networks have a variable number of neurons that is usually much less than the number of training points.

RBF Neural Network

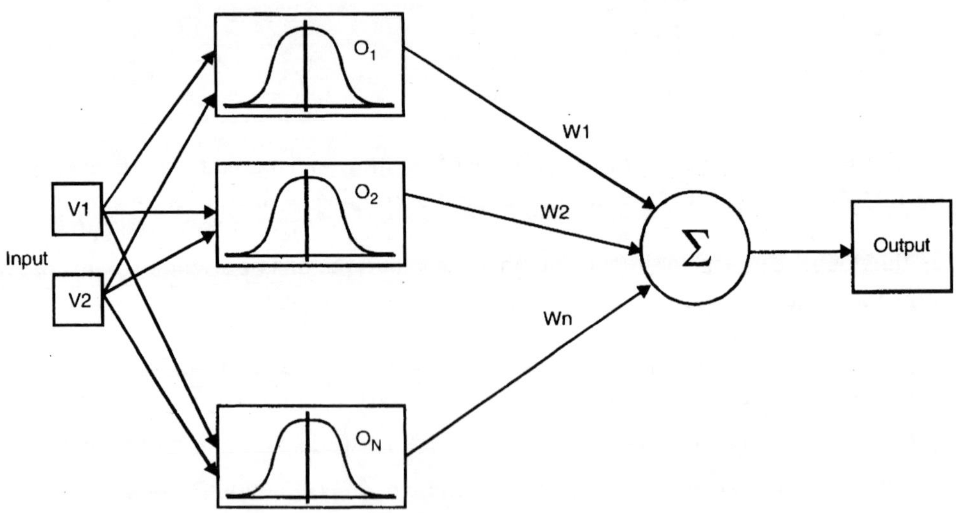

Fig. 13.

RBF networks have three layers :

1. **Input layer:** There is one neuron in the input layer for each predictor variable. In the case of categorical variables, $N-1$ neurons are used where N is the number of categories. The input neurons (or processing before the input layer) standardizes the range of the values by subtracting the median and dividing by the interquartile range. The input neurons then feed the values to each of the neurons in the hidden layer.
2. **Hidden layer:** This layer has a variable number of neurons (the optimal number is determined by the training process). Each neuron consists of a radial basis function centered on a point with as many dimensions as there are predictor variables. The spread (radius) of the RBF function may be different for each dimension. The centers and spreads are determined by the training process. When presented with the x vector of input values from the input layer, a hidden neuron computes the Euclidean distance of the test case from the neuron's center point and then applies the RBF kernel function to this distance using the spread values. The resulting value is passed to the the summation layer.
3. **Summation layer:** The value coming out of a neuron in the hidden layer is multiplied by a weight associated with the neuron (W1, W2, ...,Wn in this figure) and passed to the summation which adds up the weighted values and presents this sum as the output of the network. Not shown in this figure is a bias value of 1.0 that is multiplied by a weight W0 and fed into the summation layer. For classification problems, there is one output (and a separate set of weights and summation unit) for each target category. The value output for a category is the probability that the case being evaluated has that category.

Training RBF Networks

The following parameters are determined by the training process:

1. The number of neurons in the hidden layer.
2. The coordinates of the center of each hidden-layer RBF function.
3. The radius (spread) of each RBF function in each dimension.
4. The weights applied to the RBF function outputs as they are passed to the summation layer.

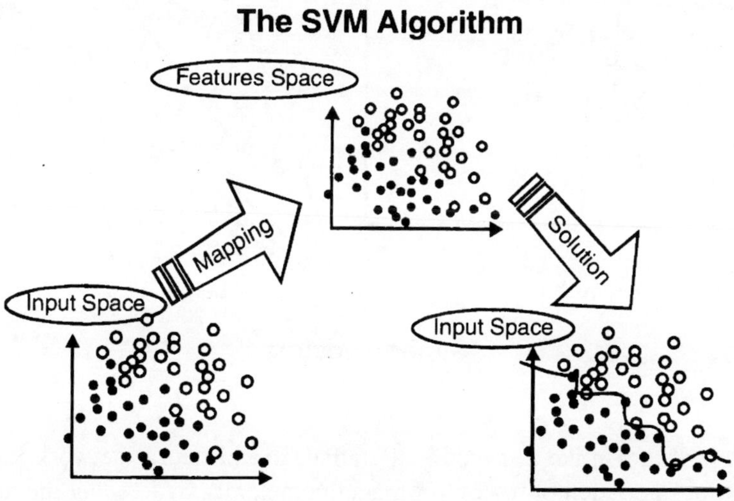

Fig. 14.

1. **Support vector machine:** Support Vector Machine (SVM) performs classification by constructing an N-dimensional hyperplane that optimally separates the data into two categories. A SVM model using a sigmoid kernel function is equivalent to a two-layer, perceptron neural network. SVM's are an alternative training method for polynomial, radial basis function and multi-layer perceptron classifiers in which the weights of the network are found by solving a quadratic programming problem with linear constraints, rather than by solving a non-convex, unconstrained minimization problem as in standard neural network training.

The goal of SVM modeling is to find the optimal hyperplane that separates clusters of vector in such a way that cases with one category of the target variable are on one side of the plane and cases with the other category are on the other size of the plane. The vectors near the hyperplane are the *support vectors*. The figure below presents an overview of the SVM process.

Classifying data is a common task in machine learning. Suppose some given data points each belong to one of two classes, and the goal is to decide which class a *new* data point will be in. In the case of support vector machines, a data point is viewed as a p-dimensional vector (a list of p numbers), and we want to know whether we can separate such points with a $(p - 1)$-dimensional hyperplane. This is called a linear classifier. There are many hyperplanes that might classify the data. One reasonable choice as the best hyperplane is the one that represents the largest separation, or margin, between the two classes. So we choose the hyperplane so that the distance from it to the nearest data point on each side is maximized. If such a hyperplane exists, it is known as the *maximum-margin hyperplane* and the linear classifier it defines is known as a *maximum margin classifier*, or - equivalently - the *perceptron of optimal stability*. The distance between the dashed lines is called the *margin*. The vectors (points) that constrain the width of the margin are the *support vectors*.

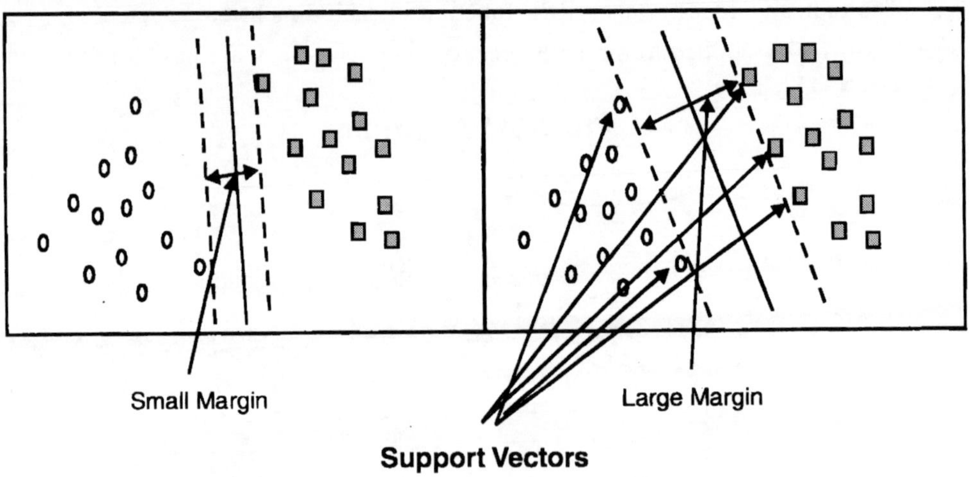

Fig. 23.

(*ii*) **SVM vs. ANN :** Unlike conventional statistical and neural network methods, the SVM approach does not attempt to control model complexity by keeping the number of features small.

"Classical learning systems like neural networks suffer from their theoretical weakness, e.g. back-propagation usually converges only to locally optimal solutions. Here SVMs can provide a significant improvement." Rychetsky (2001)

"In contrast to neural networks SVMs automatically select their model size (by selecting the Support vectors)." Rychetsky (2001)

"The absence of local minima from the above algorithms marks a major departure from traditional systems such as neural networks,..." Shawe-Taylor and Cristianini (2004)

"While the weight decay term is an important aspect for obtaining good generalization in the context of neural networks for regression, the margin plays a somewhat similar role in classification problems." Suykens et al. (2002), page 29

"In comparison with traditional multilayer perceptron neural networks that suffer from the existence of multiple local minima solutions, convexity is an important and interesting property of nonlinear SVM classifiers. [more]" Suykens et al. (2002)

B.E.
7TH SEMESTER EXAMINATION, DEC. 2011
NEURAL NETWORKS

Time : Three hours *Maximum Marks : 100*

Before answering the question, candidates should ensure that they have been supplied the correct and complete question paper. No complain in this regard, will be entertained after examination.

Note : Attempt any *five* questions.

Q. 1. (*a*) **Explain the structure of biological neuron in detail.** **(10)**

(*b*) **Differentiate between artificial and biological neuron in detail. Also discuss various applications of neural network.** **(10)**

Ans. (*a*) Refer to Chapter 1, Art. 1.3.1, pages 4-5.

(*b*) Refer to Chapter 1, pages 9-10.

Q. 2. Explain the various models of artificial neural network with their corresponding advantages and disadvantages. **(20)**

Ans. Refer to Chapter 2, Art. 2.4, pages 25-33.

Q. 3. (*a*) **What do you mean by discrete perceptron ? Explain the use of discrete perceptron in single layer perceptron model.** **(10)**

(*b*) **Explain the linearly separable classification with suitable example ?** **(10)**

Ans. (*a*) Refer to Chapter 3, Art. 3.6, pages 70-73.

(*b*) Refer to Chapter 1, Art. 1.3.1, pages 4-5.

Q. 4. (*a*) **Explain the delta learning rule for multi-perceptron neural network with its equations.** **(10)**

(*b*) **Define the following terms :** **(10)**

Error Back Propagation Training, Learning Factors, Hoff Learning Rule.

Ans. (*a*) Refer to Chapter 4, Art. 4.3, pages 80-86.

(*b*) Refer to Chapter 4, Art. 4.3, pages 80-86.

Q. 5. Explain the various architectures of Hopfield network in detail. How learning process does occurs Hopfield network ? **(20)**

Ans. Refer to Chapter 5, Art. 5.21, pages 106-112.

Q. 6. (*a*) **Explain the architecture of associative memory in detail. Also discuss its retrieval algorithm.** **(10)**

(*b*) **Explain how encoding and decoding takes place in associative memory.** **(10)**

Ans. (*a*) Refer to Chapter 6, Art. 6.1, pages 121-123.

(*b*) Refer to Chapter 6, Art. 6.2, pages 126-127.

Q. 7. Describe unsupervised learning of clusters in detail. **(20)**

Ans. Refer to Chapter 7, Art. 7.4, pages 148-150.

Q. 8. Write short notes on the following :

(*a*) Winner Take All Learning.

(*b*) Correlation learning rule.

(*c*) Auto associative memory.

(*d*) Weight adjustment.

Ans. (*a*) Refer to Chapter 2, Art. 2.5.4.6, pages 46-47.

(*b*) Refer to Chapter 2, Art. 2.5.4.5, pages 46.

(*c*) Refer to Chapter 6.

(*d*) Refer to Chapter 4, Art. 4.6, pages 93-94.

B.E.
7TH SEMESTER EXAMINATION, MAY 2012

NEURAL NETWORKS

Time : Three hours *Maximum Marks : 100*

Note : *Attempt any five questions.*

Q. 1. (*a*) **Explain the structure of artificial neuron in detail.** **(10)**

(*b*) **Differentiate between artificial and biological neuron in detail. Also discuss various applications of neural network.** **(10)**

Ans. (*a*) Refer to Chapter 1, Art. 1.3.1, pages 4-5.

(*b*) Refer to Chapter 1, pages 9-10.

Q. 2. Explain the various models of artificial neural network with their corresponding advantages and disadvantages. **(20)**

Ans. Refer to Chapter 2, Art. 2.4, pages 25-33.

Q. 3. (*a*) **What do you mean by discrete perceptron ? Explain the use of discrete perceptron in multi layer perceptron model.** **(10)**

(*b*) **Explain the Hebbian learning rule in detail.** **(10)**

Ans. (*a*) Refer to Chapter 3, Art. 3.6, pages 70-73.

(*b*) Refer to Chapter 2, Art. 2.5.4.1, pages 36-37.

Q. 4. (*a*) **What are various learning rules for multi-perceptron neural network ? Explain these learning rules with their equations.** **(10)**

(*b*) **Define the following terms :** **(10)**

Delta learning rule, separable pattern classification. **(10)**

Ans. (*a*) Refer to Chapter 4, Art. 4.3, pages 80-86.

(*b*) Refer to Chapter 4, Art. 4.1.1, pages 76-78.

Q. 5. Explain the various architectures of Hopfield network in detail. How learning process does occurs Hopfield network ? **(20)**

Ans. Refer to Chapter 5, Art. 5.21, pages 106-112.

Q. 6. (*a*) **What is associative memory ? Explain its various types in detail with suitable example. (10)**

(*b*) **Explain storage and retrieval algorithm for associative memory.** **(10)**

Ans. (*a*) Refer to Chapter 6, Art. 6.1, pages 121-123.

(*b*) Refer to Chapter 6, Art. 6.1, pages 122-123.

Q. 7. (*a*) **Describe unsupervised learning of clusters with suitable example.** **(10)**

(*b*) **Explain separation limitation for unsupervised learning.** **(10)**

Ans. (*a*) Refer to Chapter 7, Art. 7.4, pages 148-150.

(*b*) Refer to Chapter 7, Art. 7.10, pages 157-159.

Q. 8. Write short notes on the following :

(*a*) **Correlation learning rule.** (*b*) **Bi-Directional associative memory.**

(*c*) **Biological neuron.** (*d*) **Learning factors.**

Ans. (*a*) Refer to Chapter 2, Art. 2.5.4.6, pages 46-47.

(*b*) Refer to Chapter 6, Art. 6.4, page 131-133.

(*c*) Refer to Chapter 1, Art. 1.5.2, page 8-8.

(*d*) Refer to Chapter 4, Art. 4.6, pages 93-97.

❑❑❑

INDEX

T

U

V

W

X

❑❑❑

Win Prizes

Win Prizes

ATTENTION READERS

We request for your frank assessment regarding various aspects of the book as given below:

INTRODUCTION TO ARTIFICIAL NEURAL NETWORKS

By: GUNJAN GOSWAMI

Please Fill up the Sheet in Neat and Clean Writing

(i) What topics of your syllabus, important for your examination point of view, are not covered in this book ?

(ii) What are the chapters wherein the treatment of the subject matter is not systematic or organised or updated?

(iii) Have you come across misprints/mistakes/factual inaccuracies in the book? Please specify the chapters and the page numbers.

(iv) What other books on the same subject have you found/heard better than the present book? Please specify in terms of price and quality

(v) Further suggestions and comments for the improvement of this book.

(PS: If need be please attach a separate sheet to write your views)

OTHER DETAILS

(i) Who recommended you this book? (Please tick (✓) in the box near the option relevant to you.

Teacher ☐ ☐ Friends ☐ Bookseller

....

....

....

....

(ii) Name of the recommending teacher, his designation and address:

....

....

....

....

(iii) Name and address of the bookseller you purchased the book from :

....

....

....

....

(iv) Name and address of your institution (Please mention the University or Board as the case may be)

....

....

....

....

(v) Your name and complete postal address:

....

....

....

....

(vi) Course you are enrolled in:

....

....

....

....

The best assessment will be awarded each month: The award will be in the form of our publications as decided by the Editorial Board.

Please mail the filled up coupon at your earliest to:

S.K. KATARIA & SONS®
4885/109, Prakash Mahal, Dr. Subhash Bhargav Lane,
Opposite Delhi Medical Association, Daryaganj, New Delhi–110002 (INDIA)
Phone: +91-11-43551243; Mobile: +91-9871775858, 7678425209
e-mail: katariabooks@gmail.com; katariabook@yahoo.com
Website: www.skkatariaandsons.com